Digital Logic Simulation with CPLD Programming

Second Edition

Steve Waterman

DeVry Institute of Technology
DuPage Campus

Upper Saddle River, New Jersey
Columbus, Ohio

Cataloging-in-Publication Data
Waterman, Steve.
Digital logic simulation with CPLD programming / Steve Waterman.
p. cm.
ISBN 0-13-046711-1
1. Logic circuits—Design and construction—Laboratory manuals. 2. Logic circuits—Computer simulation—Laboratory manuals. 3. Logic programming—Laboratory manuals. I. Title.

TK7868.L6 W32 2003
621.39'5-dc21 2002021819

Editor in Chief: Stephen Helba
Acquisitions Editor: Dennis Williams
Editorial Assistant: Lara Dimmick
Production Editor: Stephen C. Robb
Design Coordinator: Karrie M. Converse-Jones
Cover Designer: Rod Harris
Cover photo: CY Digital Vision
Production Manager: Pat Tonneman

This book was set in Times New Roman by Steve Waterman. It was printed and bound by Banta Book Group. The cover was printed by Phoenix Color Corp.

Pearson Education Ltd.
Pearson Education Australia Pty. Limited
Pearson Education Singapore Pte. Ltd.
Pearson Education North Asia Ltd.
Pearson Education Canada, Ltd.
Pearson Educación de Mexico, S.A. de C.V.
Pearson Education—Japan
Pearson Education Malaysia Pte. Ltd.
Pearson Education, *Upper Saddle River, New Jersey*

---

10 9 8 7 6 5 4 3 2 1

ISBN 0-13-046711-1

To my sons,
Alexander and Christopher

Table of Contents

Section 3: Library of Parameterized Modules (LPM functions)

Section 4: Appendices

Parts list

Max+plus II Version 9.xx or 10.xx software by Altera Corporation
(1) Breadboard
(1) 741 Operational amplifier
(1) 555 Timer IC
(10) Light emitting diodes
(1) DAC0808 digital-to-analog converter (National)
(1) ADC0804 analog-to-digital converter (National)
(1) 6116 memory chip (2K × 8)
(1) Thermister
(3) Quad DIP switches (SPDT)
(2) Pushbutton switches (SPDT)
(1) Byte blaster (usually is supplied with the university board)
(1) * Circuit board with the MAX7000S chip

(1) 220-Ω resistor
(10) 1-KΩ resistors
(4) 4.7-KΩ resistors
(3) 10-KΩ resistors
(5) 20-KΩ resistors
(1) 33-KΩ resistor
(2) 47-KΩ resistors
(5) 100-KΩ resistors
(1) 1-KΩ potentiometer PC mount
(1) 10-KΩ potentiometer PC mount
(1) 50-KΩ potentiometer PC mount

(1) 20-μFd capacitor
(1) 1-μFd capacitor
(2) 0.1-μFd capacitors
(2) 0.01-μFd capacitors
(2) 0.001-μFd capacitors
(1) 150-pFd capacitor

* Contact the vendor directly for prices:

DeVry University Board eSOC, Electronic System On Chip or RSR PLDT Trainer PLDT-2

Both are available from Electronix Express, A Division of R.S.R. Electronics, Inc., 365 Blair Road, Avenel, New Jersey 07001 http://www.elexp.com

Phone 1-800-972-2225 (In NJ 1-732-381-8020) Fax 1-732-381-1006; 1-732-381-1572

University Board from Altera Corporation, 101 Innovation Drive, San Jose, California 95134
(408) 544-8274 http://www.altera.com

UPXa Board from Intectra, 2629 Terminal Blvd - Mountain View, CA 94043 USA (650) 967-8818
http://www.intectra.com

Free software: Download the latest Max+plus II Student Edition software from Altera Corporation
http://www.altera.com

Preface

This lab manual may be used to accompany any well-written digital electronics textbook to teach the fundamentals of digital electronics or may be used by professionals to learn how to use the Max+plus II software by Altera Corporation. Various University boards with an interface cable connect directly to the computer's RS232 port for programming the on-board CPLDs (Complex Programmable Logic Devices). Appendix C contains a step-by-step procedure to program the MAX7000S CPLD however, the instructions will work with any CPLD supported by the software. Any successfully compiled circuit may be downloaded to the CPLD through a JTAG mounted on the University board.

The sequence of the labs may be altered; however, software features described in lower numbered labs will not be described in supplemental labs. This may pose a problem for students unless the software features used in a lab are discussed in class.

The Max+plus II software has large libraries of digital devices commonly found in data books. Your application, however, may require the CPLD to be interfaced to other devices like a microprocessor, microcontroller, memory, and A/D or D/A converters. Typically, university board I/Os are not buffered and carelessness may cause permanent damage to the CPLD. Read and follow instructions provided in the user manual supplied with the university board.

Occasionally, the labs make reference to the 74 Series Integrated Circuits. Data sheets to these ICs are available for download from Texas Instruments at *http://www.ti.com*, or National Instruments at *www.national.com*. Contents and layouts of home pages periodically change so use the site's local search engine to locate the specific data sheets.

All data sheets are typically written in Adobe format, requiring Adobe Acrobat Reader to read and print the file. The latest version of the Adobe Acrobat Reader is available for download at *http://www.adobe.com*. Check your computer directories using My Computer of Windows Explorer for the Acrobat directory. If found, your computer already has the Acrobat Reader.

The lab manual is divided into three sections: asynchronous circuits, synchronous Circuits, and LPM megafunctions. Section 1 on asynchronous circuits starts with the fundamentals of digital gates and logic control circuits and progresses to MSI devices. Latches and the 555 timer are intended for first-term digital students. These 14 labs introduce the software features needed to master the Graphic and Waveform Editors in the Max+plus II Student Edition software by Altera Corporation, as well as implementing digital concepts learned in fundamental digital courses.

Section 2 is based on synchronous clock-dependent circuits, including counters and registers, memory addressing, and A/D and D/A converters, some of which will require interfacing the University board to external components on a breadboard. Several labs will focus on the LPM_MACRO functions available that are unique to this software.

Section 3 focuses on the library of parameterized megafunctions that are unique to the Max+plus II software. It is recommended that new designs be based on these megafunctions instead of the traditional macrofunctions based on the 7400 series chip set.

It is strongly recommended that students save all working files to their floppy disks in Drive A, a Zip or LS120 diskette, or be assigned a working directory on the hard drive, and save all software-generated files to the same directory/drive. Students will quickly fill the FAT (file allocation table) for a diskette, resulting in software crashes. To free up diskette space, delete all files on the diskette except for the .gdf and .scf files. If using networked computers, you may consider assigning a section of the hard drive to each student. Students should avoid using the Maxplus2 directory as their working directory.

Students always enjoy getting something for nothing, well almost for nothing. Most students in my classes who have computers at home want their own copy of the software. In this way, they can spend considerable amount of time at home doing their labs and can verify homework assignments using this computer. What a wonderful way to extend lab hours! Of course, the software is free but they will need to obtain the software from Altera. Once the software is installed, the student is directed to Altera's home page where the student fills out a form to get an authorization code necessary to activate the software. Relatively painless! Faculty can send an e-mail request to Altera's University Program asking for XXX student version CD ROMs and they will gladly send them. I normally check these out to students overnight or for a weekend to be installed on their personal PCs.

As with all software, revisions are inevitable. If using different versions of the software in lab, at work, or at home, delete all but the .gdf and .scf files on the disk. When compiled, the software will generate the necessary files recognized by the version in use.

Introduction

Programmable Logic Devices (PLDs) have evolved since the early 1980s from containing a few hundred gates to over a million gates. As the gate density within the devices has grown in excess of 100,000 times, the physical size of the devices has doubled, taking up about the same real estate on a circuit board as two small scale (SSI) or medium scale (MSI) integrated circuits. Three types of programmable logic devices are:

Simple Programmable Logic Devices (SPLDs),
Complex Programmable Logic Devices (CPLDs), and
Field Programmable Gate Arrays (FPGAs)

Simple Programmable Logic Devices

The least-expensive programmable logic devices are Simple Programmable Logic Devices (SPLDs). An SPLD can replace a few 7400-series TTL devices. Most SPLDs are programmed using Boolean-based software, ABEL being one of the most popular. The designer writes a program using a basic text editor, assembles the file into a JEDEC file, then downloads the JEDEC file to the SPLD chip mounted on a programmer unit. SPLDs may be reprogrammed in the engineering lab many times (up to 100 for some devices), making the SPLD ideal for the designer. Several SPLDs that have evolved are:

PAL (Programmable Array Logic)
GAL (Generic Array Logic)
PLA (Programmable Logic Array)
PLD (Programmable Logic Device)

Complex Programmable Logic Device

Time to market pressures have increased the demand for higher-density PLDs. Using Complex Programmable Logic Devices (CPLDs) shortens development times, provides simpler manufacturer's development and testing, and supports system upgrades in the field; ISP allows manufacturers to get their product to market quicker. In system programmability, ISP allows the designer to modify or upgrade the design as needed without altering the circuit card design.

CPLDs have significantly higher gate densities than PLDs. Software enhancements allow the designer greater programming techniques than were available for the SPLD. Designs may be graphically entered using symbols based on TTL devices, logic macros designed by the designer, or popular software languages such as VHDL (**V**ery **H**igh Speed Application Specific Integrated Circuit **H**ardware **D**escription **L**anguage). Numerous CPLD manufacturers have proprietary software that may also be used to program their devices.

CPLDs are CMOS logic devices that use nonvolatile memory cells, using either EPROM, EEPROM, or FLASH technologies. The EPM7128SLC84 is an EEPROM that can be reprogrammed 100 times or so.

Field Programmable Gate Array

Of the three, FPGAs offer the highest logic capacity. The Flex 10K series from Altera is SRAM (Static RAM) based and comes in various pin densities, up to 240 pins. Since SRAM is volatile, it is common practice to download the program to ROM memory, which then is automatically loaded into the FPGA when power is applied to the system. FPGAs consist of an array of logic blocks, surrounded by programmable I/O blocks connected together with programmable interconnects. Software places and routes the logic on the device.

The University board by Altera contains the EPM7128SLC84 CPLD and Flex10K FPGA, thus exposing students to both technologies. However, the EPM7128SLC84 will suffice for all labs within this manual.

Note to the Instructor

This lab manual is based of the Max+plus II Student Edition software and the Max7000S family complex programmable logic device (CPLD) by Altera Corporation. The software and university board are available from Altera Corporation, 101 Innovation Drive, San Jose, CA 95134, USA. A lower-cost alternative board with fewer bells and whistles is available from Intectra, 2629 Terminal Blvd., Mountain View, CA. 94043. Intectra may be reached at *www.intectra.com*. DeVry University also has an intermediate-cost board with 16 LEDs, 2 seven-segment displays, 16 SPDT switches, and 4 pushbutton switches. Information regarding this board is available at *www.devry.edu/eSOC*.

This lab is based on the "system first, details later" concept of teaching digital electronics. The bulk of the labs in Section 1 is software based. Lab 5 is the first time students are instructed to program the hardware; although previous circuits can be programmed, programming is delayed until the students have learned the fundamentals of digital and how to use the software. Depending on the features available on the university board selected, students may need to wire external switches or use the switches on the University board.

Students will do hard-wiring of components for Lab 14: The 555 Timer; Lab 21: D/A Converters; Lab 22: A/D Converters; Lab 23: Memory Addressing; Lab 24: Analog Storage; and Lab 25: Synchronous Data Transceiver. As you can see, the labs near the end do have more intense hard-wiring of the CPLD to external devices.

Faculty are always concerned that students don't put enough emphasis on troubleshooting when using software-based labs as opposed to breadboarding TTL devices. Adding software-based labs and interfacing the CPLD to hardware is not always a cut-and-dry task guaranteeing success every time. Students now have to determine if the problem is software based or hardware based. If software based, they can simulate and play "what if" scenarios to isolate software faulty designs or verify that the software is correct. If the problem is hardware based, students should be encouraged to use test equipment to verify input and output conditions on the CPLD pins, as well as on each I/O of the external devices. Students need to learn troubleshooting at a systems level in addition to component level.

Each lab requires students to write a technical summary with embedded graphics. We are all aware that people skills, written and oral communication skills, and technical knowledge all play an important role on the job. I have students write 1- to 2-page technical summaries with embedded graphics for each lab submitted. Technical summaries should never include words such as "I," "we," or "learned"! Each summary should have an opening paragraph stating the lab objectives, an electronically embedded schematic, and a paragraph or two explaining the schematic. This is not a step-by-step procedure of what was done, but rather a technical disclosure on a segment of the lab. Students could also include waveforms to support their discussion. In this way, students can focus on a section of the lab for their summary.

I encourage students to write about problems encountered, either with the software or hardware, and to explain how they would advise a customer or fellow colleague on how to work around the problem and how to avoid the situation that may have caused the problem.

Each lab contain numerous parts. You may want to skip a section here or there in order for the student to complete the lab on a timely schedule. Maybe you have a project that you would like to assign, so I would advise students to read certain sections of predefined labs, and then implement concepts learned to complete the project.

Acknowledgments

I am very grateful to my colleagues, Gary Luechtefeld, Zoran Ulcivich, and Art Ramirez, at DeVry DuPage in Addison, Illinois, and J. S. Watson at DeVry in Tinley Park, Illinois for their patience, advice, moral support, and willingness to assign the labs in rough draft to their students as the labs were being developed.

I gratefully acknowledge the following reviewers for their insightful suggestions for both the first and second editions of this manual:

Dan Black, DeVry Institute of Technology, Oakbrook Terrace, Illinois
James Hamblen, Georgia Institute of Technology, Atlanta, Georgia
Joe Hanson, Altera Corporation, 101 Innovation Drive, San Jose, California
Mike Phipps, Altera Corporation, 101 Innovation Drive, San Jose, California
John Hull, DeVry Institute of Technology, DuPage campus, Addison, Illinois
Bill Kleitz, Tompkins Courtland Community College, Drayden, New York
Shodi Mazidi, DeVry Institute of Technology - Calgary, AB, Canada
Guillermo Jaquenod, Calle 24 # 709 - (1900), La Plata, Argentina
Carlos Puleston, Intectra, 2629 Terminal Blvd., Mountain View, California
Warren Foxwell, DeVry Institute of Technology, DuPage campus, Addison, Illinois

Many thanks go to my students who provided constructive criticisms as they completed each lab during development.

It is a pleasure working with the people at Prentice Hall, and special thanks to my editor Dennis Williams and local Prentice Hall representative Evan Girard for their continued support during the completion of this project.

About the Author

Professor Steven Waterman, M.Ed., teaches in the, Electronics and Computer Technology, Electronics Engineering, and Telecommunications programs at the DuPage Campus of DeVry Institute of Technology in Addison, Illinois. As the digital/micro/controls sequence chair, Professor Waterman is very active teaching Computer Hardware/Software, Digital Fundamentals, Introduction to Microprocessors, and Advanced Microprocessors, including the Intel 80xxx assembler or Motorola 68xxx assembler. Courses are Internet based where students can access classroom activities using a Web browser and can complete assignments online. Professor Waterman is a member of DeVry's National Digital Committee, which determines course sequence and content, as well as a member of IEEE, ISA, and a local SCUBA diving club.

waterman@dpg.devry.edu *http://www.dpg.devry.edu/~waterman*

Lab 1: Logic Gates

Objectives:

1. Learn how to use the Max+plus II^R software to create a schematic, create waveforms, compile, and simulate a circuit
2. Analyze waveforms and develop truth tables for these fundamental logic gates:
   AND OR NOT NAND
   NOR XOR XNOR
3. (Optional) Program the EPM7128SLC84-7 Programmable Logic Device and verify each gate operation

Materials list:

- Max+plus II software by Altera Corporation
- University board with Altera EPM7128SLC84 CPLD (optional)
- Computer requirements:
  Minimum 486/66 with 8 MB RAM
- Floppy disk

Discussion:

The symbols for the logic gates discussed in this lab are shown in Figure 1. The **NOT** gate, or the INVERTER, produces an output opposite of the input. The NOT symbol shows an active-HIGH input and an active-LOW output.

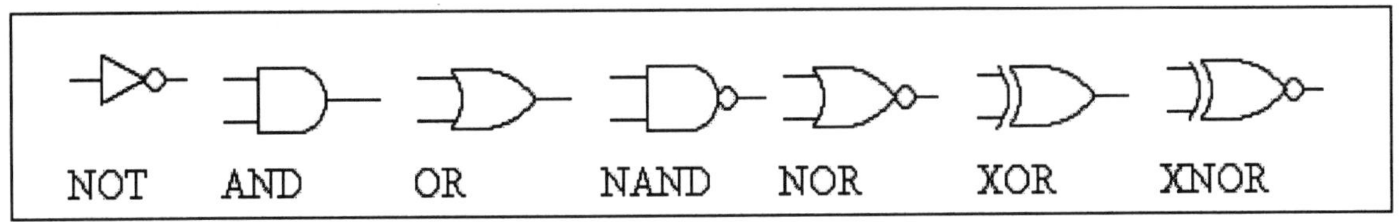

Figure 1

The **AND** gate produces a logic-HIGH output only when ALL inputs to the gate are a logic-HIGH. The AND gate has active-HIGH inputs and an active-HIGH output.

The **OR** gate produces a logic-HIGH output if either or both inputs are a logic-HIGH. The inputs and output are active-HIGH.

The **NAND** gate produces a logic-LOW output only when ALL inputs to the gate are a logic-HIGH. The NAND gate has active-HIGH inputs and an active-LOW output. The NAND gate is derived from inverting the output of the AND gate.

The **NOR** gate is the complement of the OR gate. The **NOR** gate produces a logic-LOW output if either or both inputs are a logic-HIGH. The inputs are active-HIGH whereas the output is active-LOW.

The exclusive OR gate, or **XOR**, produces a logic-HIGH output when either, but not both, input to the gate is a logic-HIGH. Both inputs and the output of the XOR gate are active-HIGH.

If the output of the XOR gate is inverted, you have an **XNOR** function. The XNOR gate produces a logic-HIGH output when both inputs to the gate are the same logic level.

The Draw and Tools tool bar icons you will be using in this lab are identified in Figure 2. The tool bar contains buttons such as: New, Open, Save, Print, Cut, Copy, Paste, and Undo. The Font and Point size can be quickly changed using the buttons on the right side of this tool bar. The vertical row of buttons on the left side of the screen contain draw functions such as Select, Text, Line, Diagonal, Arc, Circle, Zoom In, and Zoom Out. Other tool bar buttons will be discussed in the lab where appropriate.

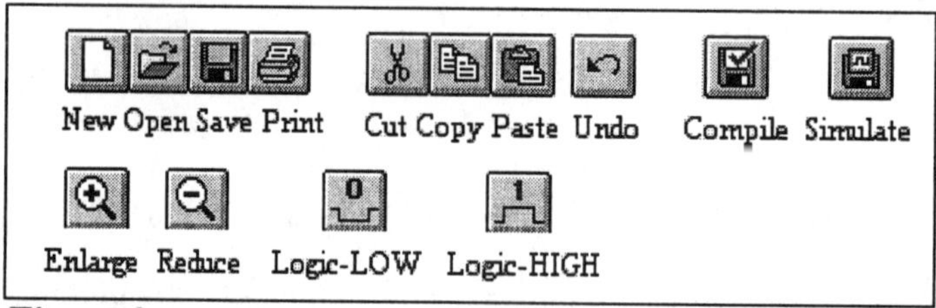

Figure 2

Part 1 Procedure

1. Open the Max+plus II software. From the desktop, select **Start - Programs - MAX+plus II**. This lab may be used with the student version or network versions 9.x or 10.x written by Altera Corporation.

   Once the software opens, you will see a menu bar at the top of the screen with names similar to those you would find in most Windows-based programs, that is, File, Edit, View, Options, Window, and Help.

2. From the main menu, select **File - Project - Name**. See Figure 3. The Project Name dialog box (see Figure 4) will appear.

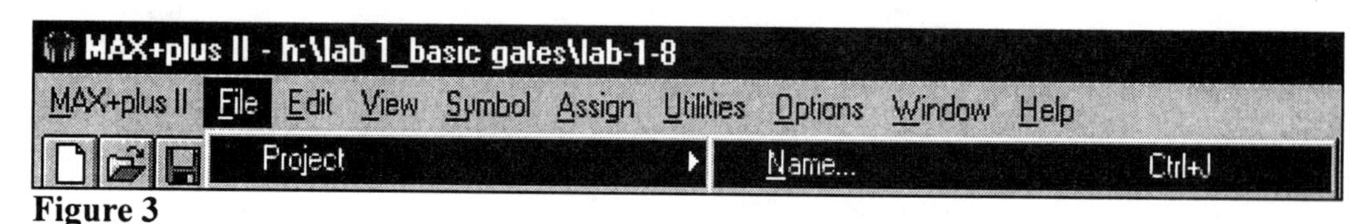

Figure 3

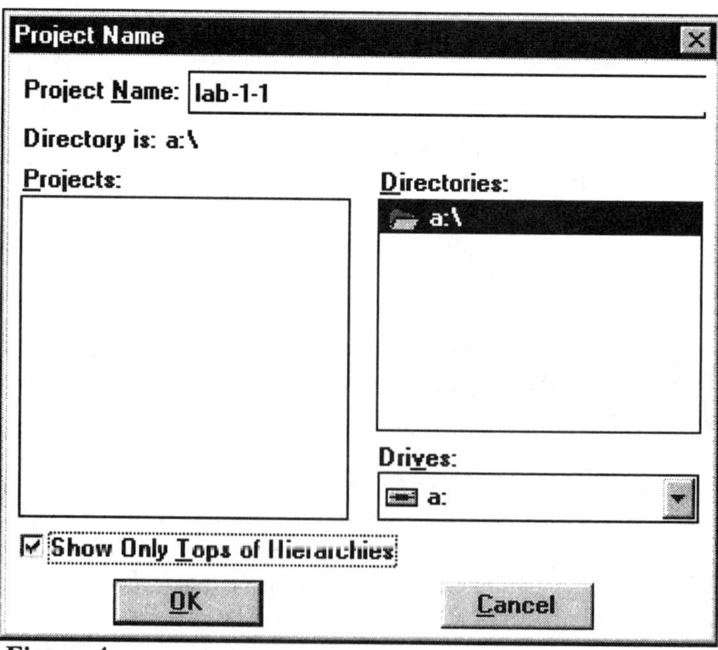

Figure 4

3. Insert a diskette into the floppy or zip drive and select the drive letter in the **Drives:** window of the Project Name dialog box. Delete all text in the Project Name window, then type **lab-1-1**. Click **OK**. Every project must have a project name that matches the design file name. Because a project name must be assigned every time the Max+plus II is run, your files must have that same project name.

4. Select **File - New - Graphic Editor File**. Click **OK**. The Graphic Editor screen will appear with **Untitled1 - Graphic Editor** as the file name.

5. The student edition allows the programming for numerous complex programmable logic devices manufactured by Altera Corporation. The IC you will program is the 84-pin EPM7128SLC84 IC. To assign this IC, select **Assign - Device**. When the Device dialog box appears (Figure 5), select the MAX7000S device family. Scroll through the list of ICs and select the EPM7128SLC84-7 device. Click **OK**. The device name need only be assigned once, when the software is first opened.

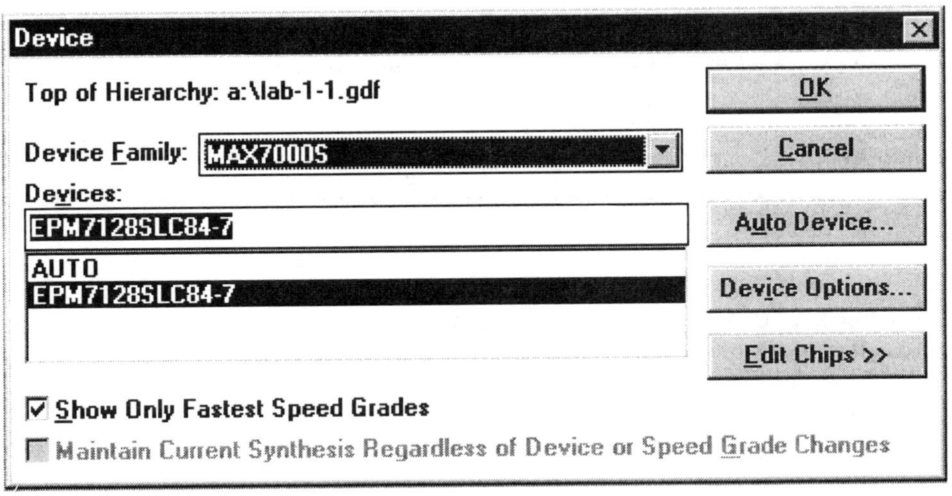

Figure 5

6. Now you are ready to enter the logic symbol. Place the cursor (arrow) at the point in the work area where you want to insert a symbol, then click on the left mouse pointer to obtain a flashing square dot. This will be called the "insertion point" throughout this lab. Select **Symbol - Enter Symbol - OK** and the Enter Symbol dialog box appears (Figure 6). Available libraries are the Altera primitives (prim) containing basic logic building blocks, macrofunctions (mf) of the 7400 family logic, library of parameterized modules (LPMs) and Megafunctions (mega_lpm) for high-level circuit functions, and the building blocks (edif) for the LPMs.

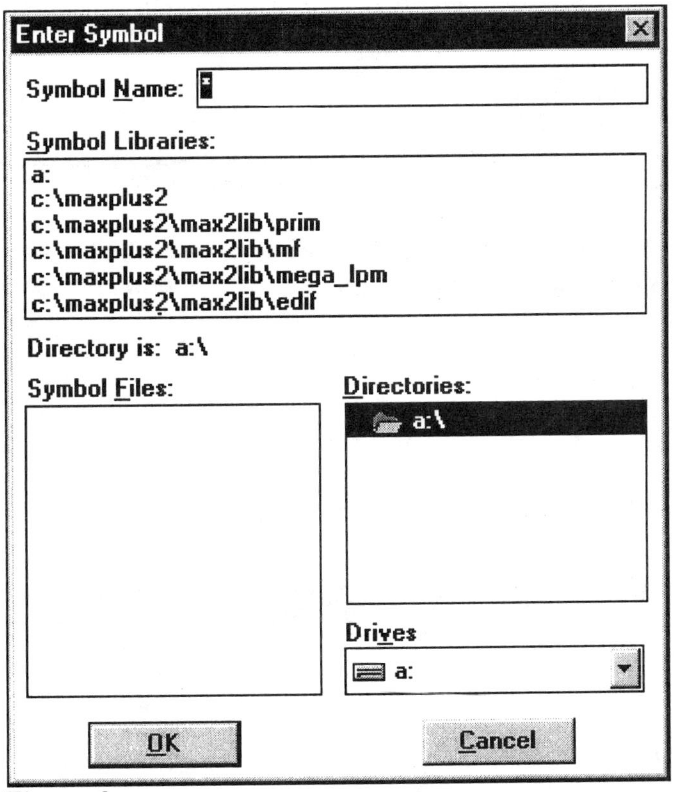

Figure 6

7. Set the mouse pointer on the **..\maxplus2\max2lib\prim** directory and double click on the left mouse button. Select **and2** in the Symbol files list and click **OK**. (You could also double click on the symbol name.) The symbol will appear in the Graphic Editor.

8. Insert input and output symbols found in the **..\maxplus2\max2lib\prim** directory and position these symbols to create the circuit shown in Figure 7. Single click the left side of each symbol to "click and drag" the symbol to the desired position.

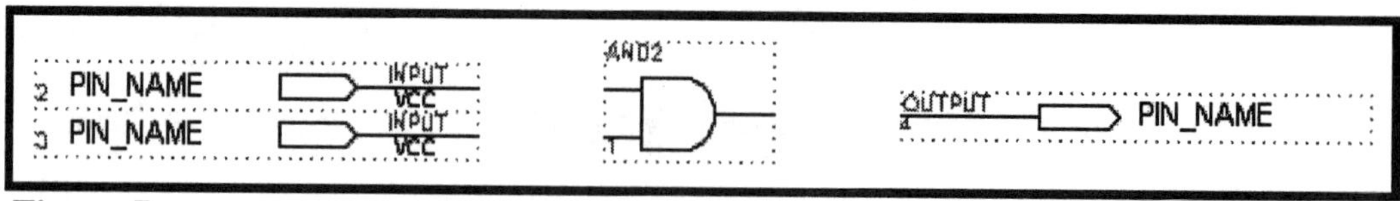

Figure 7

9. To change the name "PIN_NAME" of the input (or output) connector, double click on the current name. When the name appears in reverse text (highlighted), type the new name. Change the current names of the input connectors to A and B and change the output connector's name to AND2. Your circuit should now look like the circuit in Figure 8.

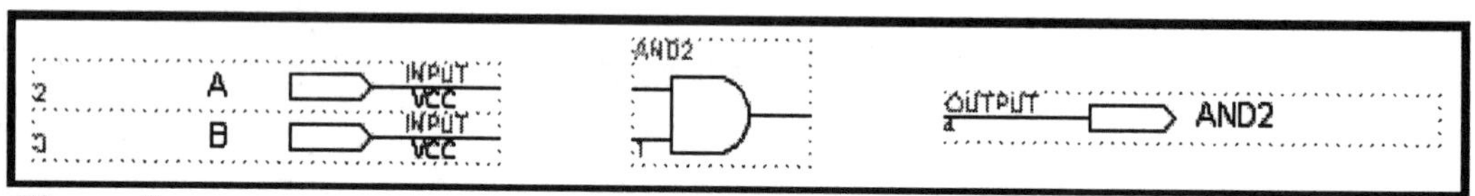

Figure 8

10. Add lines from each gate terminal to the respective input or output terminal. To make these lines, position the mouse pointer on a gate terminal. When the mouse pointer turns into a cross, click and drag the mouse so that it barely touches the input or output connector lead. Do *not* drag and release the wire inside a component. The circuit should appear as shown in Figure 9.

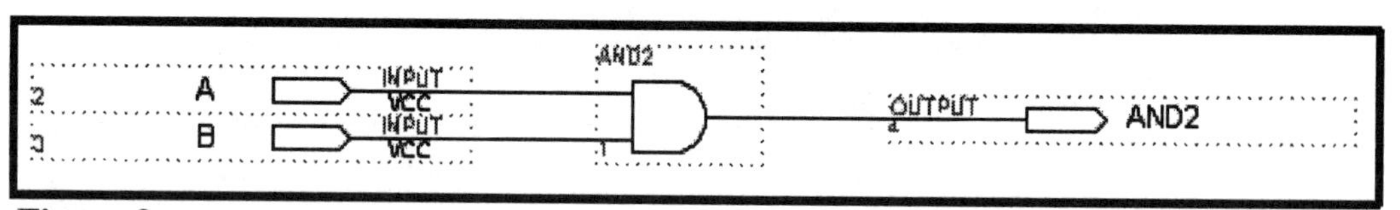

Figure 9

11. Now that the circuit is constructed, you are ready to create a set of input waveforms. Select **File - New - Waveform Editor File,** then click **OK**. The Untitled1 Waveform Editor will appear on the screen. An unlimited number of "fields" are below the Name column. The first field is shown in Figure 10, but it does not appear on your screen when the Waveform Editor file is first opened. Don't worry if your scale is different than what is shown in Figure 10.

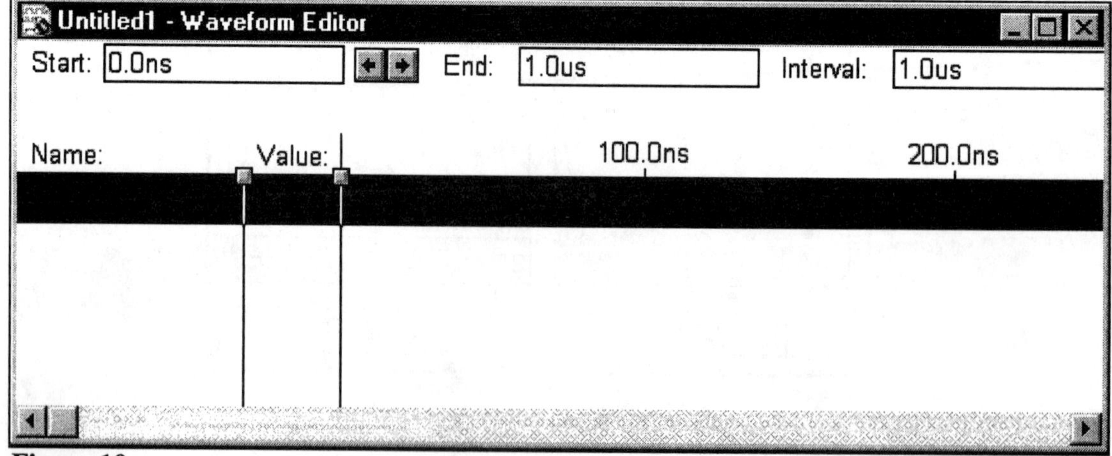

Figure 10

12. Place your mouse pointer in the first field below the Name column, then double click on the left mouse pointer. The Insert Node dialog box will appear. Enter **A** for the Node Name and select **Input Pin** in the **I/O Type** frame. Click **OK**. The input waveform "A" was created with a "value" of "0" from 0 ns to 1 µs.

13. Place your mouse pointer in the first field below Waveform A, then double click on the left mouse pointer. Enter **B** for the Node Name, then click the **OK** button. The input waveform "B" was created with a "value" of "0" from 0 ns to 1 µs.

14. Place your mouse pointer in the first field below Waveform B, then double click on the left mouse pointer. The Insert Node dialog box will appear. Enter **AND2** for the Node Name and select **Output Pin** in the I/O Type frame. Click **OK**. The output waveform "AND2" was created with a "value" of "0" from 0 ns to 1 µs. The Waveform Editor should look like Figure 11.

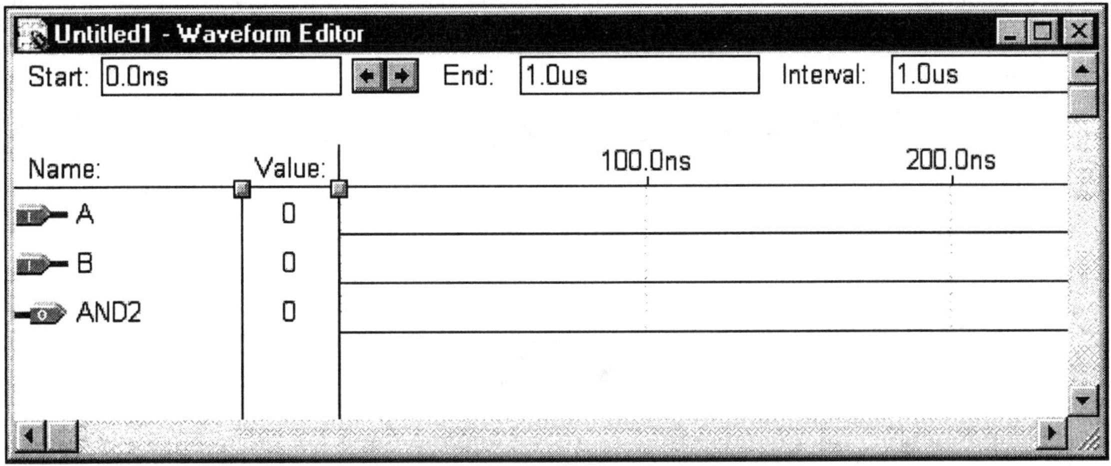

Figure 11

15. Turn on the **Snap to Grid** feature in the Options menu. If this feature already has a check mark next to it, as shown in Figure 12, the feature is already turned on.

16. Select **Grid Size** (Figure 12) in the Options menu, type **100** in the Grid Size dialog box, then press the Enter key. The Waveform Editor now has vertical timing marks every 100 ns.

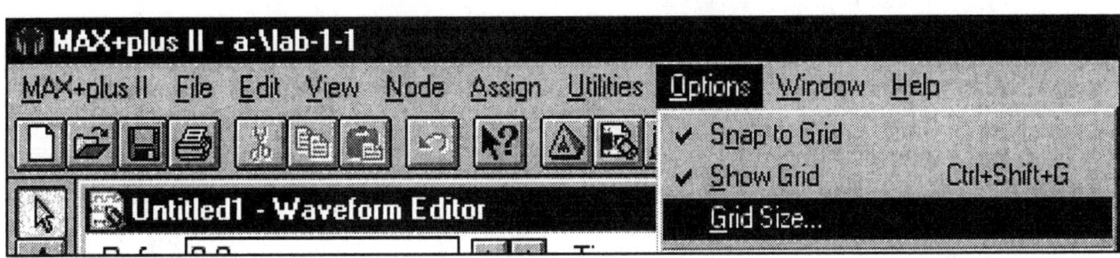

Figure 12

17. Highlight the section of Waveform A from 100 ns to 200 ns by clicking and dragging the mouse pointer, starting at the 100 ns marker. Change this highlighted section to a logic-HIGH pulse using the Logic-HIGH pulse button in the Draw tool bar (refer to Figure 2). In a similar manner, change Waveform A from 300 ns to 400 ns and Waveform B from 200 ns to 400 ns to a logic-HIGH. Your waveforms should look like those shown in Figure 13.

18. Click the Compile button in the tool bar (disk with red check mark). Save the **.GDF** and **.SCF** files when the respective Save As dialog boxes appear. You should have zero errors. Click **OK**, then click the Start button in the Compiler. At this time, you should see zero errors and one warning. Click **OK**, then close the Compiler window.

19. Click the Simulate button in the tool bar (disk with blue square wave). Assuming zero errors, click **OK** in the pop-up window, then the **Open SCF** button in the Simulator window.

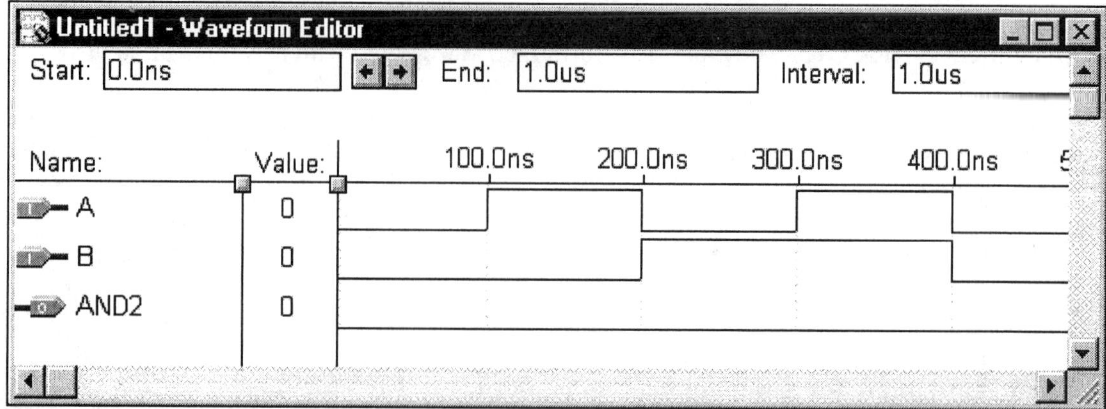

Figure 13

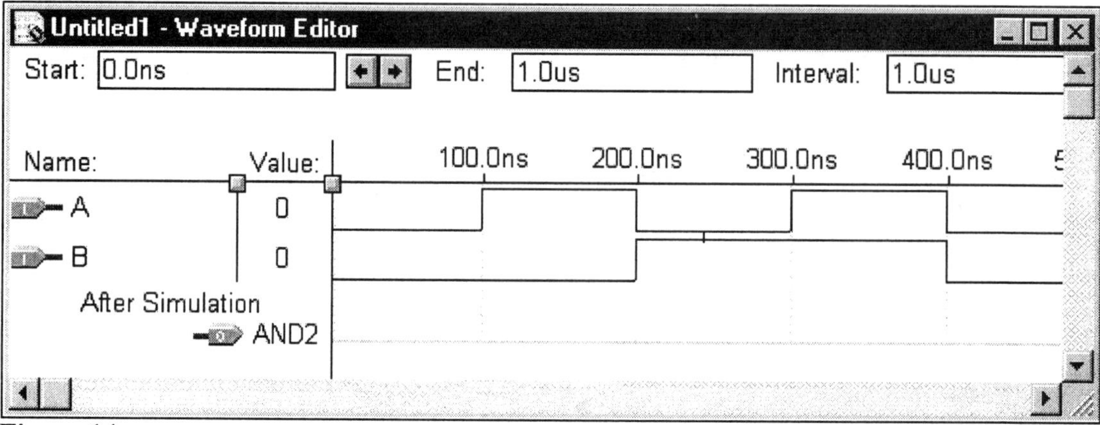

Figure 14

20. Sketch the output waveform from 0 to 400 ns in Figure 14 as displayed on the Graphic Editor.

21. Analyze the waveforms in Figure 14 and complete the function table (Table 1) for the 2 input AND gate.

B	A	X
0	0	0
0	1	0
1	0	0
1	1	1

Table 1

22. Write the Boolean expression to represent the 2-input AND gate.

X = AB

23. Explain when the output of the AND gate is a logic-HIGH, with respect to inputs A and B.

BOTH HAVE TO BE HIGH

24. Save the editor files to Drive A as **lab-1-1**, then exit the Graphic and Waveform Editors.

Part 2 Procedure

1. Open the Max+plus II software. Assign the project name **lab-1-2**, assign MAX7000S for the device family, and select the EPM7128SLC84 chip.

2. Open a new Graphic Editor file and construct the circuit shown in Figure 15.

3. Open a new Waveform Editor file, set Grid Size to 100 ns, and create the waveforms shown in Figure 15.

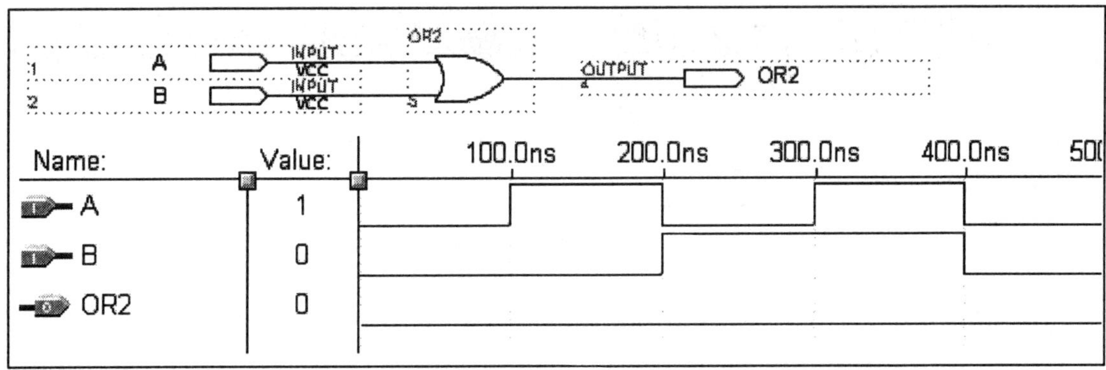

Figure 15

4. Click the Compile and Simulate buttons. Click **OK** in the Save As dialog boxes. Correct all errors before continuing.

5. Assuming zero errors, sketch the output waveform of the 2-input OR gate in Figure 16.

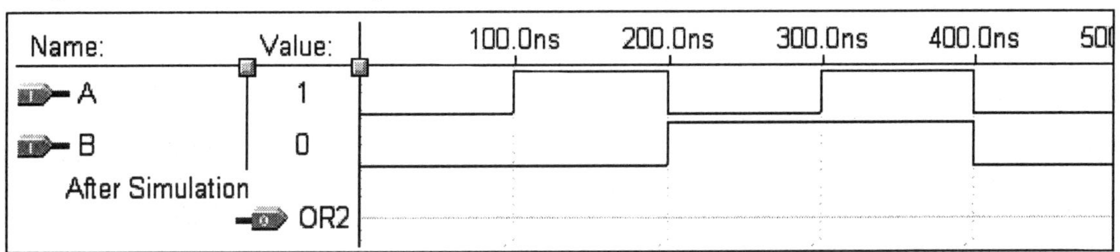

Figure 16

6. Analyze the waveforms in Figure 16 and complete the function table (Table 2) for the 2- input OR gate.

B	A	X
0	0	0
0	1	1
1	0	1
1	1	1

Table 2

7. Write the Boolean expression to represent the 2-input OR gate.

   X = A+B

8. Explain when the output of the OR gate is a logic-HIGH, with respect to inputs A and B.

   IF THERE'S A HIGH

9. Save the editor files to Drive A as **lab-1-2**, then exit the Graphic and Waveform Editors.

Part 3 Procedure

1. Open the Max+plus II software. Assign the project name **lab-1-3**, assign MAX7000S for the device family, and select the EPM7128SLC84 chip.

2. Open a new Graphic Editor file and construct the circuit shown in Figure 17.

3. Open a new Waveform Editor file, set Grid Size to 100 ns, and create the circuit and waveforms shown in Figure 17.

4. Click the Compile and Simulate buttons. Click **OK** in the Save As dialog boxes. Correct all errors before continuing.

5. Assuming zero errors, sketch the output waveform for the NOT gate in Figure 18.

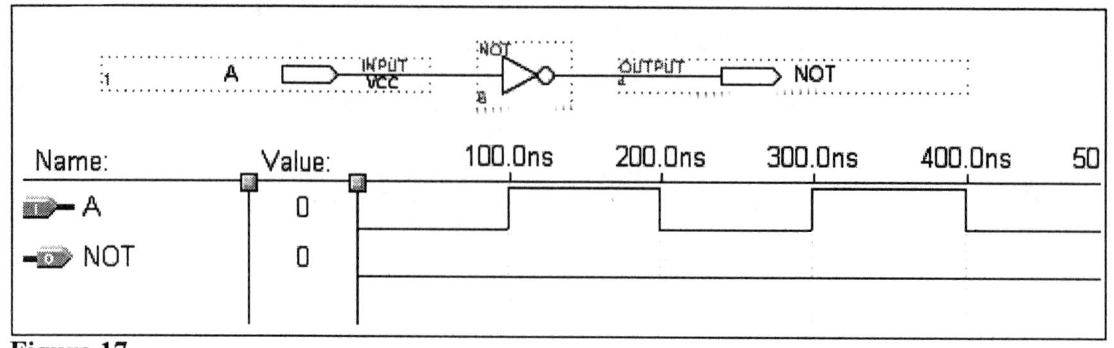

Figure 17

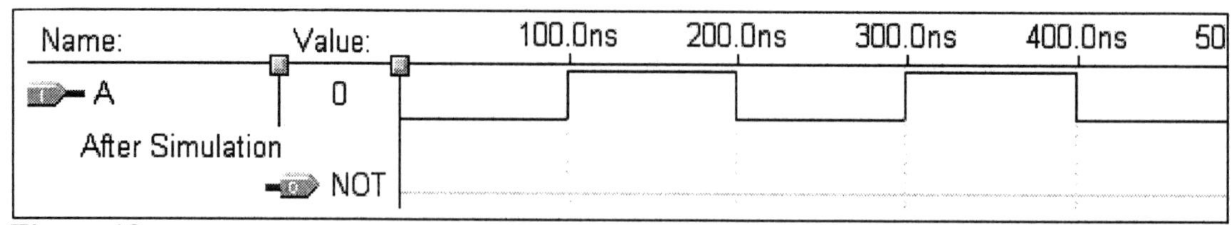

Figure 18

6. Analyze the waveforms in Figure 18 and complete the function table (Table 3) for the NOT gate.

A	X
0	1
1	0

Table 3

7. Write the Boolean expression to represent the NOT gate.

$X = \overline{A}$

8. Write a statement explaining output X with respect to input A of the NOT gate.

IF INPUT IS HIGH OUTPUT IS OPPOSITE

9. Save the editor files to Drive A as **lab-1-3**, then exit the Graphic and Waveform Editors.

Part 4 Procedure

1. Open the Max+plus II software. Assign the project name **lab-1-4**, assign MAX7000S for the device family, and select the EPM7128SLC84 chip.

2. Open a new Graphic Editor file and construct the circuit shown in Figure 19.

3. Open a new Waveform Editor file, set Grid Size to 100 ns, and create the waveforms shown in Figure 19.

4. Click the Compile and Simulate buttons. Correct all errors before continuing.

5. Assuming zero errors, draw the output waveform for the 2-input NAND gate in Figure 20.

6. Analyze the waveforms in Figure 20 and complete the function table (Table 4) for the NAND, NOR, XOR, and XNOR gates.

7. Write the Boolean expression to represent each logic gate with respect to inputs A and B.

NAND = ______________ NOR = ______________

XOR = ______________ XNOR = ______________

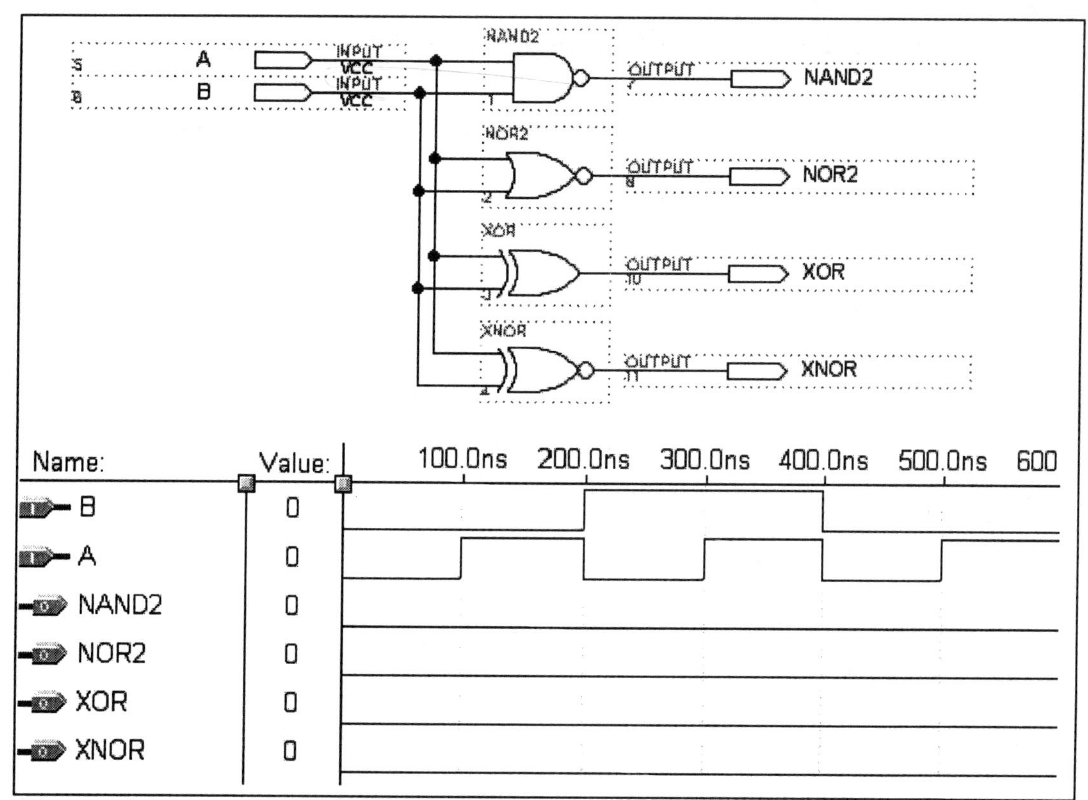

Figure 19

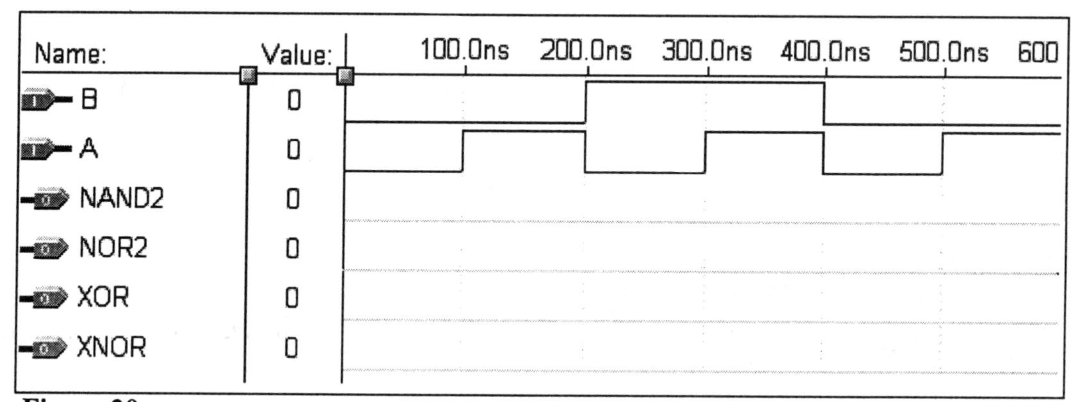

Figure 20

B	A	NAND	NOR	XOR	XNOR
0	0	____	____	____	____
0	1	____	____	____	____
1	0	____	____	____	____
1	1	____	____	____	____

Table 4

8. Write a statement explaining when the NAND output is a logic-HIGH with respect to inputs A and B.

____

9. Write a statement explaining when the NOR output is a logic-HIGH with respect to inputs A and B.

____

10. Write a statement explaining when the XOR output is a logic-HIGH with respect to inputs A and B.

____

11. Write a statement explaining when the XNOR output is a logic-HIGH with respect to inputs A and B.

____

12. Bring the Graphic Editor to the foreground. Place the mouse pointer in the upper-left side of your figure, then click and drag the mouse to the lower-right side of the figure, highlighting the entire figure. Select **File - Print** from the main menu. In the Pages block, select **Selected Area** and **Fit into 100% of one page**, then click **OK**. This method allows you to print only the selected work area.

13. An alternative method for printing a legible schematic is to select **File - Size**, then set the **Orientation** to Horizontal and select the **Sheet Size** to A: 9×11 in [59 x 77 GUs]. If you receive a "Sheet size is too small for drawing" error message, then select the next paper size listed, B:, C:, and so on. Once the paper size is selected, click the Print icon in the tool bar or select **File - Print** in the main menu.

14. Bring the Waveform Editor to the foreground. Select **File - Print Setup** in the main menu. Check **Landscape** in the Orientation box, then click **OK**. Select **File - Print - OK** from the main menu.

15. Demonstrate the waveforms printed to your instructor. Obtain the signature directly on the answer page for this lab.

16. Save the editor files to Drive A as **../work/lab-1-4**, then exit the Max+plus II software.

17. Create a cover page for the lab in the following format.
    - Your name and section in upper-right corner
    - Title of lab centered on page with instructor's name below the title
    - Today's date in the lower-right corner

18. Write a 1- to 2-page summary using Times New Roman 12-point font pertaining to the results obtained from this lab. Include the lab objectives in the opening paragraph of your summary, a single function table with logic gates AND, OR, NOT A, NAND, NOR, XOR, and XNOR as output column headings, embedded graphics of each logic gate and a brief logic statement for each gate. Include the modified VHDL code for Part 6 of the lab with a paragraph explaining the modifications.

19. Staple all papers for this lab in the following sequence, then submit the lab to your instructor for grading.
    - Cover page
    - Typed summary
    - The completed answer page for this lab
    - The hard copy of the Graphic Editor, **Part 4, Step 12**
    - The hard copy of the Waveform Editor, **Part 4, Step 14**
    - The hard copy of the VHDL_2.vhd text file after modifications

Lab 2: Boolean: Laws, Principles, and Rules

Objectives:

1. Verify Boolean laws using waveform analysis
2. Prove the 12 Boolean rules
3. Verify DeMorgan's principle
4. Demonstrate dual-function gates

Materials List:

- ♦ Max+plus II software by Altera Corporation
- ♦ University Board with CPLD (optional)
- ♦ Computer requirements:
  Minimum 486/66 with 8 MB RAM
- ♦ Floppy disk

Discussion:

All logic circuits can be expressed mathematically using Boolean equations. Boolean rules, principles, and theorems are used to describe relationships of logic gates and as a means of expressing, analyzing, and simplifying complex logic circuits.

Since the development of sophisticated computer software, analyzing and simplifying circuits has become easy. You develop the expression for a circuit from a function table or circuit, then let the computer software simplify the circuit logic. Hence, lengthy derivations and complex circuit reduction by paper and pencil are rarely necessary. Standardized circuits become macros placed in libraries used by the design tool.

This lab utilizes the Max+plus II software to illustrate and graphically analyze the following Boolean identities, theorems, and laws.

Commulative law: Rearranging the order of the expression

$A \bullet B = B \bullet A$ and $A + B = B + A$

Associative law: Grouping terms or variables using parenthesis.

$A \bullet B \bullet C = A\,(B \bullet C) = (A \bullet B)C$
$A + B + C = A + (B + C) = (A + B) + C$

Distributive law: The process of ANDing a single variable over each term of an OR expression or factoring out a common term from an OR expression.

$A(B + C) = AB + AC$

The twelve rules of Boolean Algebra are:

$A + 0 = A$	$A \bullet 0 = 0$	$\overline{\overline{A}} = A$
$A + 1 = 1$	$A \bullet 1 = A$	$A + AB = A$
$A + A = 1$	$A \bullet A = A$	$(A + B)(A + C) = A + BC$
$A + \overline{A} = 1$	$A \bullet \overline{A} = 1$	$A + \overline{A}B = A + B$

Gate Duality

Gate		Double Negate	DeMorgans	Statement
OR	$A + B$	$\overline{\overline{A + B}}$	$\overline{\overline{A} \bullet \overline{B}}$	Active-LOW input NAND
NOR	$\overline{A + B}$		$\overline{A} \bullet \overline{B}$	Active-LOW input AND
AND	$A \bullet B$	$\overline{\overline{A + B}}$	$\overline{\overline{A} + \overline{B}}$	Active-LOW input NOR
NAND	$\overline{A \bullet B}$		$\overline{A} + \overline{B}$	Active-LOW input OR

Part 1 Procedure

1. Open the Max+plus II software. Assign the project name **boolean1** and assign EPM7128SLC84-7 as the device.

2. Open a new Graphic Editor file and construct the circuits shown in Figure 1.

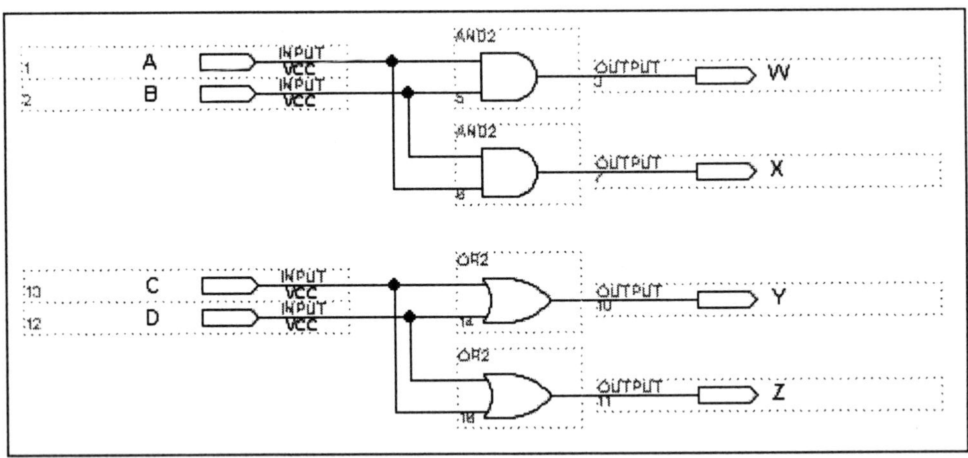

Figure 1

3. Open a new Waveform Editor file and create the waveforms shown in Figure 2.

4. Click the Compile and Simulate buttons. Correct all errors before continuing.

5. Sketch output waveforms, W, X, Y, and Z in Figure 3 that were created by the software for Figure 1.

6. Write the Boolean expression to represent the output of each AND gate in Figure 1.

   W = __________ X = __________

7. Based on waveforms W and X, does the order of the variables ANDed together alter the output of the AND gate? (Yes/No)

8. Which law, theorem, or rule is illustrated by the circuits in Figure 1? ______________

9. Write the Boolean expression to represent the output of each OR gate in Figure 1.

   Y = __________ Z = __________

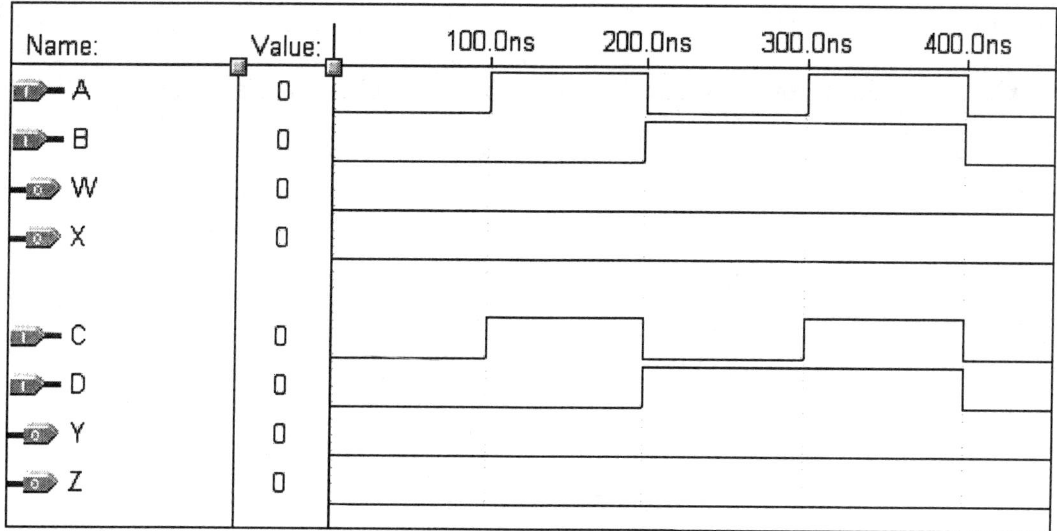

Figure 2

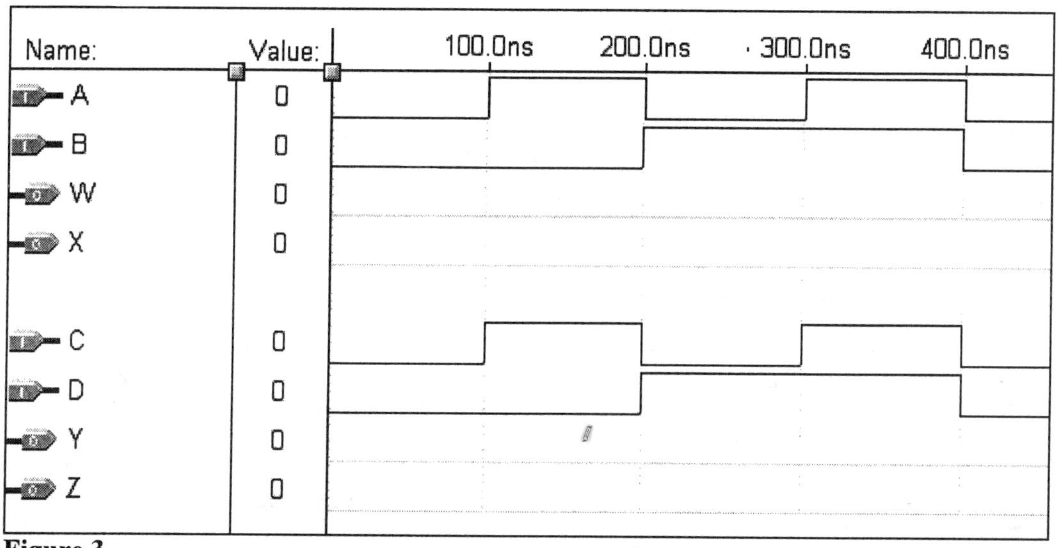

Figure 3

10. Based on waveforms Y and Z, does the order of the variables ORed together alter the output of the OR gate? (Yes/No)

11. Complete the folowwing function tables (Tables 1 and 2) for outputs W, X, Y, and Z based on Figure 3.

B	A	W	X
0	0	___	___
0	1	___	___
1	0	___	___
1	1	___	___

Table 1

D	C	Y	Z
0	0	___	___
0	1	___	___
1	0	___	___
1	1	___	___

Table 2

12. Save the editor files to Drive A as **boolean1**, then exit the Graphic and Waveform Editors.

Part 2 Procedure

1. Open the Max+plus II software. Assign the project name as **boolean2** and assign EPM7128SLC84-7 as the device.

2. Open a new Graphic Editor file and construct the circuits shown in Figure 4.

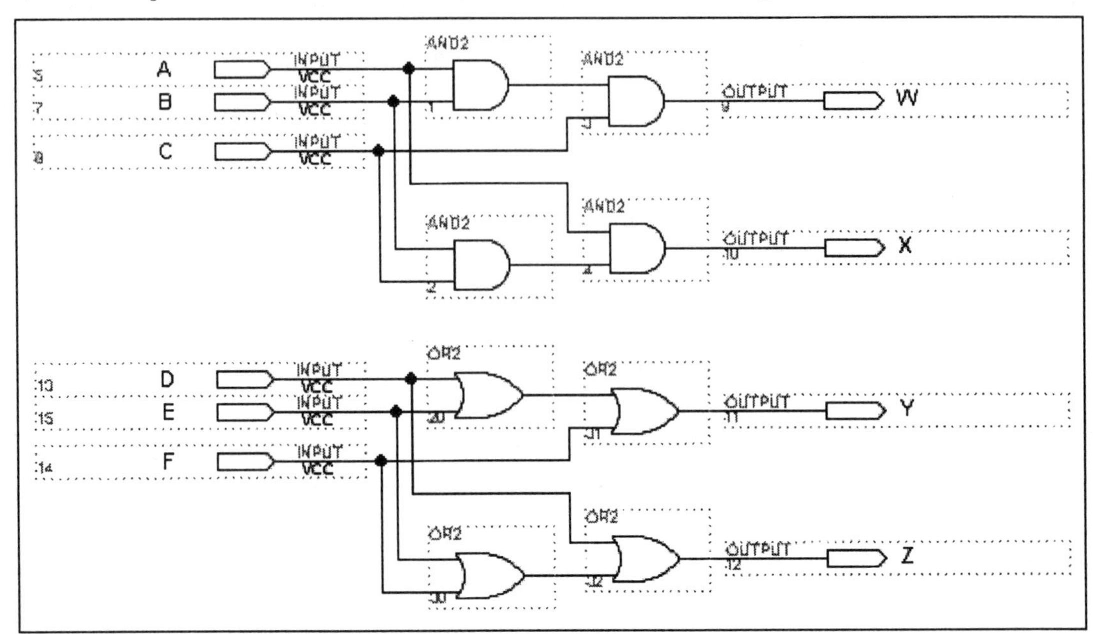

Figure 4

3. Open a new Waveform Editor file and create the waveforms shown in Figure 5.

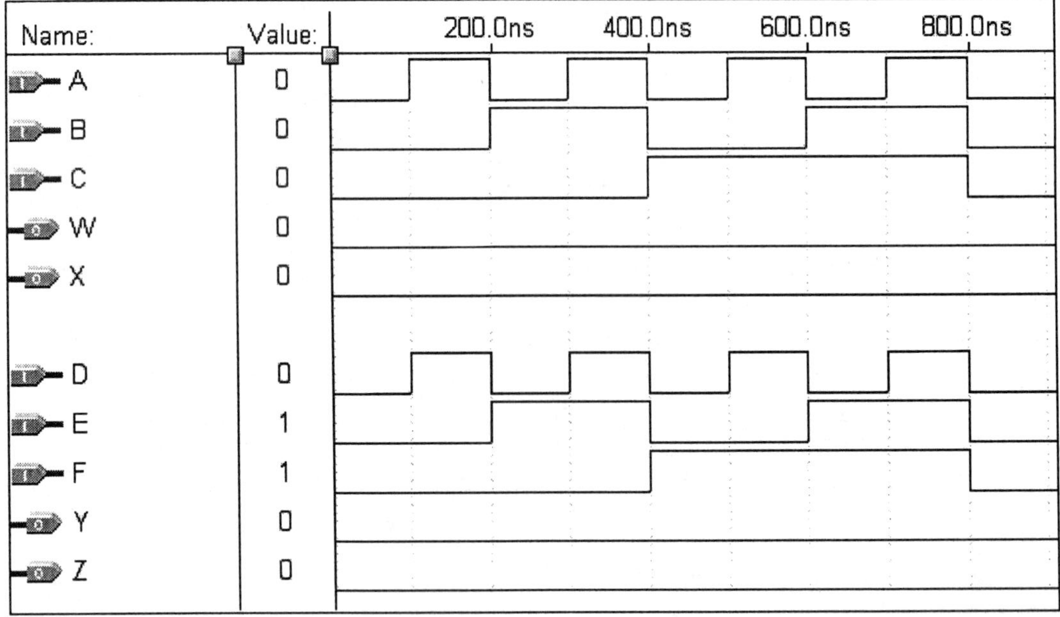

Figure 5

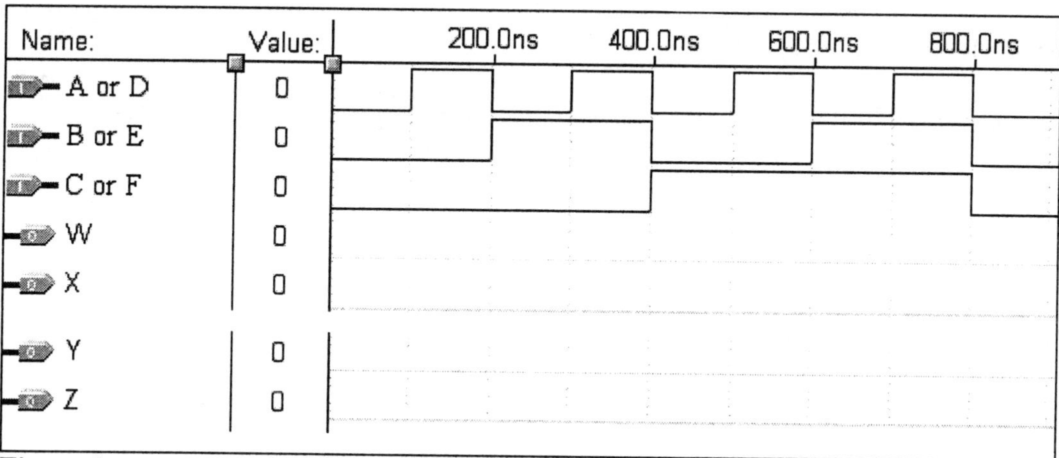

Figure 6

4. Click the Compile and Simulate buttons. Correct all mistakes before continuing.

5. Sketch output waveforms W, X, Y, and Z in Figure 6 that were created by the software for Figure 4.

6. Does output W = output X in Figure 4? (Yes/No)

7. Does output Y = output Z in Figure 4? (Yes/No)

8. Write the Boolean expression to represent output W and output X in Figure 4.

   W = __________ X = __________

9. Write the Boolean expression to represent output Y and output Z in Figure 4.

   Y = __________ Z = __________

10. Which law, theorem, or rule is illustrated by the circuits in Figure 4? ____________________

11. Complete the following function tables (Tables 3 and 4) for outputs W, X, Y, and Z of Figure 4.

C	B	A	W	X
0	0	0	___	___
0	0	1	___	___
0	1	0	___	___
0	1	1	___	___
1	0	0	___	___
1	0	1	___	___
1	1	0	___	___
1	1	1	___	___

Table 3

F	E	D	Y	Z
0	0	0	___	___
0	0	1	___	___
0	1	0	___	___
0	1	1	___	___
1	0	0	___	___
1	0	1	___	___
1	1	0	___	___
1	1	1	___	___

Table 4

12. Save the editor files to Drive A as **boolean2**, then exit the Graphic and Waveform Editors.

Part 3 Procedure

1. Open the Max+plus II software. Assign the project name as **boolean3** and assign EPM7128SLC84-7 as the device.

2. Open a new Graphic Editor file and construct the circuits shown in Figure 7.

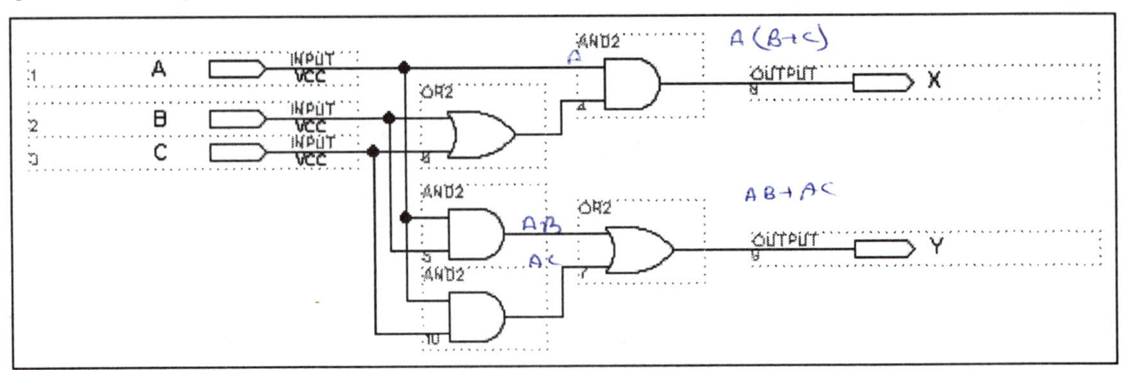

Figure 7

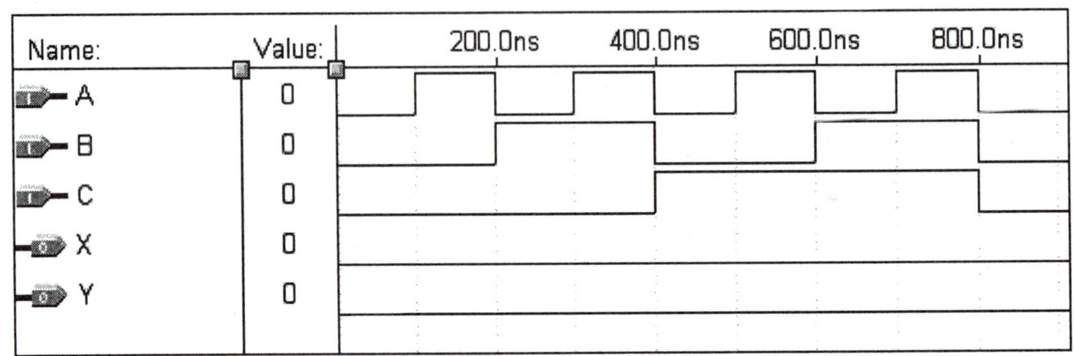

Figure 8

3. Open a new Waveform Editor file and create the waveforms shown in Figure 8.

4. Click the Compile and Simulate buttons. Correct all mistakes before continuing.

5. Sketch output waveforms X and Y in Figure 9 that were created by the software for Figure 7.

6. Write the Boolean expression for each output in Figure 7.

   X = ______________________   Y = ______________________

7. Which law, theorem, or rule is illustrated by the circuits in Figure 7? ____________________

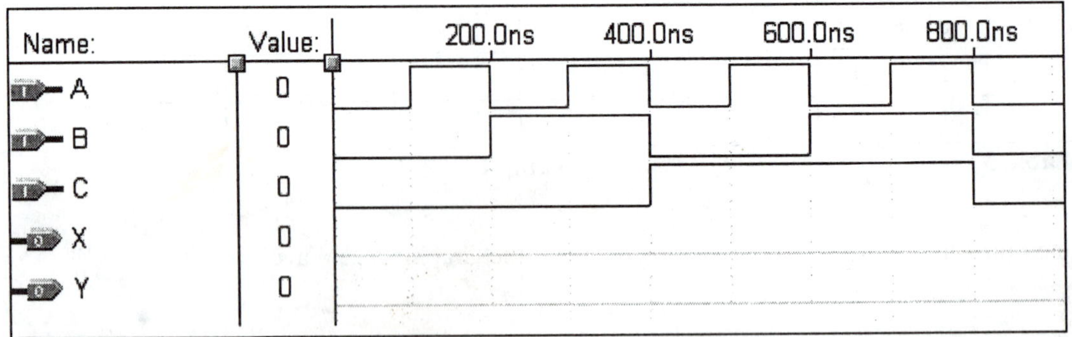

Figure 9

8. Complete the following function tables (Tables 5a and 5b) for outputs X and Y of Figure 7.

C	B	A	X	Y
0	0	0	___	___
0	0	1	___	___
0	1	0	___	___
0	1	1	___	___

Table 5a

C	B	A	X	Y
1	0	0	___	___
1	0	1	___	___
1	1	0	___	___
1	1	1	___	___

Table 5b

9. Save the editor files to Drive A as **boolean3**, then exit the Graphic and Waveform Editors.

Part 4 Procedure

1. Open the Max+plus II software. Assign the project name as **boolean4** and assign EPM7128SLC84-7 as the device.

2. Open a new Graphic Editor file and construct the circuits shown in Figure 10.

3. Open a new Waveform Editor file and create the waveforms shown in Figure 11.

4. Click the Compile and Simulate buttons. Correct all mistakes before continuing.

5. Sketch output waveforms W and Y in Figure 12 that were created by the software for Figure 10.

6. Examine waveforms W and X. Does output W = output X? (Yes/No)

7. Write the Boolean expressions for output W and output X with respect to inputs A and B.

   W = ________________ X = ________________

8. What law, theorem, or rule was demonstrated by the circuit consisting of inputs A and B and outputs W and X? (Proper name) ______________________________

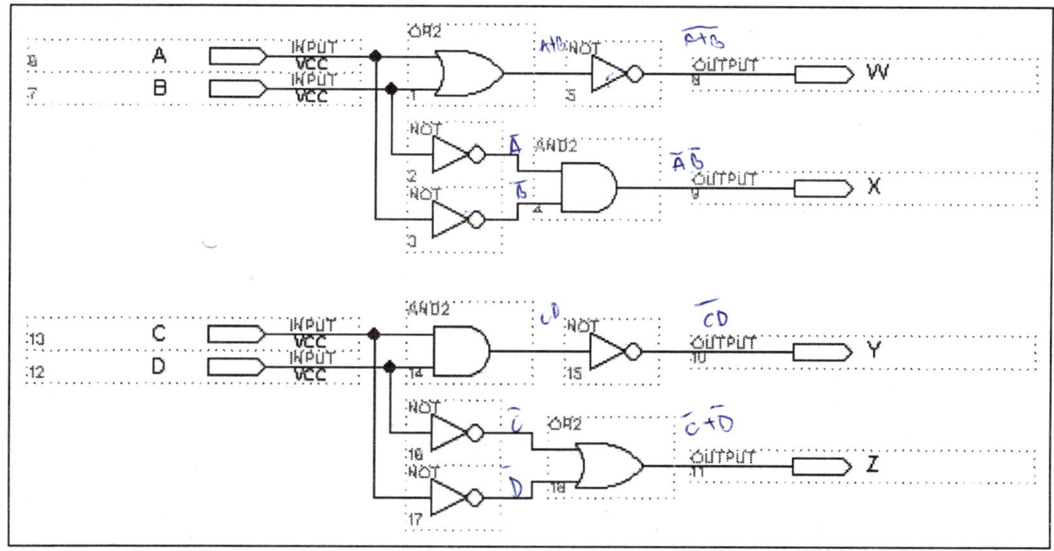

Figure 10

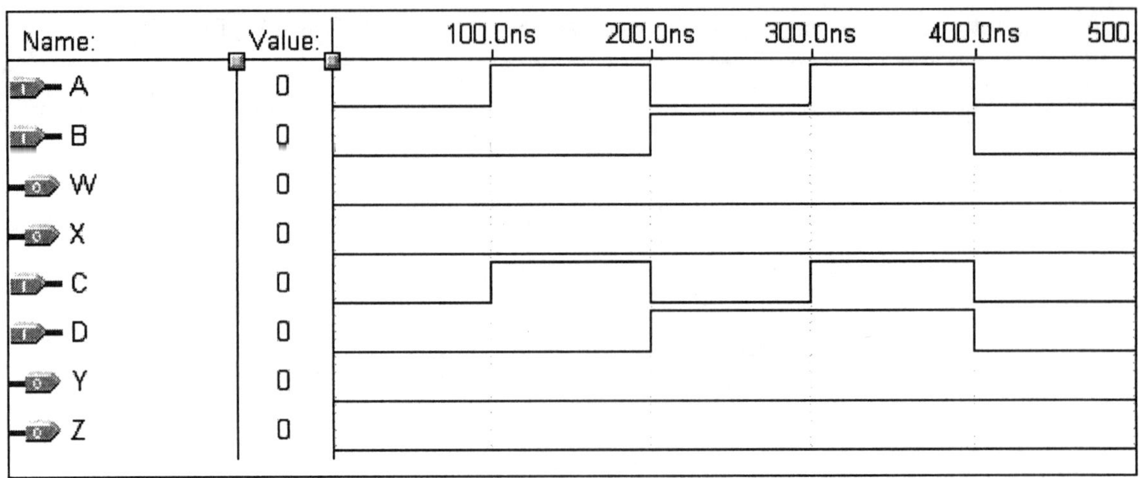

Figure 11

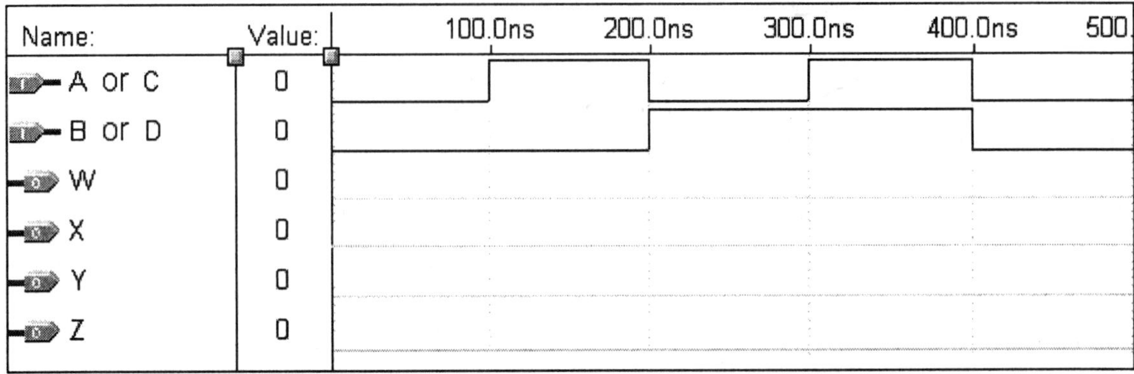

Figure 12

9. Examine waveforms Y and Z. Does output Y = output Z? (Yes/No)

10. Write the Boolean expressions for output Y and output Z with respect to inputs A and B.

 Y = ________________ Z = ________________

11. What law, theorem, or rule was demonstrated by the circuit consisting of inputs C and D and outputs Y and Z? (Proper name) ____________________________

12. Complete the following function tables (Tables 6 and 7) for the circuits in Figure 10.

B	A	W	X
0	0	___	___
0	1	___	___
1	0	___	___
1	1	___	___

Table 6

D	C	Y	Z
0	0	___	___
0	1	___	___
1	0	___	___
1	1	___	___

Table 7

13. Save the editor files to Drive A as **boolean4**, then exit the Graphic and Waveform Editors.

Part 5 Procedure

1. Open the Max+plus II software. Assign the project name as **boolean5** and assign EPM7128SLC84-7 as the device.

2. Open a new Graphic Editor file and construct the circuits shown in Figure 13.

3. Open a new Waveform Editor file and create Waveform A but leave Waveform A a logic-LOW.

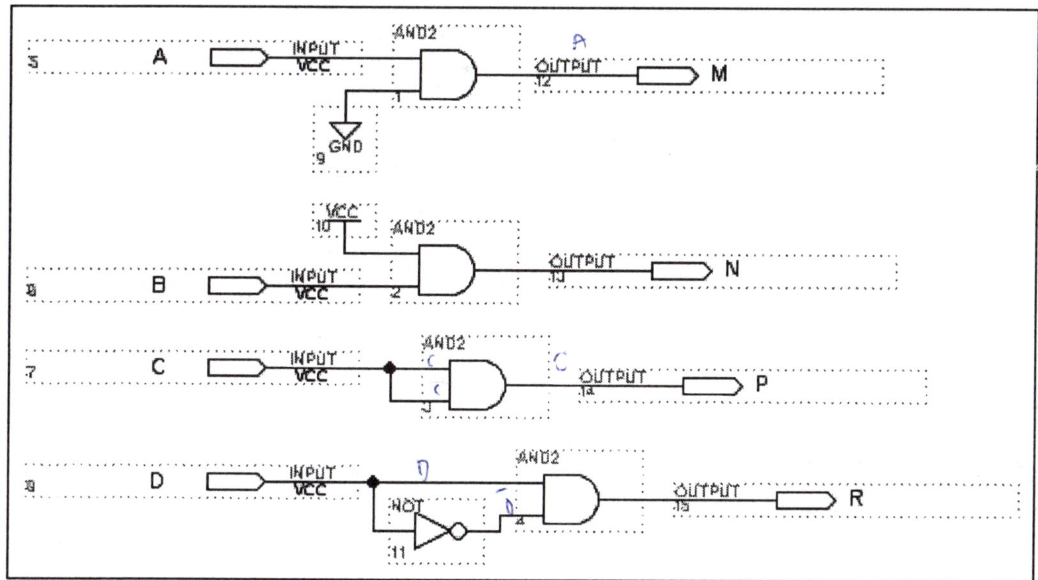

Figure 13

4. Highlight Waveform A shown in Figure 14 by single clicking on the left mouse button. Set the grid size in the **Options** menu to 100 ns.

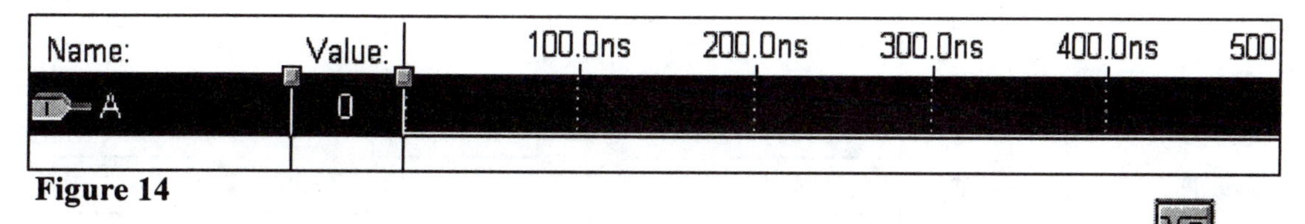

Figure 14

5. Click the "Clock" button in the Draw tool bar. The Clock button is shown in Figure 15 (Right).

Figure 15

6. Note the Overwrite Count Value dialog box (Figure 16) identifies the pulse width as 100 ns (Count Every). Click **OK**. The software generates a square wave from 0.0 ns to 1.0 μs. Figure 17 shows what Waveform A should look like.

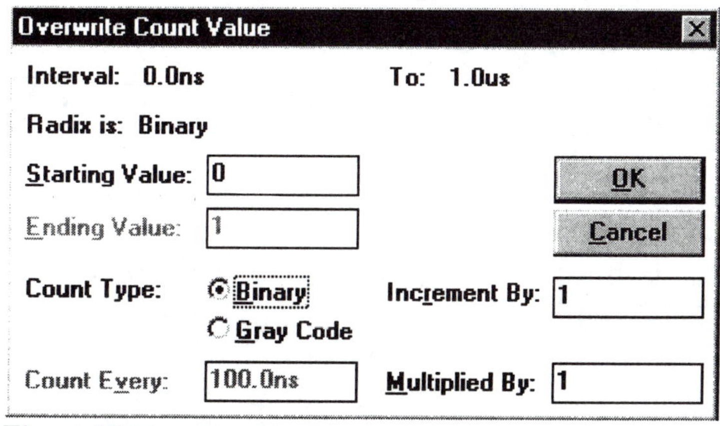

Figure 16

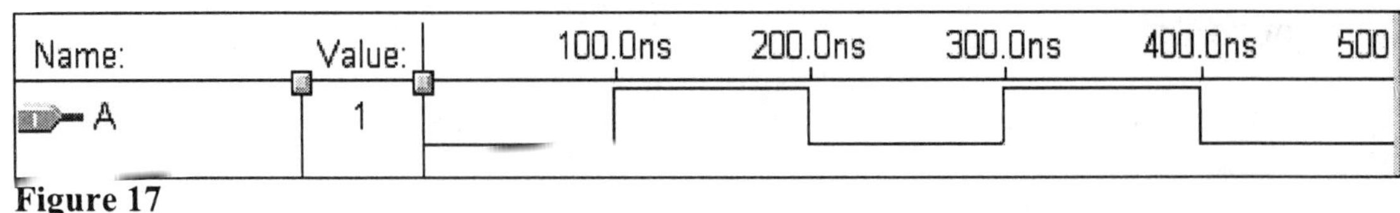

Figure 17

7. Create Waveform B, then with Waveform B highlighted, click on the Clock button (Figure 15) in the Draw tool bar. Set **Multiplied By:** in the Overwrite Count Value dialog box (see Figure 16) to 2. Click **OK**.

8. Create the rest of the waveforms shown in Figure 18. Be sure to create Waveforms C and D using the Clock button in the Draw tool bar. Waveforms M, N, P, and R are outputs.

9. Click the Compile and Simulate buttons. Correct all mistakes before continuing.

10. In Figure 19, sketch output waveforms M, N, P, and R of the respective gates in Figure 14.

11. Inspect the output waveforms of each gate with respect to the corresponding inputs. Write the Boolean identity that represents output M with respect to input A.

M = ______________________________

12. Write the Boolean identity that represents output N with respect to input B. N = ____________

13. Write the Boolean identity that represents output P with respect to input C. P = ____________

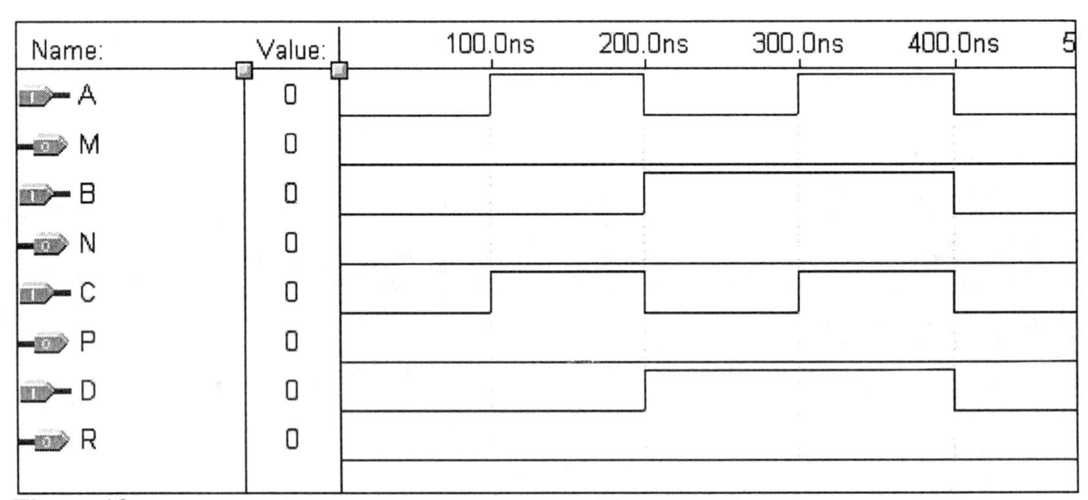

Figure 18

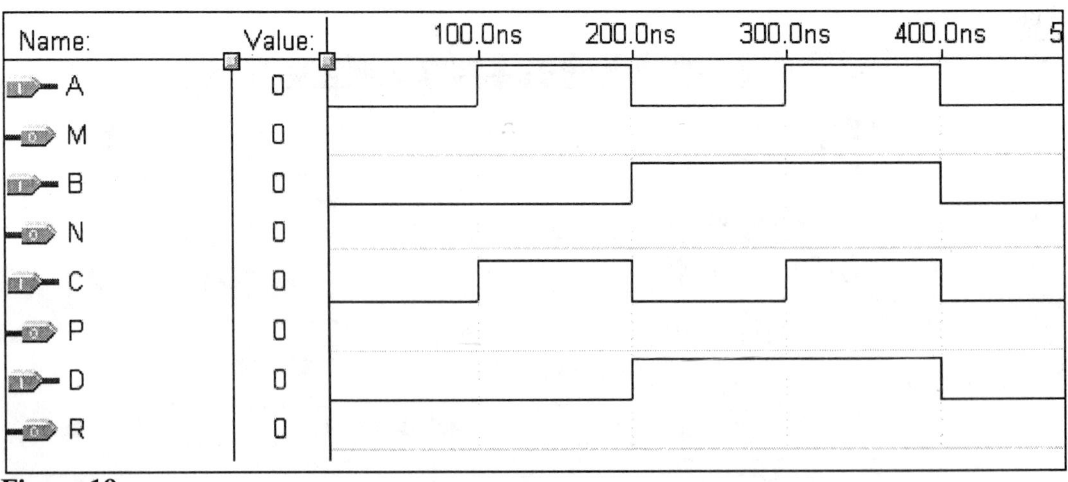

Figure 19

14. Write the Boolean identity that represents output R with respect to input D. R =_______________

15. Complete the following function tables (Tables 8 through 11) for the circuits shown in Figure 14.

A	0v	M
0	0	___
1	0	___

Table 8

B	5v	N
0	1	___
1	1	___

Table 9

C	C	P
0	0	___
1	1	___

Table 10

D	$\overline{D}$	R
0	___	___
1	___	___

Table 11

16. Add the circuits shown in Figure 20 to your Graphics Editor file, aligning input E below input D.

17. Add the waveforms shown in Figure 21 to the Waveform Editor below Waveform R.

18. Click the Compile and Simulate buttons. Correct all errors before continuing.

19. Sketch output waveforms S, T, U, and V in Figure 22 that were generated by the software for Figure 20.

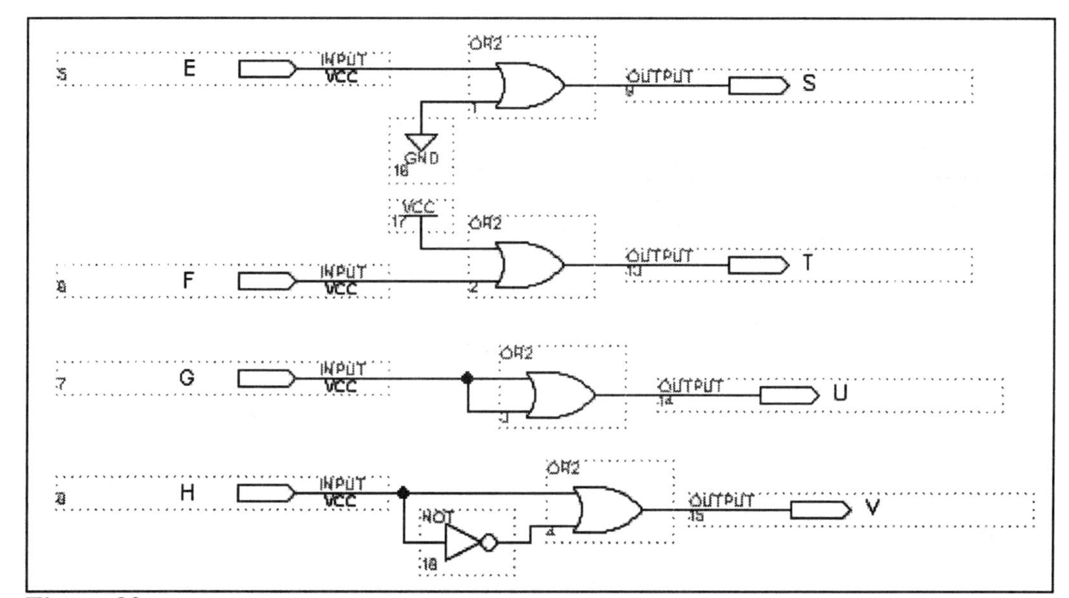

Figure 20

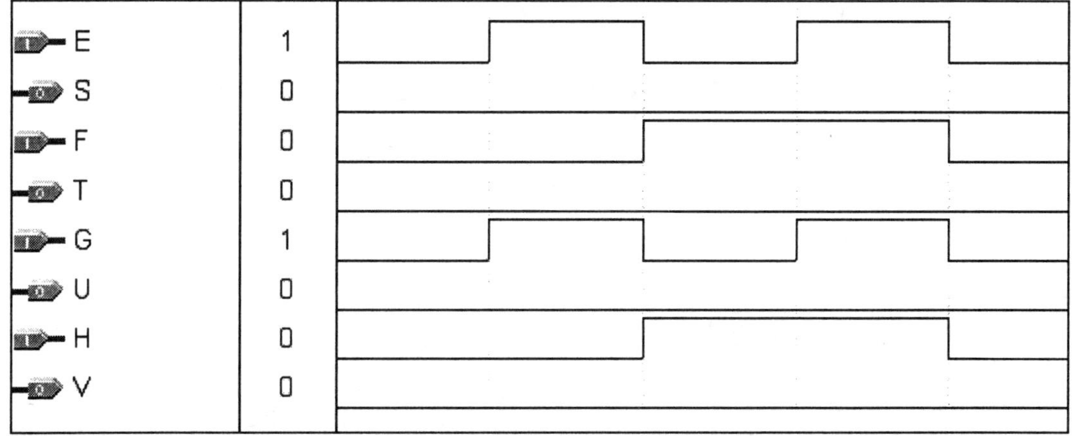

Figure 21

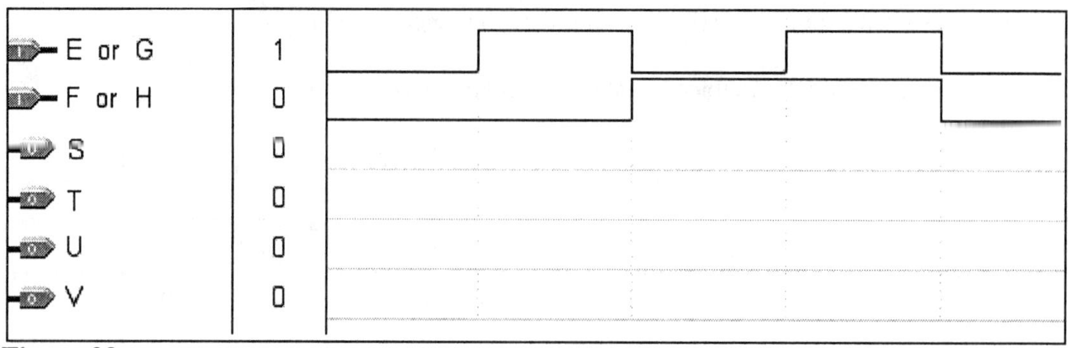

Figure 22

20. Inspect the output waveforms of each gate with respect to the respective inputs. Write the Boolean identity that represents output S with respect to input E.

S = ________________

21. Write the Boolean identity that represents output T with respect to input F. T = ________________

22. Write the Boolean identity that represents output U with respect to input G. U = ________________

23. Write the Boolean identity that represents output V with respect to input H. V = ________________

24. Complete the following function tables (Tables 12 through 15) for the circuits shown in Figure 20.

E	0v	S
0	0	___
1	0	___

Table 12

F	5v	T
0	1	___
1	1	___

Table 13

G	G	U
0	0	___
1	1	___

Table 14

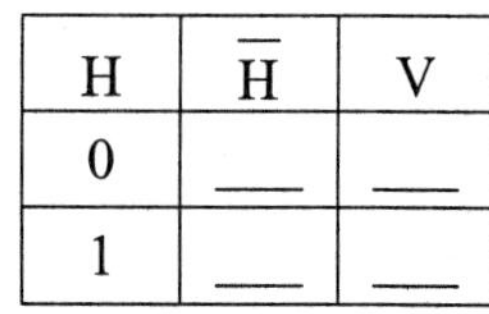

H	$\overline{H}$	V
0	___	___
1	___	___

Table 15

25. Save the editor files to Drive A as **boolean5**, then exit the Graphic and Waveform Editors.

Part 6 Procedure

1. Open the Max+plus II software. Assign the project name as **boolean6** and assign EPM7128SLC84-7 as the device.

2. Open a new Graphic Editor file and construct the circuits shown in Figure 23.

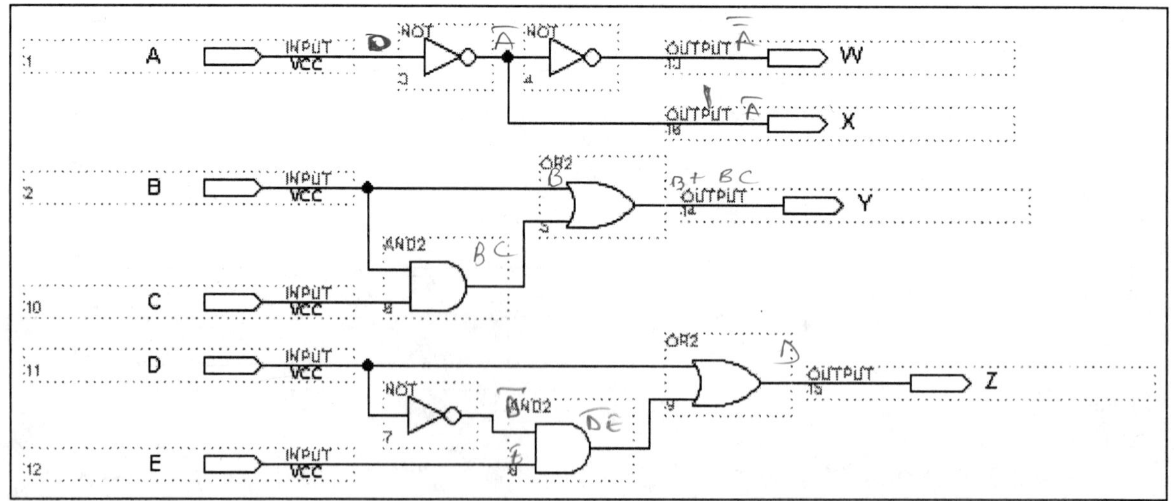

Figure 23

3. Open a new Waveform Editor file and create the waveforms shown in Figure 24.

4. Click the Compile and Simulate buttons. Correct all errors before continuing.

5. For Figure 25, sketch output waveforms W, X, Y, and Z in Figure 23.

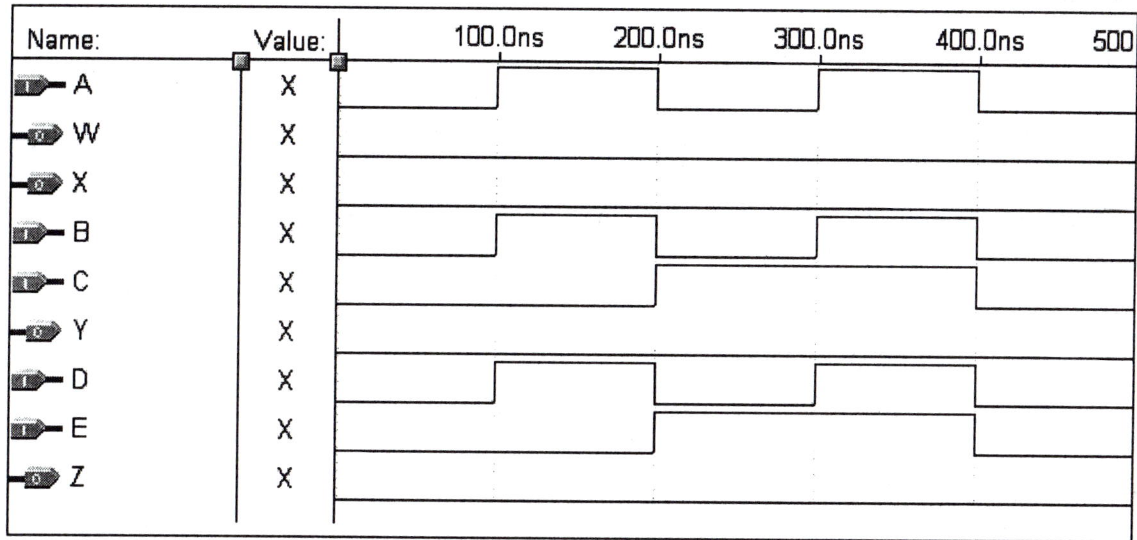

Figure 24

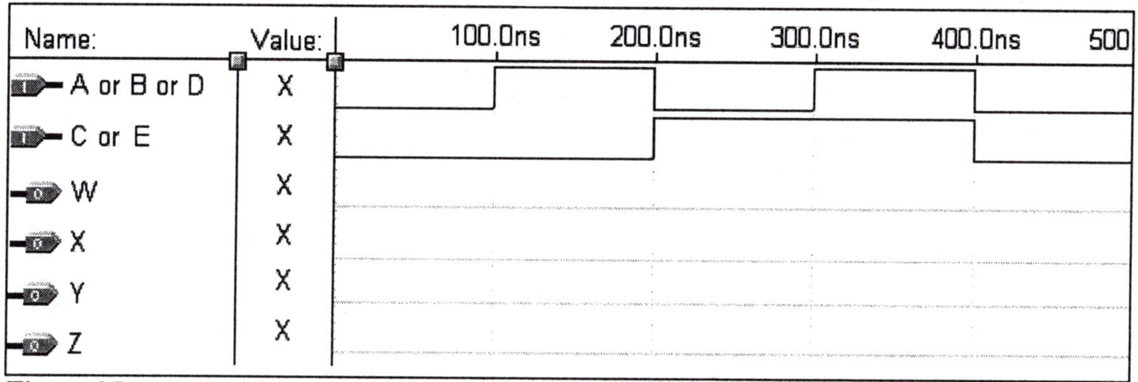

Figure 25

6. Write the Boolean identity that represents output W with respect to input A. W = A

7. Write the Boolean identity that represents output X with respect to input A. X = $\overline{A}$

8. Write the Boolean identity that represents output Y with respect to inputs B and C. Y = $B + BC$

   $1 + 1 \cdot 0$

9. Write the Boolean identity that represents output Z with respect to inputs D and E. Z = $D + \overline{D}E$

10. Complete the following function tables (Tables 16 through 18) for the circuits shown in Figure 23.

C	X	W
0	1	0
1	0	1

Table 16

B	C	Y
0	0	0
0	1	0
1	0	1
1	1	1

Table 17

D	E	Z
0	0	0
0	1	1
1	0	1
1	1	1

Table 18

11. Save the editor files to Drive A as **boolean6**, then exit the Graphic and Waveform Editors.

Part 7 Procedure

1. Open the II software. Assign the project name **boolean7** and assign EPM7128SLC84-7 as the device.

2. Open a new Graphic Editor file and construct the circuit shown in Figure 26 using the BAND2, BNAND2, BOR2, and BNOR2 gate symbols in the **prim** directory.

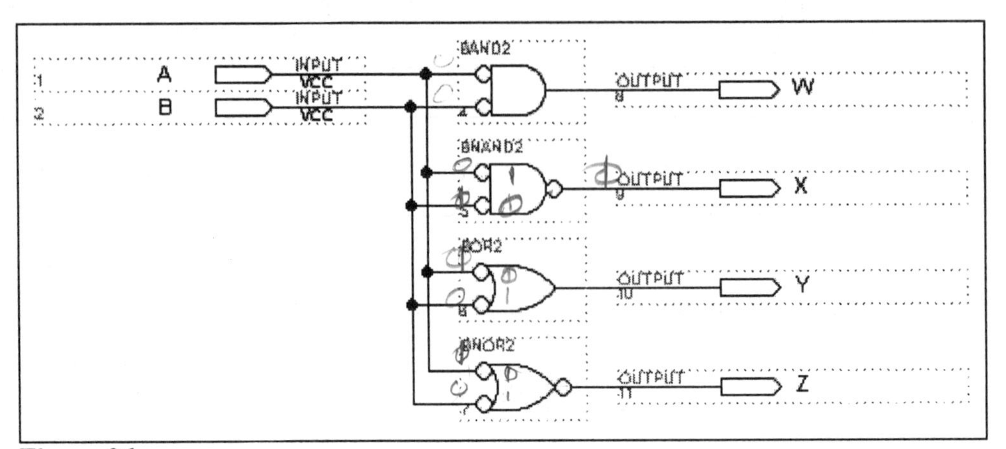

Figure 26

Name:	Value:
A	X
B	X
W	X
X	X
Y	X
Z	X

100.0ns 200.0ns 300.0ns 400.0ns 500

Figure 27

3. Open a new Waveform Editor file and create the waveforms shown in Figure 27.

4. Click the Compile and Simulate buttons. Correct all errors before continuing.

5. In Figure 28, sketch output waveforms W, X, Y, and Z for the circuit in Figure 26.

6. Examine output W with respect to inputs A and B. The active-LOW input AND gate is equivalent to what basic logic gate? Circle one: AND OR NAND NOR

7. Examine output X with respect to inputs A and B. The active-LOW input NAND gate is equivalent to what basic logic gate? Circle one: AND OR NAND NOR

   Note: The active-LOW input NAND gate may also be called the active-LOW input, active-LOW output AND gate.

8. Examine output Y with respect to inputs A and B. The active-LOW input OR gate is equivalent to what basic logic gate? Circle one: AND OR NAND NOR

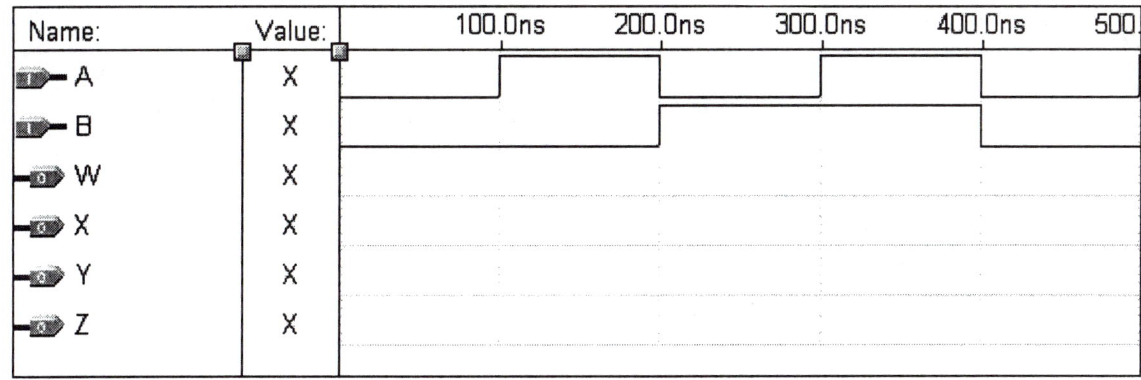

Figure 28

9. Examine output Z with respect to inputs A and B. The active-LOW input NOR gate is equivalent to what basic logic gate? Circle one: AND OR NAND NOR

   Note: The active-LOW input NOR gate may also be called the active-LOW input, active-LOW output OR gate.

10. Complete the following function tables (Tables 19 through 22) for the gates shown in Figure 26.

B	A	W
0	0	1
0	1	0
1	0	0
1	1	0

Table 19

B	A	X
0	0	0
0	1	1
1	0	1
1	1	1

Table 20

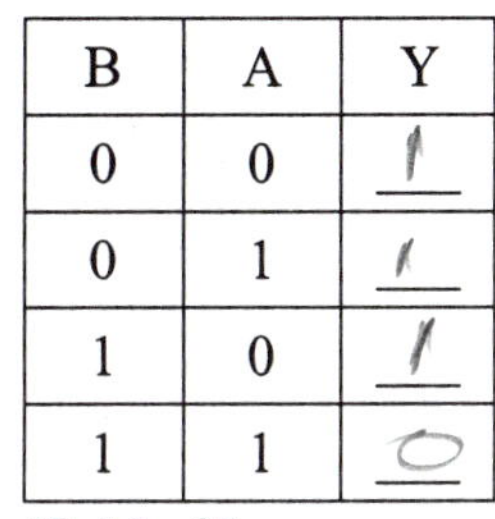

B	A	Y
0	0	1
0	1	1
1	0	1
1	1	0

Table 21

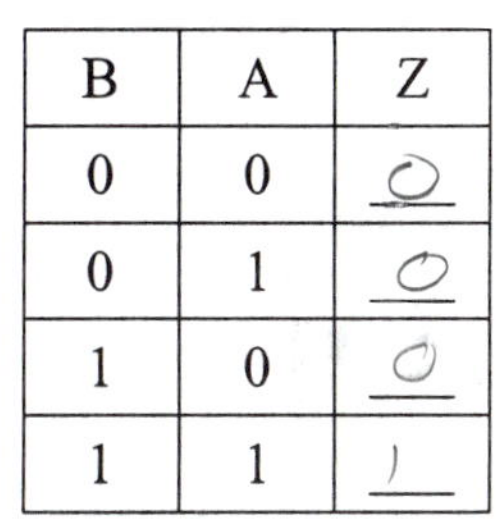

B	A	Z
0	0	0
0	1	0
1	0	0
1	1	1

Table 22

11. Obtain a hard copy of the Graphic and Waveform Editor displays. Print the waveforms in landscape mode. Label these hard copies **Part 7, Step 11A** and **Part 7, Step 11B**.

12. Demonstrate the waveforms printed to your instructor. Obtain the signature of approval on the answer page.

13. Save the editor files to Drive A as **boolean7**, then exit the Graphic and Waveform Editors.

14. Create a cover page for the lab in the following format.
    - Name and section in upper-right corner
    - Title of lab centered on the page with instructor's name below the title
    - Today's date in the lower-right corner

15. Write a 1- to 2-page technical summary pertaining to the results obtained from this lab. Paragraph 1 should stage the objectives of the lab. Paragraphs two and three should provide examples of the three laws and define enable and inhibit as applied to basic logic gates. Paragraphs 4 and 5 should include Figure 26 and Figure 28 (after simulation) as embedded graphics in your summary. Write a paragraph explaining each of these figures.

16. Staple all papers for this lab in the following sequence, then submit the lab to your instructor for grading.
    - Cover page
    - Typed summary
    - The completed answer page for this lab
    - Hard copy of the Graphic Editor, **Part 7, Step 11A**
    - Hard copy of the Waveform Editor, **Part 7, Step 11B**

Figure 22

20. S = ____________

21. T = ____________

22. U = ____________

23. V = ____________

Part 6

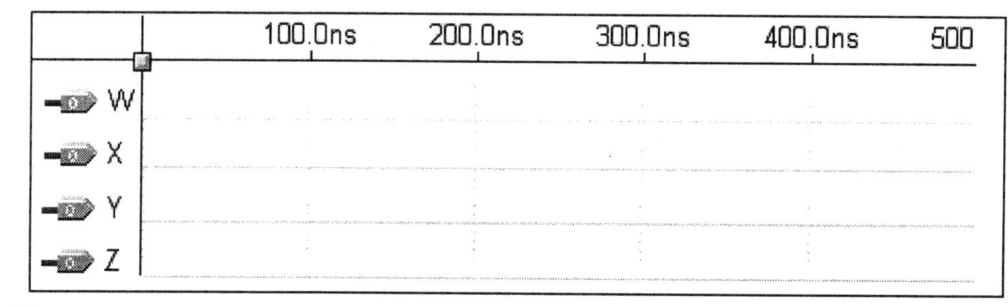

Figure 25

6. W = ____________

7. X = ____________

8. Y = ____________

9. Z = ____________

C	X	W
0	___	___
1	___	___

Table 17

B	C	Y
0	0	___
0	1	___
1	0	___
1	1	___

Table 18

D	E	Z
0	0	___
0	1	___
1	0	___
1	1	___

Table 19

Part 7

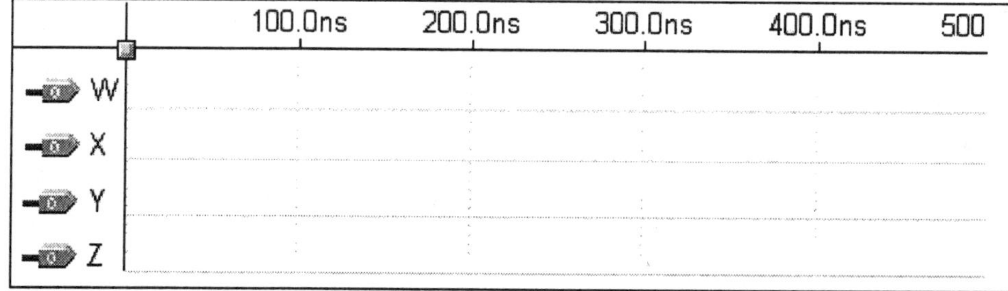

Figure 28

6. ____________

7. ____________

8. ____________

9. ____________

B	A	W
0	0	___
0	1	___
1	0	___
1	1	___

Table 20

B	A	X
0	0	___
0	1	___
1	0	___
1	1	___

Table 21

B	A	Y
0	0	___
0	1	___
1	0	___
1	1	___

Table 22

B	A	Z
0	0	___
0	1	___
1	0	___
1	1	___

Table 23

15. Demonstrated to: ______________________ Date: ____________

Grade: ________

Lab 3: Combinational Logic Circuits

Objectives:

1. Analyze output waveforms of different combinational logic circuits
2. Complete function tables for these combinational logic circuits
3. Write Boolean expressions for each combinational logic circuit

Materials List:

- Max+plus II software by Altera Corporation
- University Board with CPLD (optional)
- Computer requirements:
  - Minimum 486/66 with 8 MB RAM
- Floppy disk

Discussion:

All circuits, no matter how complex, will have multiple inputs and at least one output. The inputs and outputs are either "Data" or "Control." The Data is information in binary code, such as the data typed while using a word processor or data that is printed on a page. The Control inputs will direct the data flow through a circuit.

The most fundamental control circuit is the 2-input (AND, OR, NAND, NOR, XOR) gate shown in Figure 1A. The Control input will inhibit the data flow (Figure 1B) or enable the data flow (Figure 1C). As the circuit gets more complex (Figure 1D) the basic function of the control inputs remains the same to control the data flow through the circuit.

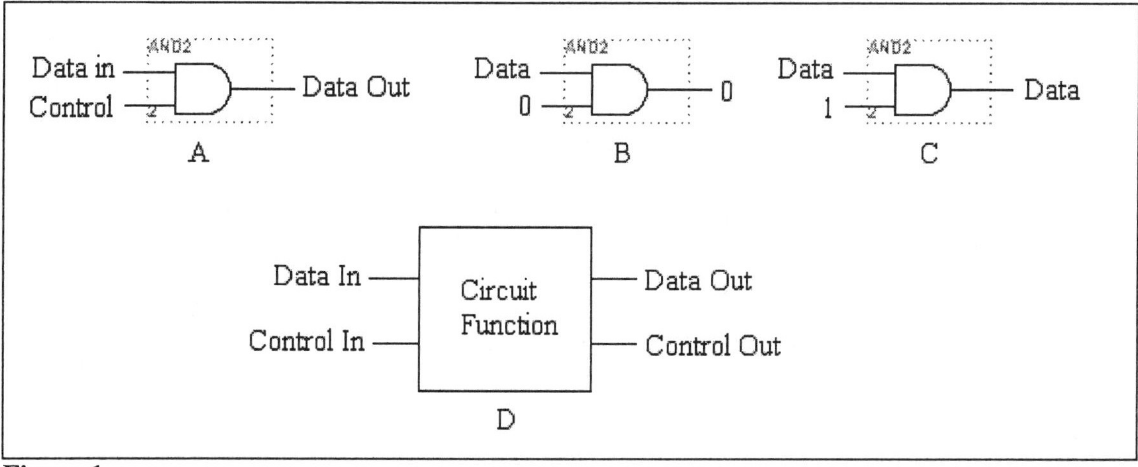

Figure 1

Several commonly used control circuits will be demonstrated in this lab. Each circuit will be constructed and simulated using the Max+plus II software. Waveforms will be analyzed, function tables will be created, and Boolean expressions will be developed from the circuits in this lab.

Part 1 Procedure

1. Open the Max+plus II software. Assign the project name **comb1**, MAX7000S as the device family, and EPM7128SLC84 as the device.

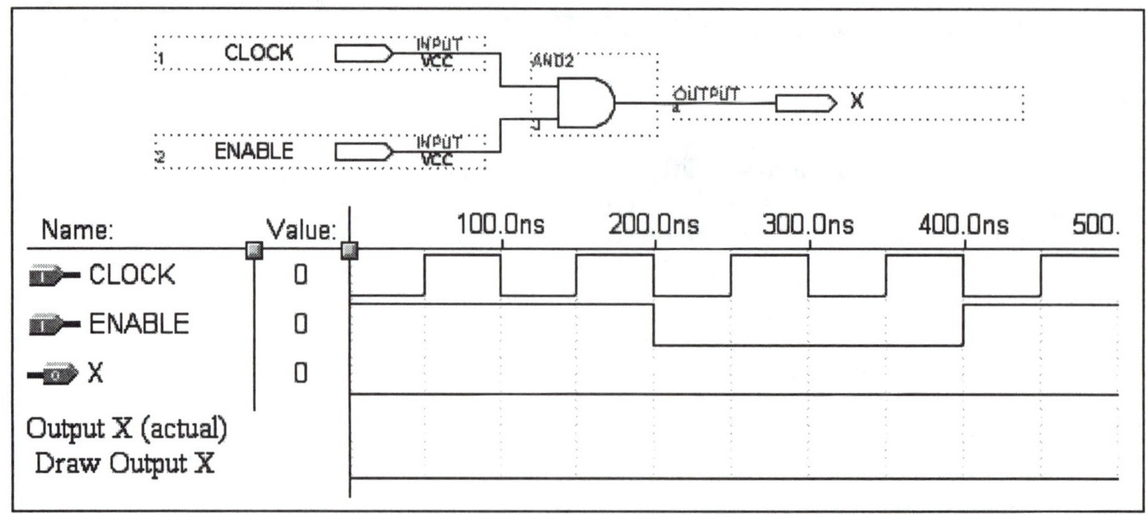

Figure 2

2. Open a new Graphic Editor file and construct the circuit shown in Figure 2.

3. Open a new Waveform Editor file, set the grid size in the Options menu to 50 ns, and create the waveforms shown in Figure 2.

4. Click the Compile and Simulate buttons. Correct all errors before continuing.

5. Click the Open SCF button in the Simulator dialog box, then sketch output X at the bottom of Figure 2 as shown in the Waveform Editor.

6. Complete the function table (Table 1) for the Clock Enable circuit of Figure 2.

Enable	Output
0	______
1	______

Table 1

Enter "Clock" if the signal is passed, otherwise enter the logic level for output X.

7. When the enable input is a logic-LOW, the gate is (enabled/inhibited) and output X is (a logic-LOW/ a logic-HIGH/ the clock). When the enable input is a logic-HIGH, the gate is (enabled/inhibited) and output X is (a logic-LOW/ a logic-HIGH/ the clock).

8. Write the Boolean expression for output X with respect to inputs Clock and Enable in Figure 2.

   X = ______________________

9. Save all files to Drive A as **comb1**, then exit the Waveform and Graphic Editors.

Part 2 Procedure

1. Open the Max+plus II software. Assign the project name **comb2**, MAX7000S as the device family, and EPM7128SLC84 as the device.

2. Open a new Graphic Editor file and construct the circuit shown in Figure 3.

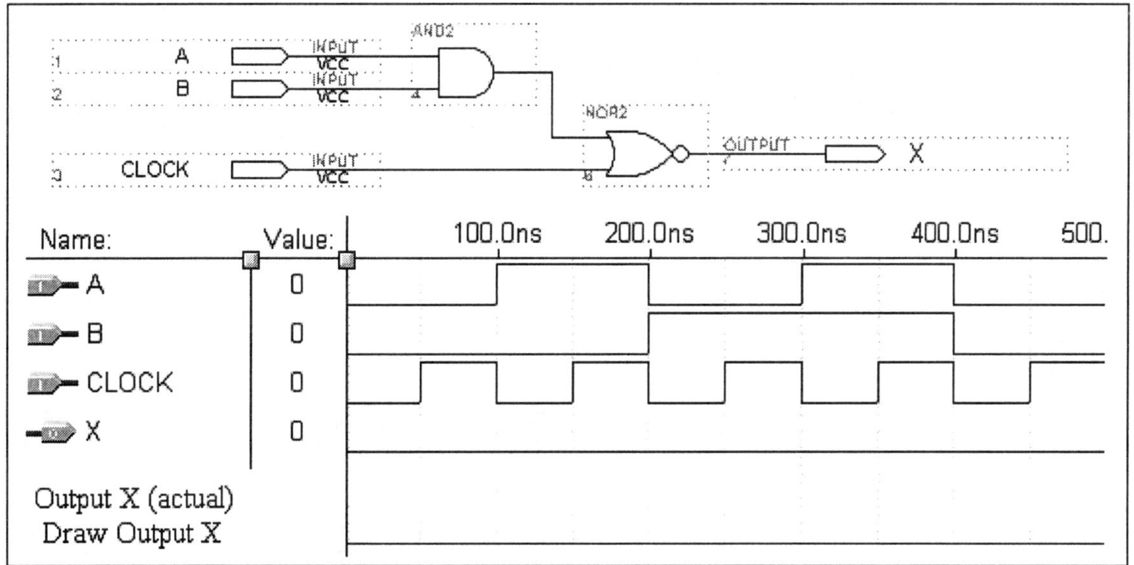

Figure 3

3. Open a new Waveform Editor file, set the grid size under the Options menu to 50 ns, and create the waveforms as shown in Figure 3.

4. Click the Compile and Simulate buttons. Correct all errors before continuing.

5. Sketch output X at the bottom of Figure 3 as shown in the Waveform Editor.

6. Complete the function table (Table 2) for the Clock Enable circuit of Figure 3.

B	A	X
0	0	____
0	1	____
1	0	____
1	1	____

Table 2

Enter "Clock" if the signal is passed, otherwise enter the logic level for output X.

7. In order to inhibit the clock from passing through the NOR gate, inputs A and B must be a logic-(LOW/HIGH). If either input A or B is a logic-(LOW/HIGH), output X is the clock signal.

8. Write the Boolean expression for output X with respect to inputs A, B, and Clock in Figure 3.

X = ________________________________

9. Save the files to Drive A as **comb2**, then exit the Waveform and Graphic Editors.

Part 3 Procedure

1. Open the Max+plus II software. Assign the project name **comb3**, MAX7000S as the device family, and EPM7128SLC84 as the device.

2. Open a new Graphic Editor file and construct the circuit shown in Figure 4.

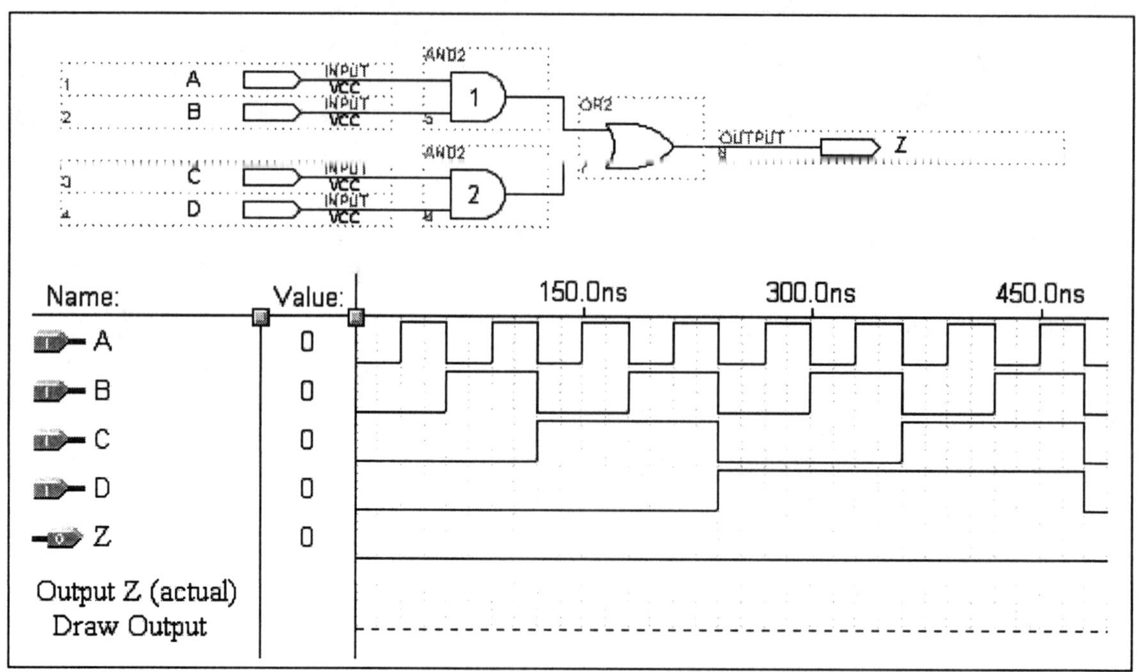

Figure 4

3. Open a new Waveform Editor file, set the grid size under the Options menu to 30 ns, and create the waveforms as shown in Figure 4.

4. Click the Compile and Simulate buttons. Correct all errors before continuing.

5. Sketch output Z at the bottom of Figure 4 as shown in the Waveform Editor.

6. Complete the function table (Table 3) for the 2-wide, 2-input, AND OR circuit of Figure 4.

7. Write the Boolean equation for output Z of the circuit in Figure 4.

   Z = ______________________________

8. Analyze the function table and Boolean expression for Figure 4. Write a statement explaining when output Z is a logic-HIGH.

D	C	B	A	Z
0	0	0	0	___
0	0	0	1	___
0	0	1	0	___
0	0	1	1	___
0	1	0	0	___
0	1	0	1	___
0	1	1	0	___
0	1	1	1	___
1	0	0	0	___
1	0	0	1	___
1	0	1	0	___
1	0	1	1	___
1	1	0	0	___
1	1	0	1	___
1	1	1	0	___
1	1	1	1	___

Table 3

____

9. Modify the circuit and waveforms (Grid Size = 20 ns) as shown in Figure 5.

10. Click the Compile and Simulate buttons. Correct all errors before continuing.

11. Click the **Open SCF** button in the Simulator dialog box, then sketch output Y at the bottom of Figure 5 as shown in the Waveform Editor.

12. Analyze the waveforms and identify which data waveform, Data A or Data B, appears on output Y when control input C is:

    a logic-LOW ________   a logic-HIGH ________

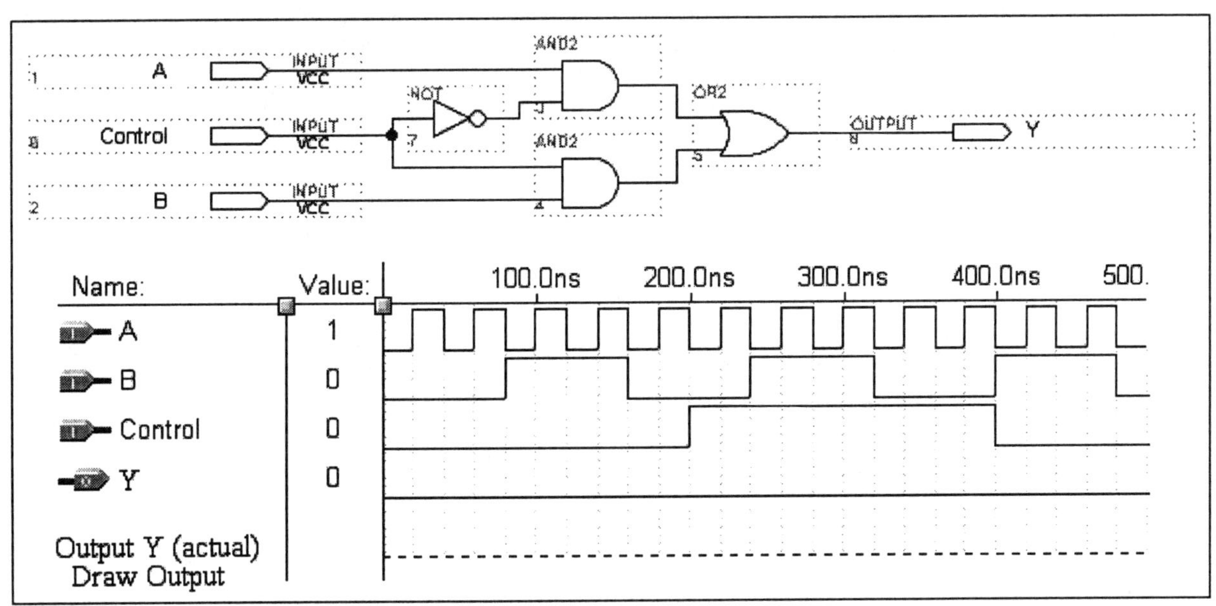

Figure 5

13. Complete the function table (Table 4) for the AND OR circuit with control input of Figure 5. Note the function table does not show all eight possible combinations of inputs C, B, and A. Since C is the control input, we are only concerned what data, either Data A or Data B, is passed through the circuit to output Y when the control input is a logic-LOW or a logic-HIGH.

Control	B	A	Y
0	Data B	Data A	_____
1	Data B	Data A	_____

Enter Data A or Data B for output Y.

Table 4

14. Complete Table 5, a simplified version of Table 4, by identifying which data, Data A or Data B, appears on output Y.

Control Input	Output Y
0	_____
1	_____

Enter Data A or Data B for output Y.

Table 5

15. When the control input is a logic-LOW, (Data A/Data B) appears on output Y. If the control input is a logic-HIGH, (Data A/Data B) appears on output Y.

16. Write the Boolean expression for output Y with respect to the inputs, Data A, Data B, and Control, in Figure 5.

Y = ______________________________________

17. Save the files to Drive A as **comb3**, then exit the Waveform and Graphic Editors.

Part 4 Procedure

1. Open the Max+plus II software. Assign the project name **comb4**, MAX7000S as the device family, and EPM7128SLC84 as the device.

2. Open a new Graphic Editor file and construct the circuit shown in Figure 6.

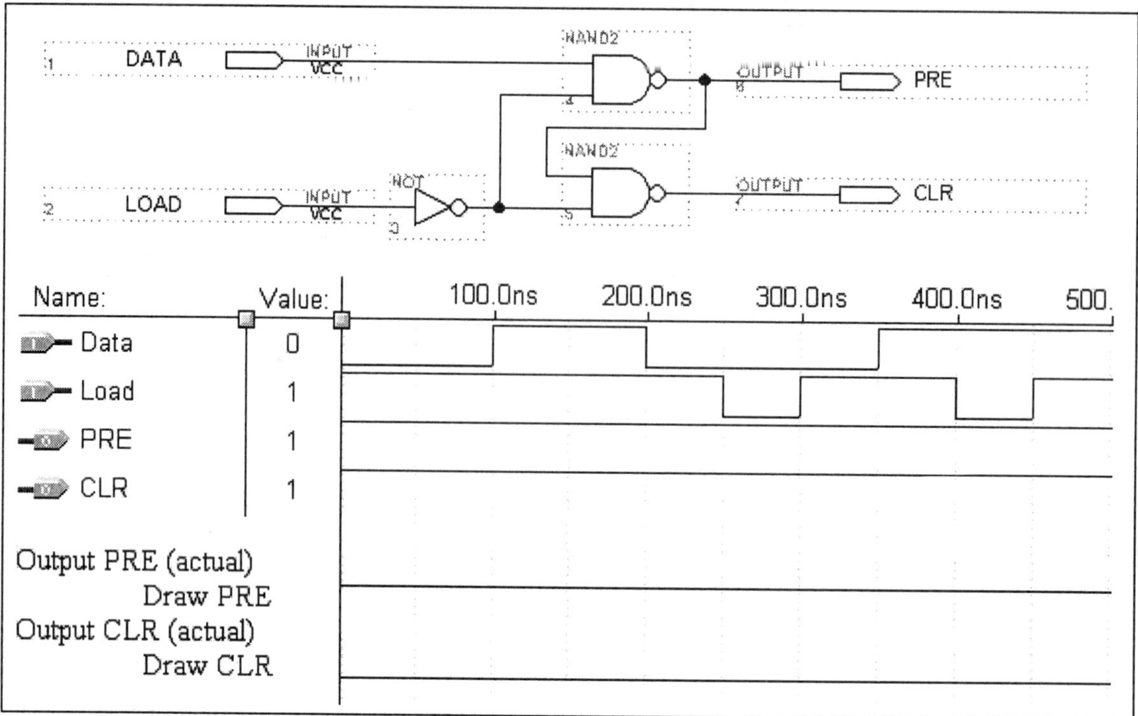

Figure 6

3. Open a new Waveform Editor file, set the Grid Size under the Options menu to 50 ns, and create the circuit and waveforms as shown in Figure 6.

4. Click the Compile and Simulate buttons. Correct all errors before continuing.

5. Sketch the PRE and CLR outputs at the bottom of Figure 6 as shown in the Waveform Editor.

6. Complete the function table (Table 6) for the **load-data** circuit of Figure 6.

LOAD	DATA	PRE	CLR
0	0	____	____
0	1	____	____
1	0	____	____
1	1	____	____

Table 6

7. To cause output PRE to become a logic-LOW, LOAD must be a logic-(LOW/HIGH) and data must be a logic-(LOW/HIGH).

8. Explain what input conditions to the circuit of Figure 6 will cause the CLR output to be a logic-LOW.

____

9. Write the Boolean expression for the PRE and CLR outputs with respect to the inputs in Figure 6.

PRE = ______________________________

CLR = ______________________________

10. Obtain a hard copy of the Graphic and Waveform Editors and label these hard copies **Part 4, Step 10A** and **Part 4, Step 10B**, respectively.

11. Demonstrate Step 10 to your instructor. Obtain the signature on the answer page for this lab.

12. Save the files to Drive A as **comb4**, then exit the Waveform and Text Editors.

13. Create a cover page for this lab.

14. Write a 1- to 2-page technical summary, based on the Combinational Logic Circuits discussed in this lab. Include an embedded schematic and a function table for that schematic along with a technical discussion describing the circuit operation.

15. Staple the pages in the following sequence and submit the completed lab to your professor for review.
    - Cover page
    - Typed summary
    - Completed answer page for this lab
    - Hard copy of the Graphic Editor, **Part 4 Step 10A**
    - Hard copy of the Waveform Editor, **Part 4, Step 10B**

Lab 3: Combinational Logic Circuits Answer Pages

Name: ______________________________

Part 1

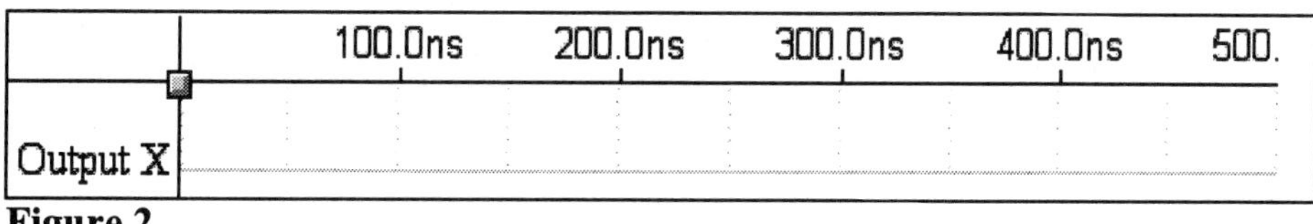

Figure 2

Enable	Output
0	____
1	____

Table 1

7. (enable/inhibit) (a logic-LOW/a logic-HIGH/ the clock)

   (enable/inhibit) (a logic-LOW/a logic-HIGH/ the clock)

8. X = ______________________________

Part 2

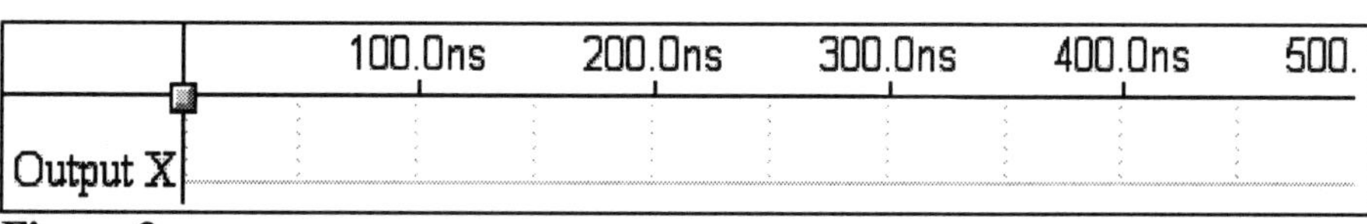

Figure 3

B	A	X
0	0	____
0	1	____
1	0	____
1	1	____

Table 2

7. (LOW/HIGH) (LOW/HIGH)

8. X = ____

Part 3

DCBA	Z
0	____
1	____
2	____
3	____
4	____
5	____
6	____
7	____
8	____
9	____
A	____
B	____
C	____
D	____
E	____
F	____

Table 3

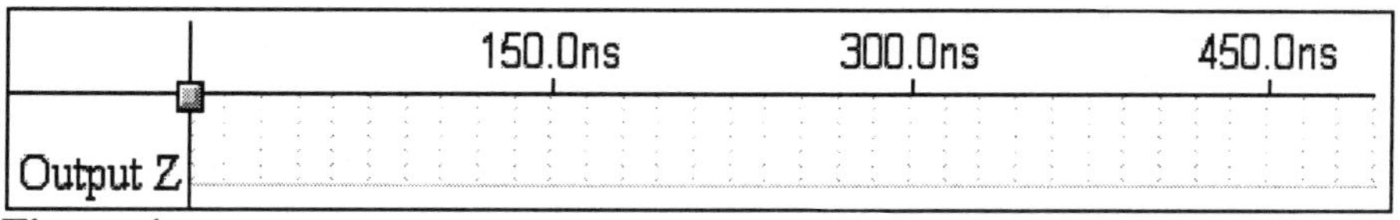

150.0ns 300.0ns 450.0ns

Output Z

Figure 4

7. Z = ____

8. ____

12. a logic-LOW__________ a logic-HIGH __________

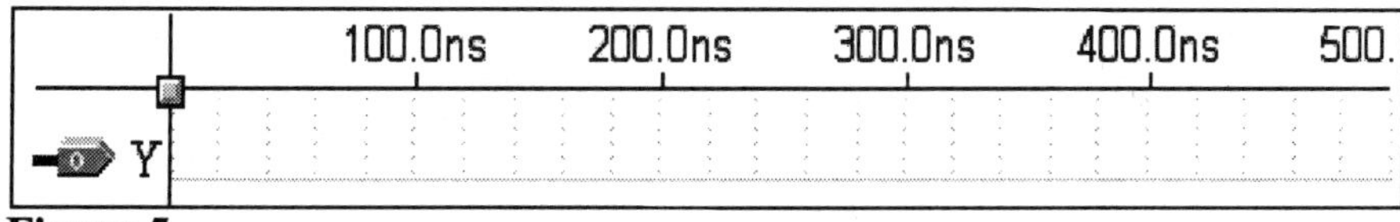

Figure 5

Control	B	A	Y
0	Data B	Data A	_____
1	Data B	Data A	_____

Table 4

Control Input	Output Y
0	_____
1	_____

Table 5

15. (Data A/Data B) (Data A/Data B)

16. Y = ____

Part 4

PRE	CLR
____	____
____	____
____	____
____	____

Table 6

	100.0ns	200.0ns	300.0ns	400.0ns	500.
PRE					
CLR					

Figure 6

7. (LOW/HIGH) (LOW/HIGH)

8. ____

9. PRE = ____

CLR = ____

11. Demonstrated to: ______________________________ Date: ______________

Grade: ____________

Lab 4: Implementing Logic Designs

Objectives:

1. Develop a function table from a statement of conditions
2. Extract the Boolean equation from the function table
3. Construct the circuit from a Boolean equation
4. Verify the circuit constructed matches the function table

Materials List:

- Max+plus II software by Altera Corporation
- University board by Altera Corporation (optional)
- Computer requirements:
  Minimum 486/66 with 8 MB RAM
- Floppy disk

Discussion:

Let us assume a circuit is to be designed that produces a logic-HIGH output (X) when one input (B) is a logic-LOW and the other input (A) is a logic-HIGH. The inputs and output are arbitrarily assigned variables such as A, B, and X. Set up a function table to represent the design statement, listing all possible combinations for the input variables, A and B. Show a logic-HIGH output for X that satisfies the statement and a logic-LOW for all other input combinations. (See Table 1.)

Once a function table has been developed, each logic-HIGH output for a row can be expressed as a multi-input AND gate. The number of inputs to the AND gate is determined by the number of inputs from the logic statement.

B	A	X
0	0	0
0	**1**	**1**
1	0	0
1	1	0

Table 1

Start by drawing the 2-input AND gate with inputs and output at a logic-HIGH as shown in Figure 1. The output of the AND gate is a logic-HIGH only when all inputs are a logic-HIGH. What must be done to input A to provide a logic-HIGH state? For row 2 of the table, when output X is a logic-HIGH, input A is already a logic-HIGH, so merely wire input A to one of the gate inputs.

However, note input B for row 2 of the table is a logic-LOW. Input B must be inverted before being wired to the AND gate. Output X for row 2 can now be expressed as a function of the inputs. The column equation for the output is derived by ORing the individual row expressions. The individual row expressions represented by the AND gates are called product terms. The column equation ORing the AND terms is called the sum of products, SOP.

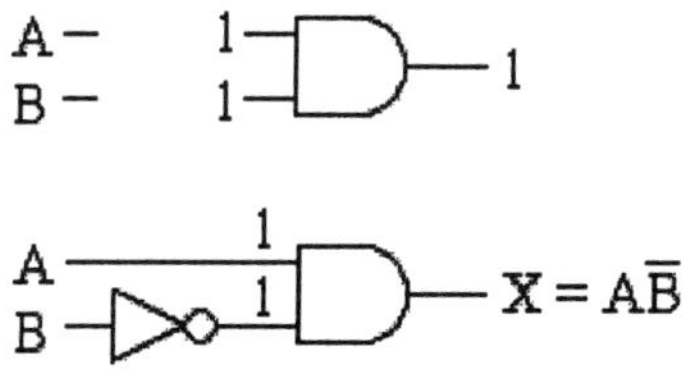

A
B
1
1
1
A
B
1
1
$X = A\overline{B}$

Figure 1

Part 1 Procedure

Statement: The output (X) of a circuit is a logic-HIGH only when inputs A and B are a logic-HIGH, when input C is a logic-LOW, **or** when inputs A and B are a logic-HIGH when input C is a logic-HIGH.

1. Complete the function table (Table 2) for the logic statement.

2. Write the Boolean expression for each product term where X = 1 with respect to the corresponding input conditions for that row.

1st product term: ______________________

2nd product term: ______________________

C	B	A	X
0	0	0	____
0	0	1	____
0	1	0	____
0	1	1	____
1	0	0	____
1	0	1	____
1	1	0	____
1	1	1	____

Table 2

3. Combine the two product terms from Part 1, Step 2 into a single OR expression to represent the output X column of the function table.

X = ______________________________________

4. Using logic gates, sketch the circuit to represent output X with respect to inputs A, B, and C.

5. Open the Max+plus II software. Assign the project name **Design1** and MAX7000S as the device family.

6. Open a new Graphic Editor file and construct the circuit you created for Part 1, Step 4.

7. Open a new Waveform Editor file and create the waveforms shown in Figure 2 following these steps:

- Set the grid size to 100 ns
- Set **Multiplied By:** in the Overwrite Count Value for Waveform A to 1.
- Set **Multiplied By:** in the Overwrite Count Value for Waveform B to 2.
- Set **Multiplied By:** in the Overwrite Count Value for Waveform C to 4.

8. At the bottom of Figure 2, sketch the output signal created by the software after simulation.

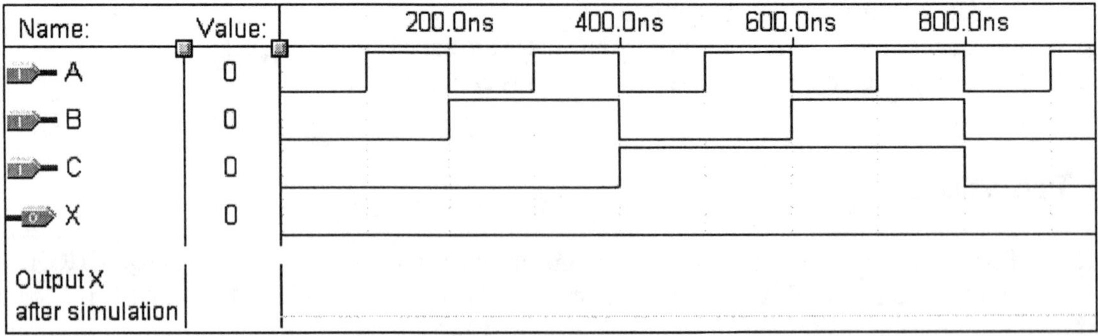

Figure 2

Lab 4: Implementing Logic Designs

Objectives:

1. Develop a function table from a statement of conditions
2. Extract the Boolean equation from the function table
3. Construct the circuit from a Boolean equation
4. Verify the circuit constructed matches the function table

Materials List:

- Max+plus II software by Altera Corporation
- University board by Altera Corporation (optional)
- Computer requirements:
  Minimum 486/66 with 8 MB RAM
- Floppy disk

Discussion:

Let us assume a circuit is to be designed that produces a logic-HIGH output (X) when one input (B) is a logic-LOW and the other input (A) is a logic-HIGH. The inputs and output are arbitrarily assigned variables such as A, B, and X. Set up a function table to represent the design statement, listing all possible combinations for the input variables, A and B. Show a logic-HIGH output for X that satisfies the statement and a logic-LOW for all other input combinations. (See Table 1.)

Once a function table has been developed, each logic-HIGH output for a row can be expressed as a multi-input AND gate. The number of inputs to the AND gate is determined by the number of inputs from the logic statement.

B	A	X
0	0	0
0	**1**	**1**
1	0	0
1	1	0

Table 1

Start by drawing the 2-input AND gate with inputs and output at a logic-HIGH as shown in Figure 1. The output of the AND gate is a logic-HIGH only when all inputs are a logic-HIGH. What must be done to input A to provide a logic-HIGH state? For row 2 of the table, when output X is a logic-HIGH, input A is already a logic-HIGH, so merely wire input A to one of the gate inputs.

However, note input B for row 2 of the table is a logic-LOW. Input B must be inverted before being wired to the AND gate. Output X for row 2 can now be expressed as a function of the inputs. The column equation for the output is derived by ORing the individual row expressions. The individual row expressions represented by the AND gates are called product terms. The column equation ORing the AND terms is called the sum of products, SOP.

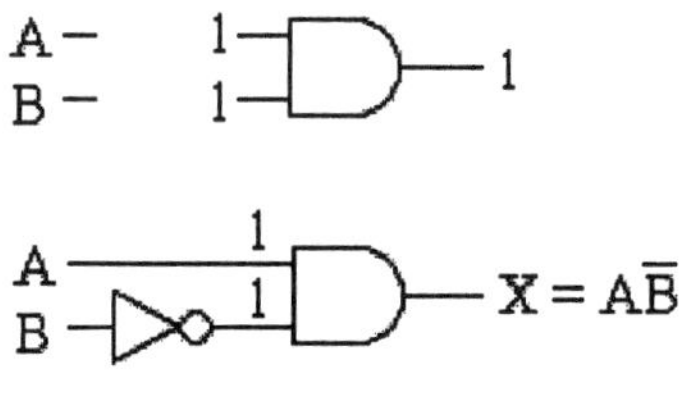

A – 1
B – 1
1

A 1
B 1
$X = A\overline{B}$

Figure 1

Part 1 Procedure

Statement: The output (X) of a circuit is a logic-HIGH only when inputs A and B are a logic-HIGH, when input C is a logic-LOW, **or** when inputs A and B are a logic-HIGH when input C is a logic-HIGH.

1. Complete the function table (Table 2) for the logic statement.

2. Write the Boolean expression for each product term where X = 1 with respect to the corresponding input conditions for that row.

   1st product term: ____________________

   2nd product term: ____________________

C	B	A	X
0	0	0	___
0	0	1	___
0	1	0	___
0	1	1	___
1	0	0	___
1	0	1	___
1	1	0	___
1	1	1	___

Table 2

3. Combine the two product terms from Part 1, Step 2 into a single OR expression to represent the output X column of the function table.

   X = ______________________________

4. Using logic gates, sketch the circuit to represent output X with respect to inputs A, B, and C.

5. Open the Max+plus II software. Assign the project name **Design1** and MAX7000S as the device family.

6. Open a new Graphic Editor file and construct the circuit you created for Part 1, Step 4.

7. Open a new Waveform Editor file and create the waveforms shown in Figure 2 following these steps:

   - Set the grid size to 100 ns
   - Set **Multiplied By:** in the Overwrite Count Value for Waveform A to 1.
   - Set **Multiplied By:** in the Overwrite Count Value for Waveform B to 2.
   - Set **Multiplied By:** in the Overwrite Count Value for Waveform C to 4.

8. At the bottom of Figure 2, sketch the output signal created by the software after simulation.

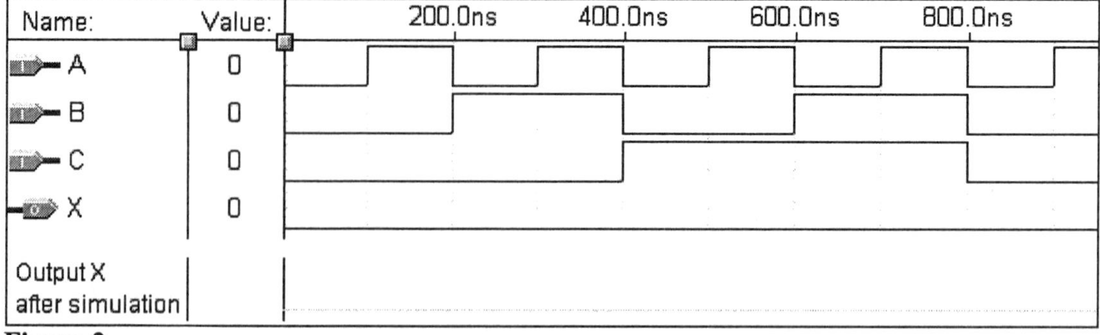

Figure 2

9. Examine the simulated output X waveform. Verify that output X is a logic-HIGH only under the two conditions mentioned in the statement. Correct all errors before continuing.

10. Save the files to Drive A as **design1**, then exit the Graphic and Waveform Editors.

Part 2 Procedure

Statement: The output of a circuit is to be a logic-HIGH when the majority of the three inputs are logic-LOW.

1. Complete the function table (Table 3) for the logic statement.

2. Write the Boolean expression for each product term where X = 1 with respect to the corresponding input conditions for that row.

   1st product term: ______________________

   2nd product term: ______________________

   3rd product term: ______________________

   4th product term: ______________________

C	B	A	X
0	0	0	____
0	0	1	____
0	1	0	____
0	1	1	____
1	0	0	____
1	0	1	____
1	1	0	____
1	1	1	____

Table 3

3. Combine the four product terms from Part 2, Step 2 into a single OR expression to represent the output X column of the function table.

   X = ____

4. Using logic gates, sketch the circuit to represent output X .

5. Open the Max+plus II software. Assign the project name **design2**.

6. Open a new Graphic Editor file and construct the circuit you created for Part 2, Step 4.

7. Open a new Waveform Editor file and create the waveforms shown in Figure 3 following these steps:

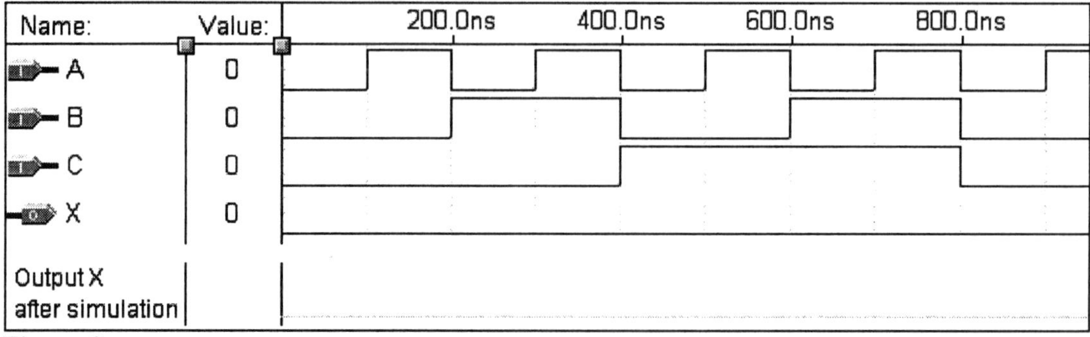

Figure 3

- Set the grid size to 100 ns
- Set **Multiplied By:** in the Overwrite Count Value for Waveform A to 1
- Set **Multiplied By:** in the Overwrite Count Value for Waveform B to 2
- Set **Multiplied By:** in the Overwrite Count Value for Waveform C to 4

8. At the bottom of Figure 3, draw the output signal created by the software after simulation.

9. Examine the simulated output X waveform. Verify that output X is a logic-HIGH only under the conditions mentioned in the statement. Correct all errors before continuing.

10. The Boolean equation may be simplified using Boolean algebra, Karnaugh mapping techniques, or computer software to reduce the number of gates; however, expression simplification is not the objective of this lab.

11. Save the files to Drive A as **design2**, then exit the Graphic and Waveform Editors.

Part 3 Procedure

Statement: A circuit with four variables will produce a logic-HIGH output only when three or four of the four variables are a logic-HIGH.

1. Complete the function table (Table 4) for the logic statement.

2. Write the Boolean expression for each product term where X = 1 with respect to the corresponding input conditions for that row.

   1st product term: ______________________

   2nd product term: ______________________

   3rd product term: ______________________

   4th product term: ______________________

   5th product term: ______________________

D	C	B	A	X
0	0	0	0	____
0	0	0	1	____
0	0	1	0	____
0	0	1	1	____
0	1	0	0	____
0	1	0	1	____
0	1	1	0	____
0	1	1	1	____
1	0	0	0	____
1	0	0	1	____
1	0	1	0	____
1	0	1	1	____
1	1	0	0	____
1	1	0	1	____
1	1	1	0	____
1	1	1	1	____

Table 4

3. Combine the five product terms from Part 3, Step 2 into a single OR expression to represent the sum of products function of output X column of the function table.

   X = ____

4. Using logic gates, sketch the circuit to represent output X.

5. Open the Max+plus II software. Assign the project name **design3**.

6. Open a new Graphic Editor file and construct the circuit you created for Part 3, Step 4.

7. Open a new Waveform Editor file and create the waveforms shown in Figure 4 following these steps:

   - Set the grid size to 50 ns
   - Set **Multiplied By:** in the Overwrite Count Value for Waveform A to 1
   - Set **Multiplied By:** in the Overwrite Count Value for Waveform B to 2
   - Set **Multiplied By:** in the Overwrite Count Value for Waveform C to 4
   - Set **Multiplied By:** in the Overwrite Count Value for Waveform D to 8

8. At the bottom of Figure 4, sketch the output signal created by the software after simulation.

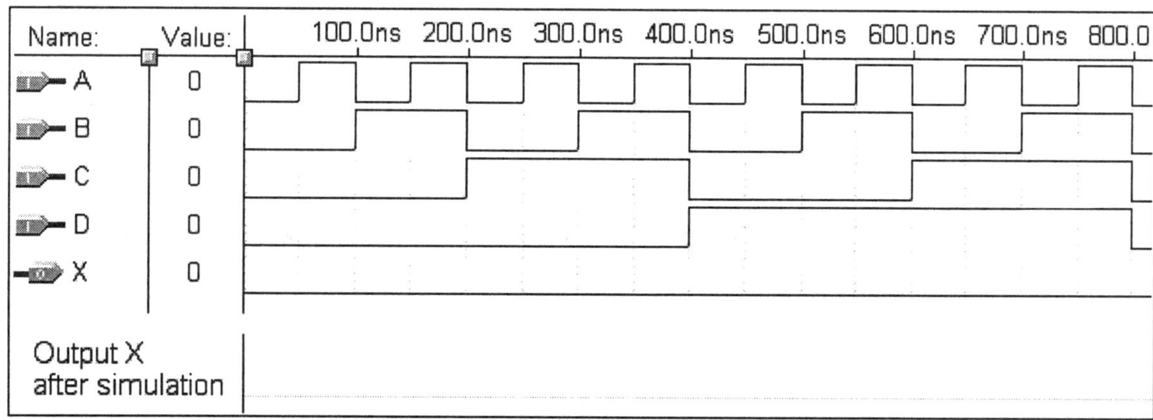

Figure 4

9. Examine the simulated output X waveform. Verify that output X is a logic-HIGH only under the two conditions mentioned in the statement. Correct all errors before continuing.

10. Save the files to Drive A as **design3**, then exit the Graphics and Waveform Editors.

Part 4 Procedure

Statement: The function table for a circuit to drive a common anode seven-segment display is shown in Table 5. A logic-LOW on the **g, f, e, d, c, b,** and **a** outputs will cause the segment to light. For instance, if the PONM inputs are 1000_2, the low states on **g, f, e, d, c, b,** and **a** will cause the corresponding segments to light, displaying an eight. (Figure 5)

Symbol	P O N M	Shape	$\overline{g}$	$\overline{f}$	$\overline{e}$	$\overline{d}$	$\overline{c}$	$\overline{b}$	$\overline{a}$
0	0 0 0 0	0	1	0	0	0	0	0	0
1	0 0 0 1	1	1	1	1	1	0	0	1
2	0 0 1 0	2	0	1	0	0	1	0	0
3	0 0 1 1	3	0	1	1	0	0	0	0
4	0 1 0 0	4	0	0	1	1	0	0	1
5	0 1 0 1	5	0	0	1	0	0	1	0
6	0 1 1 0	6	0	0	0	0	0	1	0
7	0 1 1 1	7	1	1	1	1	0	0	0
8	1 0 0 0	8	0	0	0	0	0	0	0
9	1 0 0 1	9	0	0	1	0	0	0	0
A	1 0 1 0	A	0	0	0	1	0	0	0
B	1 0 1 1	b	0	0	0	0	0	1	1
C	1 1 0 0	C	1	0	0	0	1	1	0
D	1 1 0 1	d	0	1	0	0	0	0	1
E	1 1 1 0	E	0	0	0	0	1	1	0
F	1 1 1 1	F	0	0	0	1	1	1	0

Table 5

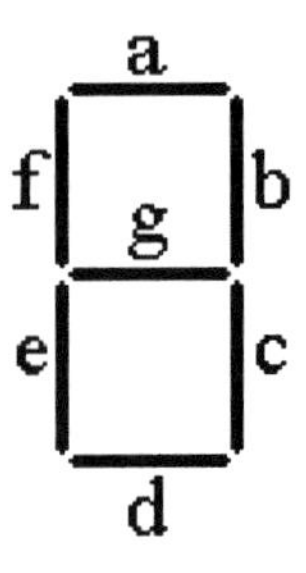

Figure 5

When developing a circuit from the function table, either the Sum of Products (SOP) method based on logic-HIGHs or the Product of Sums (POS) method based on logic-LOWs may be used since both methods will yield equivalent expressions. Table 5 shows 34 output logic-HIGHs and 78 output logic-LOWs. When driving the segments, a logic-LOW will turn the segment on, making these outputs active-LOW.

For the following example, use the Sum of Products method to extract the expressions based on logic-HIGHs, then construct the circuit. For instance, to find $\overline{g}$, write a sum expression based on the DCBA codes 0000_2, 0001_2, 0111_2, and 1100_2.

1. Write the product terms as a sum of products (SOP) Boolean expression based on the logic-HIGHs for each output, $\overline{g}$ to $\overline{a}$

$\overline{g}$ = ____

$\overline{\mathbf{f}}$ = ____________________

$\overline{\mathbf{e}}$ = ____________________

$\overline{\mathbf{d}}$ = ____________________

$\overline{\mathbf{c}}$ = ____________________

$\overline{\mathbf{b}}$ = ____________________

$\overline{\mathbf{a}}$ = ____________________

2. Using logic gates, draw the circuit to represent outputs $\overline{\mathbf{g}}, \overline{\mathbf{f}}, \overline{\mathbf{e}}, \overline{\mathbf{d}}, \overline{\mathbf{c}}, \overline{\mathbf{b}}$, and $\overline{\mathbf{a}}$ based on the Sum of Products developed for Step 1.

3. Open the Max+plus II software. Assign the project name **design4**.

4. Open a new Graphic Editor file and construct the circuit you created for Part 4, Step 2.

5. Open a new Waveform Editor file and create the waveforms shown in Figure 6 following these steps:

   - Set the grid size to 50 ns
   - Set **Multiplied By:** in the Overwrite Count Value for Waveform P to 8.
   - Set **Multiplied By:** in the Overwrite Count Value for Waveform O to 4.
   - Set **Multiplied By:** in the Overwrite Count Value for Waveform N to 2.
   - Set **Multiplied By:** in the Overwrite Count Value for Waveform M to 1.

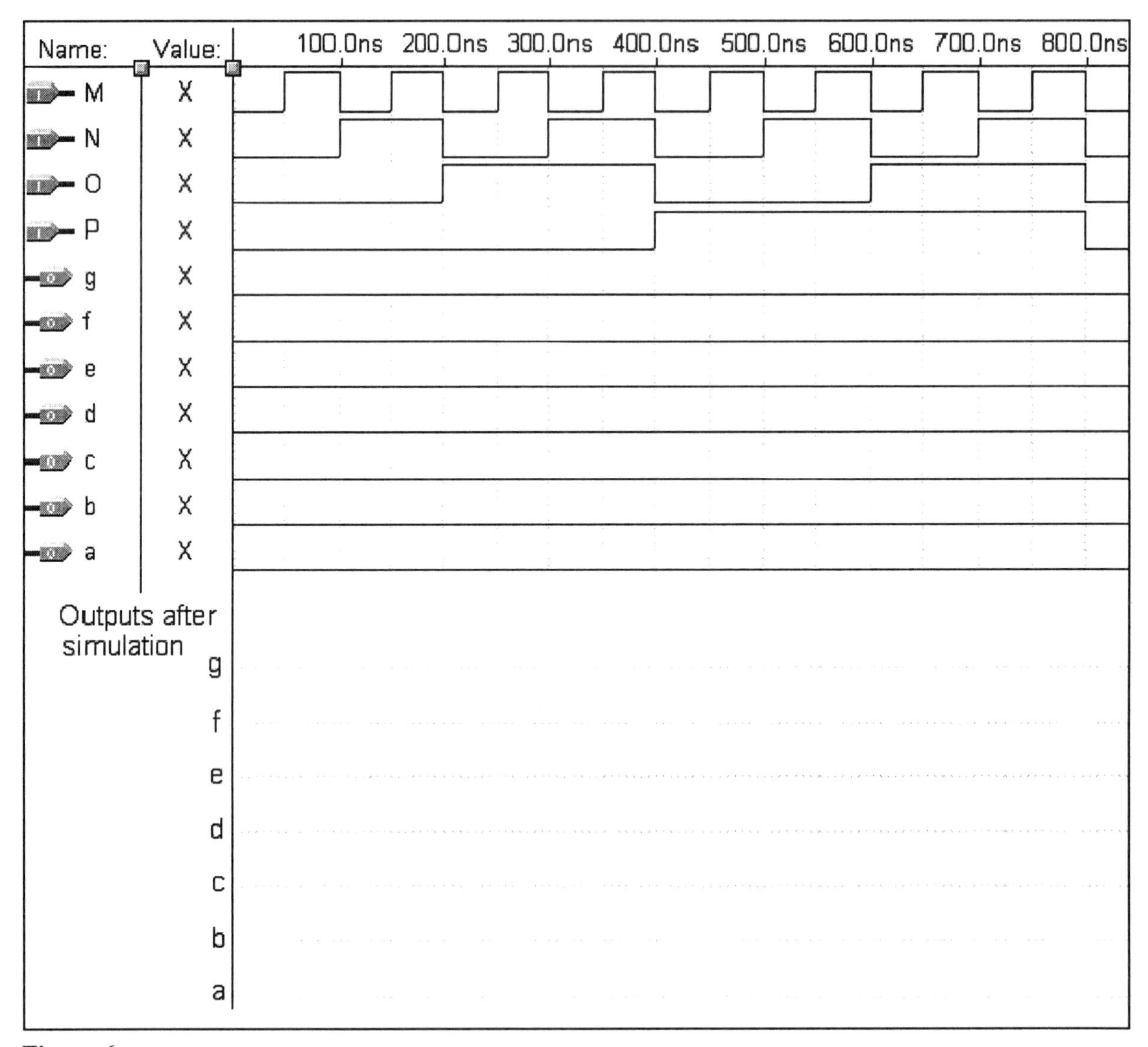

Figure 6

6. At the bottom of Figure 6, draw the output signal created by the software after simulation.

7. Examine the simulated output X waveform. Verify that output X is a logic-HIGH only under the two conditions mentioned in the statement. Correct all errors before continuing.

8. Obtain hard copies of the Graphic and Waveform Editors. Mark these pages **Part 4, Step 8A** and **Part 4, Step 8B**, respectively.

9. Demonstrate the Graphic and Waveform Editor files to the instructor. Obtain the signature of approval on the answer page.

10. Save the files to the temporary directory on Drive C as **design4**, then exit the Graphic and Waveform Editors.

11. Two integrated circuits may be used to drive the seven-segment display devices. The 7447 may be used to drive a common anode display. The outputs are active-LOW and require resistors in series between the chip and the display inputs. The 7448 outputs are active-HIGH and may drive common cathode displays. External resistors are not necessary since the 7448 has internal pull-up resistors.

12. Create a cover page and write a 1-to 2-page summary pertaining to the results obtained from this lab. Your typed summary must include Table 5 from Part 4, the sum of product expressions for outputs **g** through **a**, and a discussion explaining how the equations were extracted from the table. Use the apostrophe (') to indicate the NOT function in your equations; that is: P'O'N'M' represents 0000_2.

13. Staple all papers for this lab in the following sequence, then submit the lab to your instructor for grading.

    - Cover page
    - 1 to 2 page typed summary
    - The completed answer page for the lab
    - Printout of the Graphics Editor, **Part 4, Step 8A**
    - Printout of the Waveform Editor, **Part 4, Step 8B**

Lab 4: Implementing Logic Designs Answer Pages

Name: ____________________

Part 1

1. 1st product term: ________________________________

   2nd product term: ________________________________

2. X = ________________________________

4.

C	B	A	X
0	0	0	____
0	0	1	____
0	1	0	____
0	1	1	____
1	0	0	____
1	0	1	____
1	1	0	____
1	1	1	____

Table 2

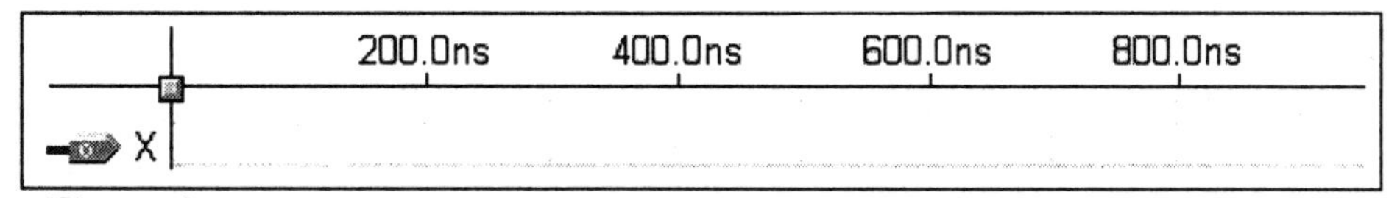

Figure 2

Part 2

2. 1st product term: ________________________________

   2nd product term: ________________________________

   3rd product term: ________________________________

   4th product term: ________________________________

C	B	A	X
0	0	0	____
0	0	1	____
0	1	0	____
0	1	1	____
1	0	0	____
1	0	1	____
1	1	0	____
1	1	1	____

Table 3

3. X = ____

4.

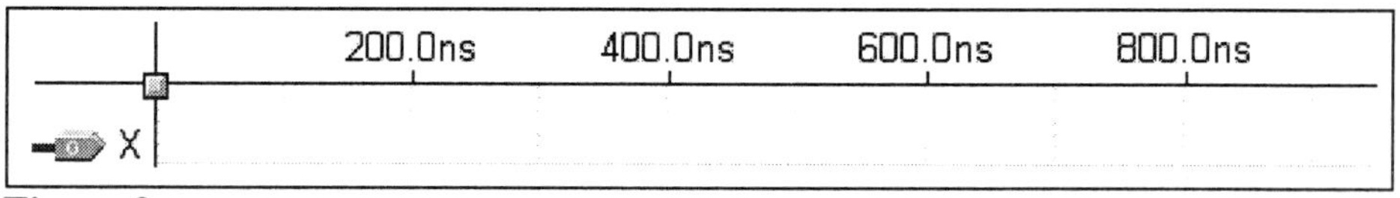

200.0ns 400.0ns 600.0ns 800.0ns

X

Figure 3

Part 3

2. 1st product term: ____________________

   2nd product term: ____________________

   3rd product term: ____________________

   4th product term: ____________________

   5th product term: ____________________

3. X = ____________________

____

____

____

D	C	B	A	X
0	0	0	0	____
0	0	0	1	____
0	0	1	0	____
0	0	1	1	____
0	1	0	0	____
0	1	0	1	____
0	1	1	0	____
0	1	1	1	____
1	0	0	0	____
1	0	0	1	____
1	0	1	0	____
1	0	1	1	____
1	1	0	0	____
1	1	0	1	____
1	1	1	0	____
1	1	1	1	____

Table 4

4.

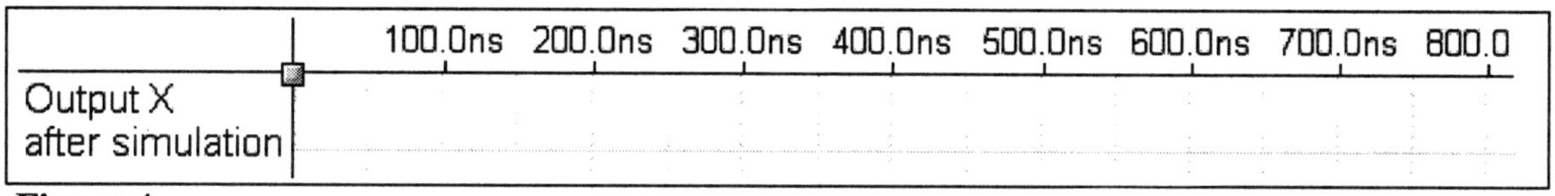

	100.0ns	200.0ns	300.0ns	400.0ns	500.0ns	600.0ns	700.0ns	800.0
Output X after simulation								

Figure 4

Part 4

$\bar{g}$ = ______________________________

$\bar{f}$ = ______________________________

$\bar{e}$ = ______________________________

$\bar{d}$ = ______________________________

$\bar{c}$ = ______________________________

$\bar{b}$ = ______________________________

$\bar{a}$ = ______________________________

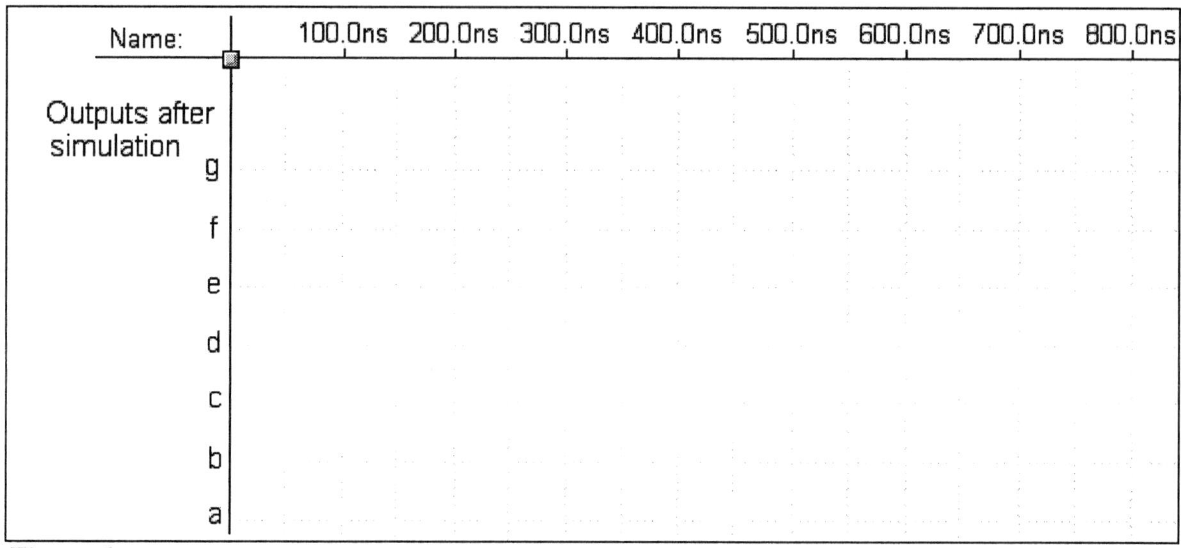

Figure 6

9. Demonstrated to: ____Date: _______________

Grade: ______________

Lab 5: Adders

Objectives:

1. Design a half adder by extracting the Boolean equation from a function table
2. Design a full adder by extracting the Boolean equation from a function table
3. Construct the half adder and full adder circuits from a Boolean equation
4. Verify the half adder and full adder circuits constructed match the function table
5. Construct and demonstrate adder circuits using the 7483 integrated circuit

Materials List:

- Max+plus II software by Altera Corporation
- Circuit card containing the EPM7128SLC84 chip
- Computer requirements:
  Minimum 486/66 with 8 MB RAM
- Floppy disk
- (9) 1 K resistors
- (9) LEDs
- 8 STDP switches
- 5-V DC power supply

Discussion:

Digital Electronics is based on two logic states, a logic-HIGH and a logic-LOW. As a result, the number system of choice is base 2, using the symbols 1 and 0 to represent the two logic states. When applied to adders, 5 + 7 = 12, right? No! The correct answer is C. The computer or microprocessor that handles the math treats all data (5 and 7) as binary numbers, performs binary addition, and gives a binary answer. For 5 + 7, the computer will add 0101_2 + 0111_2 to get 1100_2. The nibbles 5 and 7 are treated by the computer as hexadecimal symbols.

It is left up to the computer engineer or software programmer to convert the binary answer to proper form for display. This lab will treat all numbers as hex, require hex-to-binary conversions, perform all math in binary, and require binary-to-hex conversions for answers.

The fundamental building block of addition is the half adder, whose function table is shown to the right. By observation, you may recognize that the Carry output is the AND gate with respect to inputs A and B. The Sum output is the exclusive OR gate with respect to inputs A and B.

B	A	Carry	Sum
0	0	0	0
0	1	0	1
1	0	0	1
1	1	1	0

Half adder function table

The full adder is a circuit with three inputs, A, B, and C_{IN}, and two outputs, Sum and Carry. The function table for the full adder is shown below.

Cin	B	A	Carry	Sum
0	0	0	0	0
0	0	1	0	1
0	1	0	0	1
0	1	1	1	0
1	0	0	0	1
1	0	1	1	0
1	1	0	1	0
1	1	1	1	1

Full adder function table

The half adder is used to add the units column bits of a multi-digit number. Full adders are used to add bits in the 2s column, 4s column and 8s column of a nibble of data.

FULL ADDER	FULL ADDER	FULL ADDER	HALF ADDER
C3 C2 C1 C0	C3 C2 C1 C0	C3 C2 C1 C0	C3 C2 C1 C0
A3 A2 A1 A0	A3 A2 A1 A0	A3 A2 A1 A0	A3 A2 A1 A0
+ B3 B2 B1 B0	+ B3 B2 B1 B0	+ B3 B2 B1 B0	+ B3 B2 B1 B0
C3 S3 S2 S1 S0	C3 S3 S2 S1 S0	C3 S3 S2 S1 S0	C3 S3 S2 S1 S0

The 4-bit binary adder, 7483, contains four full adders cascaded together to perform hex nibble addition. This adder has a carry input, C_{IN}, to the least significant adder, and a C_{OUT} from the most significant adder allowing one to cascade several 7483 chips to perform multi-nibble addition.

The symbols for the adders used in this lab are shown in Figure 1.

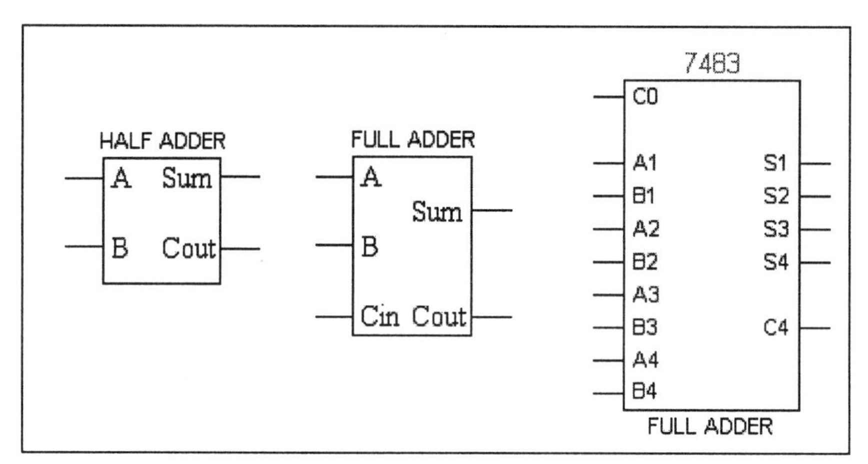

Figure 1

Part 1 Procedure

1. Complete the function table (Table 1) for the half adder.

2. Extract the Boolean equation from the function table that represents the Carry output with respect to inputs A and B.

   Carry = AND

3. Write the Boolean equation that represents the Sum output with respect to inputs A and B.

   Sum = XOR

4. Sketch the Sum and Carry circuits based on your equations from Step 2 and Step 3.

B	A	Carry	Sum
0	0	0	0
0	1	0	1
1	0	0	1
1	1	1	0

Table 1

5. Open the Max+plus II software. Assign the project name **adder1** and MAX7000S as the device family.

6. Open a new Graphic Editor file and construct the circuit from Step 4.

7. Open a new Waveform Editor file, set Grid Size to 100 ns, then construct the waveforms shown in Figure 2.

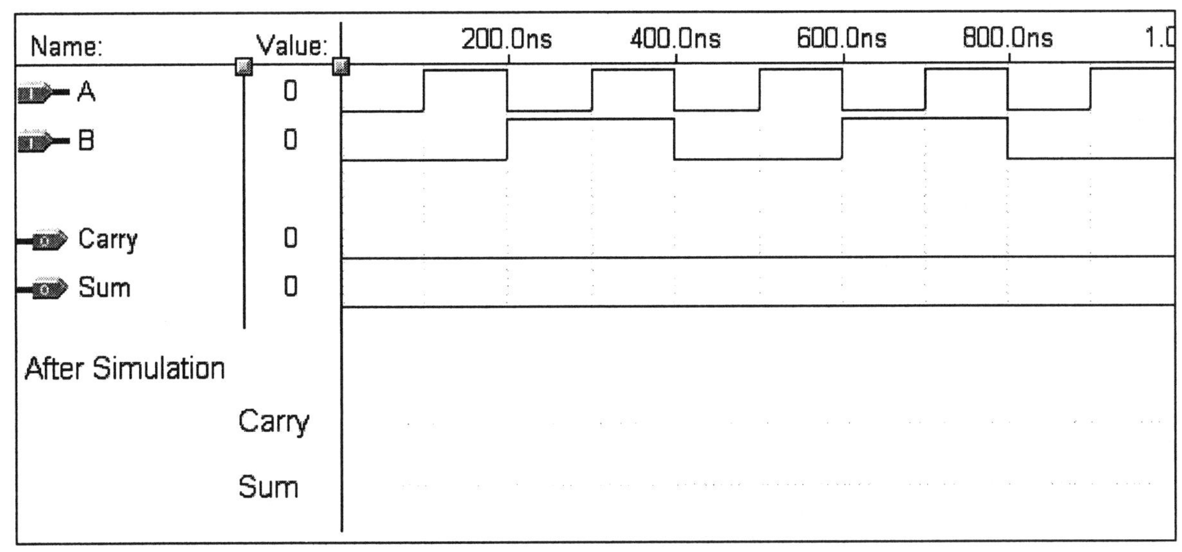

Figure 2

8. Verify that the waveforms match the function table for the half adder. The Carry output should go to a logic-HIGH only when both inputs are at a logic-HIGH.

9. Sketch the simulated waveforms at the bottom of Figure 2.

10. Save the editor files to Drive A as **adder1**, then exit the Graphic and Waveform Editors.

Part 2 Procedure

1. Complete the function table (Table 2) for the full adder.

2. Extract the Boolean equation from the function table that represents the Carry output with respect to inputs A, B, and C_{IN}.

   Carry = ______________________________

3. Extract the Boolean equation from the function table that represents the Sum output with respect to inputs A, B, and C_{IN}.

   Sum = ______________________________

C_{IN}	B	A	Carry	Sum
0	0	0		
0	0	1		
0	1	0		
0	1	1		
1	0	0		
1	0	1		
1	1	0		
1	1	1		

Table 2

4. Sketch the Sum and Carry circuits based on your equations from Step 2 and Step 3.

5. Open the Max+plus II software. Assign the project name **adder2**.

6. Open a new Graphic Editor file and construct the Sum and Carry circuits of Part 2, Step 4.

7. Open a new Waveform Editor file, set Grid Size to 100 ns, then construct the waveforms shown in Figure 3.

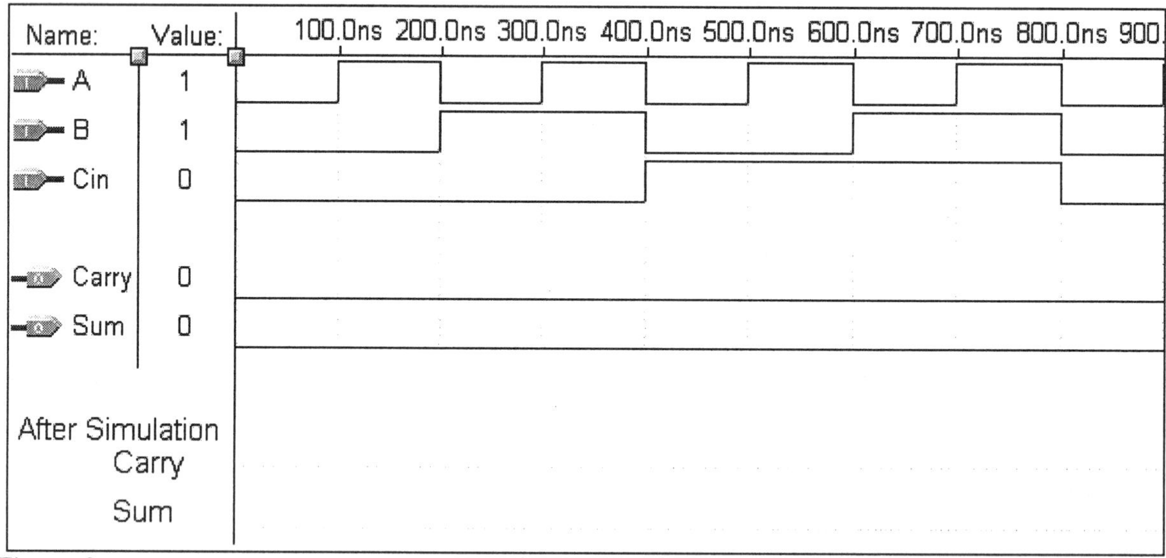

Figure 3

8. Verify that the waveforms match the function table for the full adder. The Carry output should go to a logic-HIGH only when two or three of the inputs, A, B, and C_{IN}, are at a logic-HIGH.

9. Draw the simulated waveforms at the bottom of Figure 3.

10. Save the editor files to Drive A as **adder2**, then exit the Graphic and Waveform Editors.

Part 3 Procedure

1. Open the Max+plus II software. Assign the project name **adder3**.

2. Open a new Graphic Editor file and construct the circuit shown in Figure 4, referring to the following instructions to create BUS lines and to label wires.

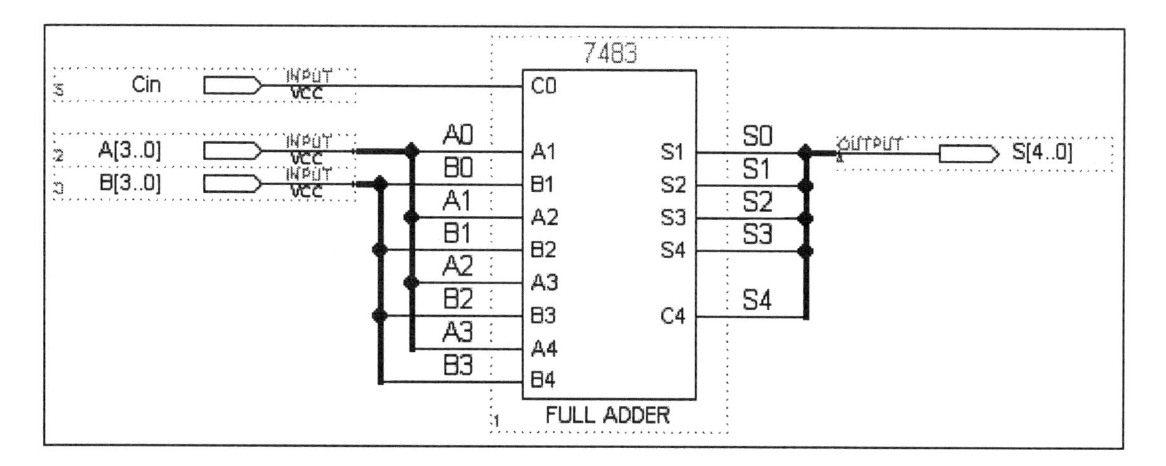

Figure 4

3. After placing the input symbol to the left of the 7483 symbol (Figure 5), place the mouse pointer on the right end of the input symbol, click and drag the mouse right 1 inch, then down 1 inch. Leave the line highlighted. Select **Options - Line Style** and click on the thick line (Figure 6) to select the **BUS** option.

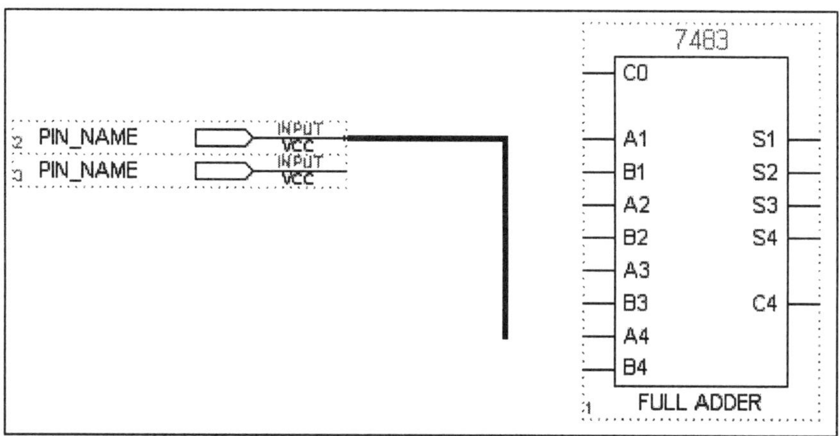

Figure 5

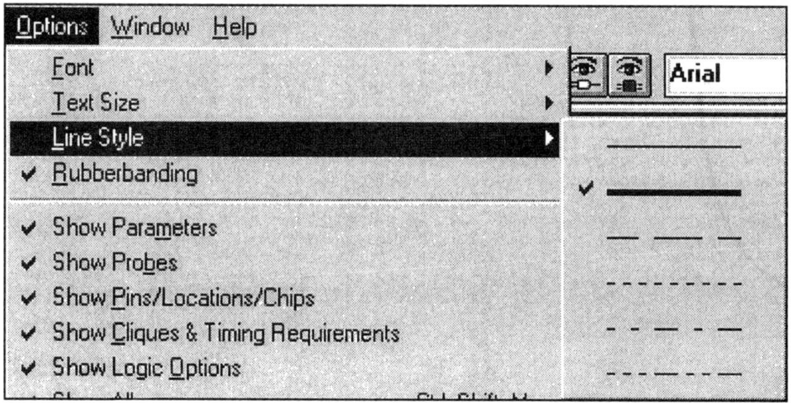

Figure 6

4. To label a line, place the mouse pointer on the line, then click on the left mouse button. The line should be highlighted with a flashing insertion point just below the line. Type the label.

5. The input buses, A[3..0] and B[3..0], each represent four bits: Bit 3, Bit 2, Bit 1, and Bit 0. The output bus, S[4..0], represents five bits: Bit 4, Bit 3, Bit 2, Bit 1, and Bit 0.

6. Open a new Waveform Editor file, set Grid Size to 200 ns, and create the waveforms shown in Figure 8. When the bus line is highlighted, select the Count Over-ride button (Figure 7) in the Draw tool bar.

Figure 7

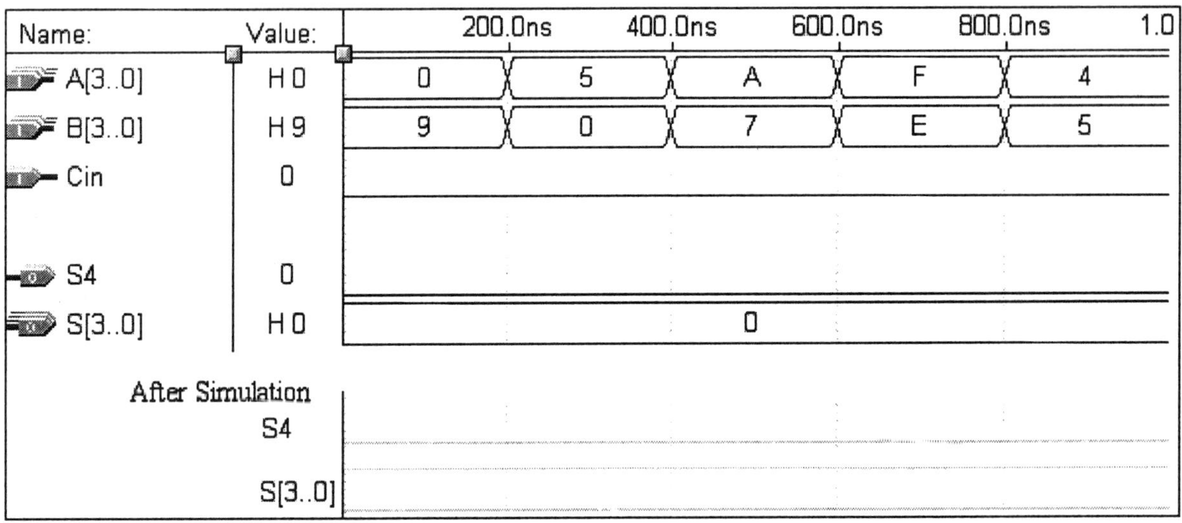

Figure 8

7. When the Overwrite Count Value box appears when defining Waveform A, set **Increment By:** to 5 before clicking the **OK** button. For Waveform B, set **Starting Value:** to 9 and **Increment By:** to 7.

8. Sketch the resulting waveforms after simulation in the space provided in Figure 8.

9. Complete each addition problem in binary. (**Show all work in binary**.)

@100 ns	@300 ns	@500 ns	@700 ns	@900 ns
0 0 0 0	0 1 0 1	1 0 1 0	1 1 1 1	0 1 0 0
+ 1 0 0 1	+ 0 0 0 0	+ 0 1 1 1	+ 1 1 1 0	+ 0 1 0 1

10. Compare your answers for each problem to the waveforms. Do your math results match the output waveforms for each time segment? (Yes/No)

11. Save the editor files to Drive A as **adder3**, then exit the Graphics and Waveform Editors.

Part 4 Procedure

1. Open the Max+plus II software. Assign the project name **adder4**.

2 Open a new Graphic Editor file.

3. Cascade two 7483 chips together to perform the following math process.

$$
\begin{array}{r}
A_7A_6A_5A_4A_3A_2A_1A_0 \\
+B_7\,B_6\,B_5B_4B_3\,B_2B_1B_0 \\
\hline
S_8S_7\,S_6\,S_5\,S_4\,S_3\,S_2\,S_1\,S_0
\end{array}
$$

4. Open a new Waveform Editor file and create a set of waveforms, A[7..0], B[7..0], C_{IN}, S_8 and S[7..0], to demonstrate that your circuit is operating properly.

5. Obtain a hard copy of the Graphic and Waveform Editors after simulation. Label these pages **Part 4, Step 5A** and **Part 4, Step 5B**.

6. Demonstrate the 8-bit adder to your instructor. Obtain the signature of approval on the answer page.

7. Save the editor files to Drive A as **adder4**, then exit the Graphic and Waveform Editors.

Part 5 Procedure

1. Connect the circuit card containing the EPM7128SLC84 chip to the computer's printer port. If the computer has a software key attached to LPT1, it may be necessary to install a parallel I/O card on the computer in order to use LPT2 as the programming port on the computer). See your instructor to determine the correct connection port.

2. Open the Max+plus II software. Assign the project name **adder4** and assign EPM7128SLC84 as the device.

3. Open the **adder4.gdf** file created in Part 4 of this lab.

4. Open the **adder4.scf** file created in Part 4 of this lab.

5. If you are programming a university board containing numerous switches and displays hard-wired to the EPM7128SLC84 CPLD, you must assign pins according to the university board user manual. Failure to adhere to the instructions in the user manual may result in damage to the CPLD. Refer to the instructions "Change the pin assignments" in Appendix A for assigning pins.

6. Run the Compiler and Simulator. Correct all errors before continuing.

7. During compilation, a report file containing technical information including pin assignments for the EPM7128SLC84 was created and stored on your disk. Use a word processor to open this **adder4.rpt** file. Highlight the section of the report file containing the symbol for the 7128S chip with pin assignments. Change the font of this section to Courier 8 point so that the symbol with the pin assignments will be easier to read.

8. Identify and record the input and output pin numbers in Table 3 for the adder4 circuit using the **adder4.rpt** file.

9. Select **Programmer** in the Max+plus II main menu item (left of the File option).

10. Proceed to program the EPM7128SLC84 chip following the instructions in Appendix C or refer to the circuit board manufacturer's programming instructions.

A Inputs	B Inputs	S Ouputs	C_{IN}
A_7: Pin ____ A_6: Pin ____ A_5: Pin ____ A_4: Pin ____ A_3: Pin ____ A_2: Pin ____ A_1: Pin ____ A_0: Pin ____	B_7: Pin ____ B_6: Pin ____ B_5: Pin ____ B_4: Pin ____ B_3: Pin ____ B_2: Pin ____ B_1: Pin ____ B_0: Pin ____	S_8: Pin ____ S_7: Pin ____ S_6: Pin ____ S_5: Pin ____ S_4: Pin ____ S_3: Pin ____ S_2: Pin ____ S_1: Pin ____ S_0: Pin ____	C_{IN}: Pin ____

Table 3

11. Assume the EPM7128SLC84 chip was successfully programmed, construct the circuit shown in Figure 9. Position the LEDs and switches in numeric sequence, with the most significant on the left side. If using a University board with at least 16 switches and 9 LEDs, demonstrate the 8-bit addition, otherwise demonstrate 4-bit addition if hard-wiring switches to the CPLD. Figure 9 shows the wiring configuration for 4-bit addition.

12. Demonstrate the 8-bit adder to the instructor. Obtain the signature of approval directly on the answer page.

13. Create a cover page and write a technical summary pertaining to the results obtained from this lab. Open the summary with stated objectives for the lab followed by a brief discussion on half and full adders. Include your graphic design file from Part 4, Step 3 with a paragraph explaining the purpose of the circuit; also, identify the inputs and outputs, and the most and least significant IC parts.

14. Staple all papers for this lab in the following sequence, then submit the lab to your instructor for grading.
    - Cover page
    - Typed summary
    - The completed answer page for this lab
    - Hard copy of the Graphic Editor, **Part 4, Step 5A**
    - Hard copy of the Waveform Editor, **Part 4, Step 5B**

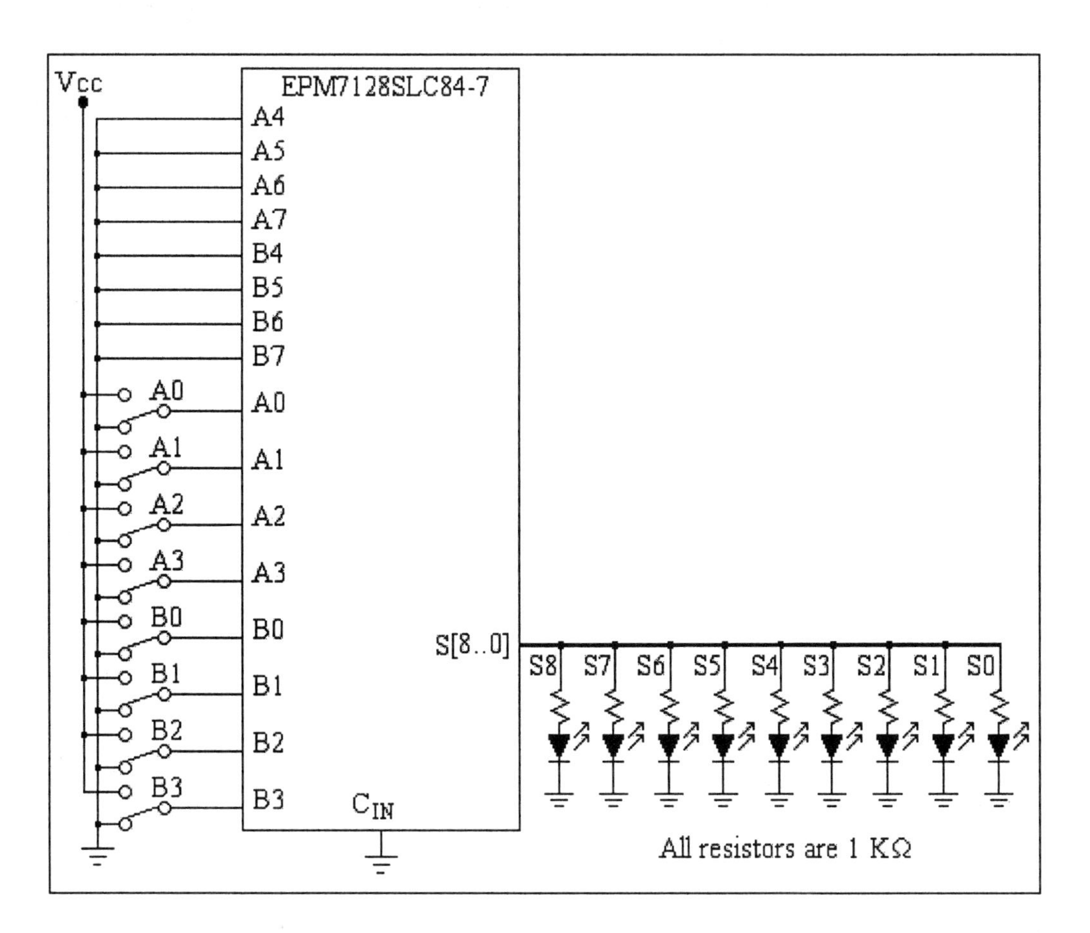
Vcc
EPM7128SLC84-7
A4
A5
A6
A7
B4
B5
B6
B7
A0
A1
A2
A3
B0
B1
B2
B3
S[8..0]
S8 S7 S6 S5 S4 S3 S2 S1 S0
CIN
All resistors are 1 KΩ

Figure 9

Lab 5: Adders Answer Pages

Name: ______________________________

Part 1

2. Carry = ______________________

3. Sum = ______________________

4. Sketch the circuit for Part 1 here.

B	A	Carry	Sum
0	0		
0	1		
1	0		
1	1		

Table 1

Part 2

2. Carry = ____

3. Sum = ____

4. Sketch the circuit for Part 2 here.

C_{IN}	B	A	Carry	Sum
0	0	0		
0	0	1		
0	1	0		
0	1	1		
1	0	0		
1	0	1		
1	1	0		
1	1	1		

Table 2

Part 3

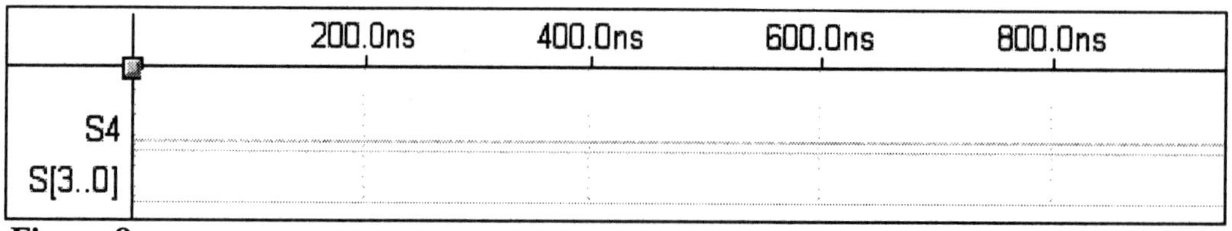

200.0ns 400.0ns 600.0ns 800.0ns

S4

S[3..0]

Figure 8

@100 ns	**@300 ns**	**@500 ns**	**@700 ns**	**@900 ns**
0 0 0 0	0 1 0 1	1 0 1 0	1 1 1 1	0 1 0 0
+ 1 0 0 1	+ 0 0 0 0	+ 0 1 1 1	+ 1 1 1 0	+ 0 1 0 1

10. Yes/No

Part 4

6. Demonstrated to: ____ Date: ____________

Part 5

11. Demonstrated to: ____ Date: ____________

Grade: _____________

Lab 6: Adding and Subtracting

Objectives:

1. Demonstrate 2's complement addition and subtraction of unsigned numbers.
2. Demonstrate 2's complement addition and subtraction of signed numbers.

Materials list:

- Max+plus II software by Altera Corporation
- University Board by Altera Corporation (optional)
- Computer requirements:
  Minimum 486/66 with 8 MB RAM
- Floppy disk

Discussion: Subtraction of unsigned numbers

The basic number system equation for 4-bit unsigned numbers is shown in Equation 1.

$$S_3B^3 + S_2B^2 + S_1B^1 + S_0B^0 \quad (1)$$

S represents the symbol and **B** represents the base of the number system. For example here, the base is 2 for the binary number system so the most significant bit, S_3, represents a weight of 8. Table 1 shows the range for 4-bit unsigned numbers and corresponding hexadecimal and decimal values. All four bits for unsigned numbers determine the magnitude of the number.

Hex Range	Binary S_3 S_2 S_1 S_0	Decimal Value
0	0 0 0 0	0
1	0 0 0 1	1
2	0 0 1 0	2
3	0 0 1 1	3
4	0 1 0 0	4
5	0 1 0 1	5
6	0 1 1 0	6
7	0 1 1 1	7
8	1 0 0 0	8
9	1 0 0 1	9
A	1 0 1 0	10
B	1 0 1 1	11
C	1 1 0 0	12
D	1 1 0 1	13
E	1 1 1 0	14
F	1 1 1 1	15

Table 1 Unsigned numbers

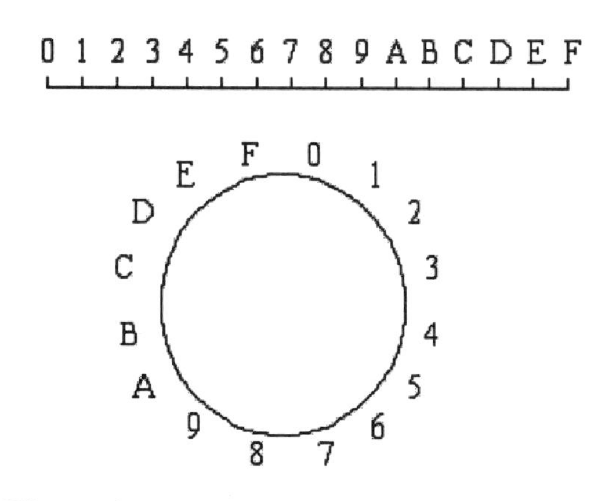

Figure 1

Number systems are often represented by a straight line, to the left of zero are negative and to the right of zero are positive. However, as shown in Figure 1, unsigned numbers are only positive and terminate due to the number of bits. A 4-bit unsigned number has a range of 0 to F_H. If 1 is added to F_H, the answer is 0 but if 1 is subtracted from 0, the answer is F_H. Therefore, the circle in Figure 1 better represents the behavior of a finite range of 4-bit numbers. Addition would be a clockwise direction and subtraction would be counterclockwise.

A circuit to perform addition, 7483, was discussed in Lab 5 so the focus here will be on subtraction. The 2's complement method shown below is used for subtraction.

$$\begin{array}{lcl} A_3A_2A_1A_0 & \rightarrow & A_3A_2A_1A_0 \\ -B_3B_2\,B_1\,B_0 & \rightarrow & +\overline{B_3}\,\overline{B_2}\,\overline{B_1}\,\overline{B_0} \\ \hline S_3\,S_2\,S_1\,S_0 & & \quad\quad +\ 1 \\ & & \overline{C_3S_3\,S_2\,S_1\,S_0} \end{array}$$

If C_3 is a logic-HIGH, the answer, $\mathbf{S_3\ S_2\ S_1\ S_0}$, is positive and in true form. If C_3 is a logic-LOW, the answer, $\mathbf{S_3\ S_2\ S_1\ S_0}$, is negative and in 2's complement form. C_3 is not part of the answer, it only determines what should be done to the answer. Examples of subtraction are shown below.

Example 1:

$$\begin{array}{ccccr} B & \rightarrow & 1011_2 & \rightarrow & 1011_2 \\ \underline{-7} & \rightarrow & \underline{-0111_2} & \rightarrow & +1000_2 \\ & & & & \underline{+1} \\ & & & & 10100_2 \end{array}$$

Since C_3 is a logic-HIGH, 0100_2 is in true form representing +4.

Example 2:

$$\begin{array}{ccccr} 7 & \rightarrow & 0111_2 & \rightarrow & 0111_2 \\ -\underline{B} & \rightarrow & \underline{-1011_2} & \rightarrow & +0100_2 \\ & & & & \underline{+1} \\ & & & & 01100 \end{array}$$

Since C_3 is a logic-LOW, 1100_2 is negative and in 2's complement form. Take the 2's complement or the answer to get the true form, −4.

Addition and subtraction of signed numbers

Examine the basic number system equation for 4-bit signed numbers shown below.

$$-S_3B^3 + S_2B^2 + S_1B^1 + S_0B^0 \qquad (2)$$

As with unsigned numbers, **S** represents the symbol and **B** represents the base (2) of the number system. For signed numbers, a logic-HIGH on $\mathbf{S_3}$, as is the case for numbers 8 to $\mathbf{F_H}$, means the number has a negative value. If $\mathbf{S_3}$ is a logic-LOW, (for 0 to 7) the number is positive. Table 2 shows the range for 4-bit signed numbers and corresponding hexadecimal and decimal values. Three bits, S2, S1, S0 determine the magnitude of the 4-bit signed number.

Hex Range	Binary $S_3\ S_2\ S_1\ S_0$	Decimal Value	Comments
0	0 0 0 0	0	Positive Numbers
1	0 0 0 1	1	
2	0 0 1 0	2	
3	0 0 1 1	3	
4	0 1 0 0	4	
5	0 1 0 1	5	
6	0 1 1 0	6	
7	0 1 1 1	7	
8	1 0 0 0	-8	Negative Numbers
9	1 0 0 1	-7	
A	1 0 1 0	-6	
B	1 0 1 1	-5	
C	1 1 0 0	-4	
D	1 1 0 1	-3	
E	1 1 1 0	-2	
F	1 1 1 1	-1	

Table 2 Signed numbers

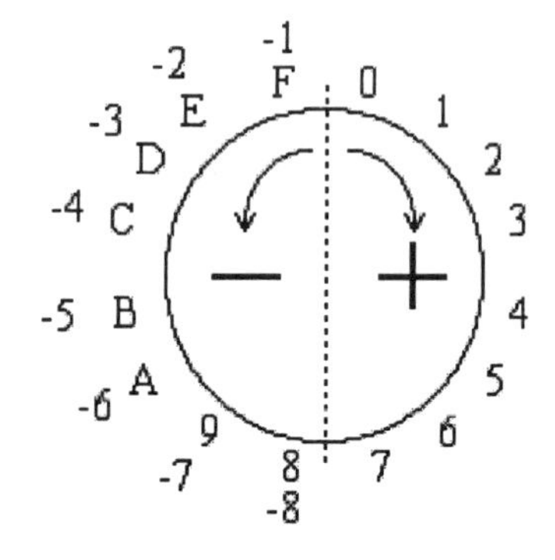

Figure 2

A circle, as shown in Figure 2, can represent the signed number system. Too far in the positive direction will give a negative number and too far in the negative direction yields a positive number. Both of these conditions will result in a false answer, are an overflow condition.

When adding two 4-bit signed numbers, one of several scenarios will result (see Table 3). If adding two positive numbers or if adding two negative numbers, an overflow could exist if the expected answer is beyond the range of the 4-bit number.

Valid processes, no overflow				Invalid processes, there is an overflow	
(+)	(+)	(-)	(-)	(+)	(-)
+(+)	+(-)	+(+)	+(-)	+(+)	+(-)
(+)	either + or −	either + or −	(-)	(-)	(+)

Table 3

For example, 2+3 = 5. Converting each number to binary, 0010 + 0011 = 0101, shows that the most significant bit, Bit 3, of each number is a logic-LOW, or positive, (+) + (+) = (+). Now add 5+4.

```
  0101
+ 0100
  1001
```

5 and 4 are both positive; however, the answer 1001 is negative! As an unsigned number (Table 1), the answer is 1001_2, or 9. But as a signed operation (Table 2), the answer is –7.

Adding two negative numbers may yield a negative answer (desirable), however just like adding two positive numbers, you may get erroneous results. (Remember, Bit 3 is the sign bit for signed numbers)

(-) + (-) = (-)		Valid		(-) + (-) = (+)		Not valid	
-4	1100	A_3	(-)	-6	1010	A_3	(-)
+-3	+1011	+B_3	(-)	+-3	+1011	+B_3	(-)
-7	11001	S_3	(-)	7	10111	S_3	(+)

Since both unsigned and signed number subtraction rely on the 2's complement method, the same circuit using a 7483 will be used for addition and subtraction of signed and unsigned numbers. Part 1 will focus on unsigned numbers, then additional circuitry will be added for signed addition and subtraction in Part 2.

Part 1 Procedure

1. Open the Max+plus II software and assign the project name **lab6_1**. Assign MAX7000S as the device family and the EPM7128SLC84 as the device.

2. Open the Graphic Editor and create the circuit shown in Figure 3. For neatness, all inputs and outputs have been placed on the left of the figure and bus lines are identified with labels, but the physical wires are missing.

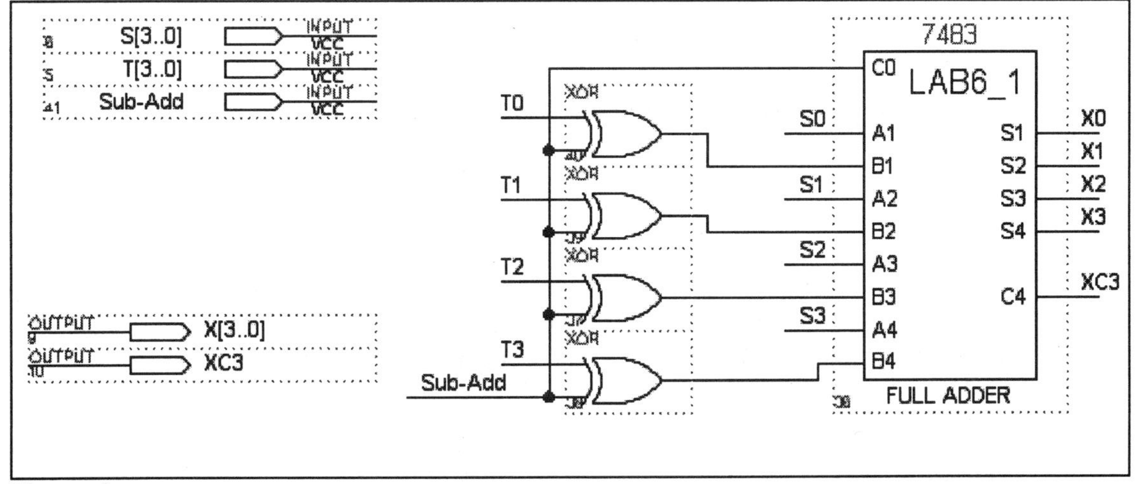

Figure 3

		200.0ns	400.0ns	600.0ns	800.0ns
Sub-Add	0				
S[3..0]	H D	7	A	D	0	3
T[3..0]	H 3	5	C	3	A	1
XC3	X				
X[3..0]	H X		X		

Figure 4

3. Open the Waveform Editor and create the waveforms shown in Figure 4.

4. Click the Compile and Simulate buttons. Correct all errors before continuing.

5. Analyze the results of the Waveform Editor after simulation. From 0 to 500 ns, the circuit (Subtracts/Adds).

6. Complete the following addition and subtraction problems, showing your work in binary, then equate the answers to hexadecimal.

7 ________$_2$
+5 + ________$_2$
= ________$_2$
= ________$_H$

A ________$_2$
+C + ________$_2$
= ________$_2$
= ________$_H$

D ________$_2$
+3 + ________$_2$
= ________$_2$
= ________$_H$

0 ________$_2$ ________$_2$
-A - ________$_2$ + ________$_2$
+____ (C_{IN})
= ________$_2$

The carry out is a logic-(LOW/HIGH).
The answer is (positive/negative).
The answer is in (true/two's complement) form.

3 ________$_2$ ________$_2$
-1 - ________$_2$ + ________$_2$
+____
= ________$_2$

The carry out is a logic-(LOW/HIGH).
The answer is (positive/negative).
The answer is in (true/two's complement) form.

7. Do the predicted answers for Step 6 match the sum and carry output waveforms displayed by the Waveform Editor for the corresponding time segments? (Yes/No)

8. Do the predicted answers for Step 8 match the sum and carry output waveforms displayed by the Waveform Editor for the corresponding time segments? (Yes/No)

9. Save all files as **Lab6_1,** then exit the Graphic and Waveform Editors.

Part 2 Procedure

1. It is desirable to have an error flag set to a logic-HIGH to indicate an overflow condition, V = 1; otherwise be at a logic-LOW state. Based on the valid and invalid conditions shown in Table 3, complete function Table 4 for the overflow bits (logic-HIGH) for errors, then extract the sum-of-products expression for the invalid states. Only write product terms for the logic-HIGH states.

A_3	B_3	S_3	$A_3 + B_3 = S_3$	V	Product term
0	0	0	(+) + (+) = (+)	___	____________
0	0	1	(+) + (+) = (−)	___	____________
0	1	0	(+) + (−) = (+)	___	____________
0	1	1	(+) + (−) = (−)	___	____________
1	0	0	(−) + (+) = (+)	___	____________
1	0	1	(−) + (+) = (−)	___	____________
1	1	0	(−) + (−) = (+)	___	____________
1	1	1	(−) + (−) = (−)	___	____________

Table 4

SOP: V = ________________________________

2. Sketch the overflow circuit based on the equation extracted from Table 4.

3. Open the Max+plus II software and assign the project name **Lab6_2**. Select the 7000S family and the EPM7128LCS84 as the device.

4. Open the Graphic Editor and open the **Lab6_1.gdf** file that was created in Part 1. Add the circuit you sketched for **Lab 6, Part 2, Step 2** to the circuit, then save the resulting circuit as **Lab6_2.gdf**.

5. Open the Waveform Editor and create waveforms shown in Figure 5.

6. Click the Compile and Simulate buttons. Correct all errors before continuing.

7. Evaluate the results as displayed by the Waveform Editor. From 0 to 400 ns the circuit is (Adding/Subtracting). From 400 ns to 1 μs the circuit is (Adding/Subtracting).

8. Complete the addition and subtraction problems at the top of page 74 assuming the A[3..0] and B[3..0] are unsigned numbers. Show all work!

		200.0ns		400.0ns		600.0ns		800.0ns
Sub-Add	0	0 (0–400 ns)		1 (400 ns–1 μs)				
S[3..0]	H 6	F	6		D		4	B
T[3..0]	H 3	C	3		A		1	8
V	0							
XC3	X							
X[3..0]	H X	X						

Figure 5

From 0 to 200 ns		From 200 to 400 ns	
F	______ $_2$	6	______ $_2$
+ C	+ ______ $_2$	+ 3	+ ______ $_2$
	= ______ $_2$		= ______ $_2$
	______ $_H$		______ $_H$

From 400 – 600 ns

	Convert to Binary	Add Using 2's Complement
D	______ $_2$	______ $_2$
- A	- ______ $_2$	+ ______ $_2$
		= ______ $_2$

When adding, S3 = (0/1), T3 = (0/1), and X3 = (0/1)

This addition process was (valid/invalid) so V = (0/1)

From 600 – 800 ns

	Convert to Binary	Add Using 2's Complement
4	______ $_2$	______ $_2$
- 1	- ______ $_2$	+ ______ $_2$
		= ______ $_2$

When adding, S3 = (0/1), T3 = (0/1), and X3 = (0/1)

This addition process was (valid/invalid) so V = (0/1)

	200.0ns	400.0ns	600.0ns	800.0ns
V				
XC3				
X[3..0]				

Figure 6

9. Sketch the output waveforms displayed by the Waveform Editor at the bottom of Figure 6.

10. Save all files as **Lab6_2**, then exit the Graphic and Waveform Editors.

Part 3 Procedure

Supplemental discussion:

Let's assume we add 1+4 with a simple calculator. The operation is either addition or subtraction. The steps are shown as superscripts in the equation to the right of the process steps.

The keystrokes are:

1. Press 1 (0001 is entered into register A.)
2. Do not press +/–
3. Press + (The operator is defined.)
4. Press 4 (0100 is entered into register B.)
5. Do not press +/–
6. Press = (The answer is displayed.)

(sign) Register A	$^{2}( )\ ^{1}0001_2$
(Operator) (sign) Register B	$^{3}+\ ^{5}( )\ ^{4}0100_2$
= Answer	$^{6}=\ 0011_2$

Since the +/– sign key was not pressed, we could consider this an unsigned addition for numbers between 0 and F.

Let's now assume we add –1+4 using a cheap 4-bit calculator.

The keystrokes are:

1. Press 1 (0001 is entered into register A)
2. Press +/– (A negative sign appears in front of 0001. Perform the 2's complement of register A)
3. Press + (The operator is defined)
4. Press 4 (0100 is entered into register B)
5. Do not press +/–
6. Press = (The answer is displayed)

(sign) Register A	$^{2}(-)\ ^{1}0001_2$
2's complement Register A	1111_2
(Operator) (sign) Register B	$^{3}+\ ^{5}(\,)\ ^{4}0100_2$
= Answer	$^{6}\ 1\ 0011_2$

Since the carry bit was a logic-HIGH, the answer 0011_2 is positive and in true form.

Now add –5 and –6.
The keystrokes are:

1. Press 5 (0101 is entered into register A.)
2. Press +/– (A negative sign appears in front of 0101. Perform the 2's complement of register A.)
3. Press + (The operator is defined.)
4. Press 6 (0110 is entered into register B.)
5. Press +/– (a negative sign appears in front of 0110. Perform the 2's complement of register B.)
6. Press = (Overflow occurs, V=1, because when adding 1011 (Step 2) + 1010 (Step 5), the answer was 0101. Two negatives cannot be a positive. Now we see an error displayed, not the correct answer.)

(sign) Register A	$^{2}(-)\ ^{1}0101_2$
2's complement Register A	1011_2
(Operator) (sign) Register B	$^{3}+\ ^{5}(-)\ ^{4}0110_2$
2's complement Register B	1010_2
= Answer	$^{6}\ 1\ 0101_2$

This last scenario is of interest. We may need to take the 2's complement of input A, take the 2's complement of input B, and may need to complement the complement of input B if this was a subtraction process, (–)5 - (–)6.

1. Open the Max+plus II software and assign the project name **Lab6_3**. Assign MAX700S as the device family and EPM7128SLC84 as the device.

2. Open the Graphic Editor, then modify **Lab6_2.gdf** to match the circuit shown in Figure 7.

3. Open the Waveform Editor and create waveforms shown in Figure 8.

4. Click the Compile and Simulate buttons. Correct all errors before continuing.

5. Predict the answers and the logic level of the overflow bit for the following problems.

1	1	1	(–)1	(–) 1	1	(–)1	(–) 1
+4	- 4	+(–)4	+ 4	+ (–)4	- (–)4	- 4	- (–)4
V = ___	V = ___	V = ___	V = ___	V = ___	V = ___	V = ___	V = ___

6. Compare your predicted results for Step 5 to the waveforms displayed by the Waveform Editor. Your predictions should match the waveforms; if not, you have errors in the graphic design file or else your predictions were wrong. Correct your mistakes.

7. Obtain a hard copy of the Graphic and Waveform Editors files. Mark these hard copies **Part 3, Step 7A** and **Part 3, Step 7B**, respectively.

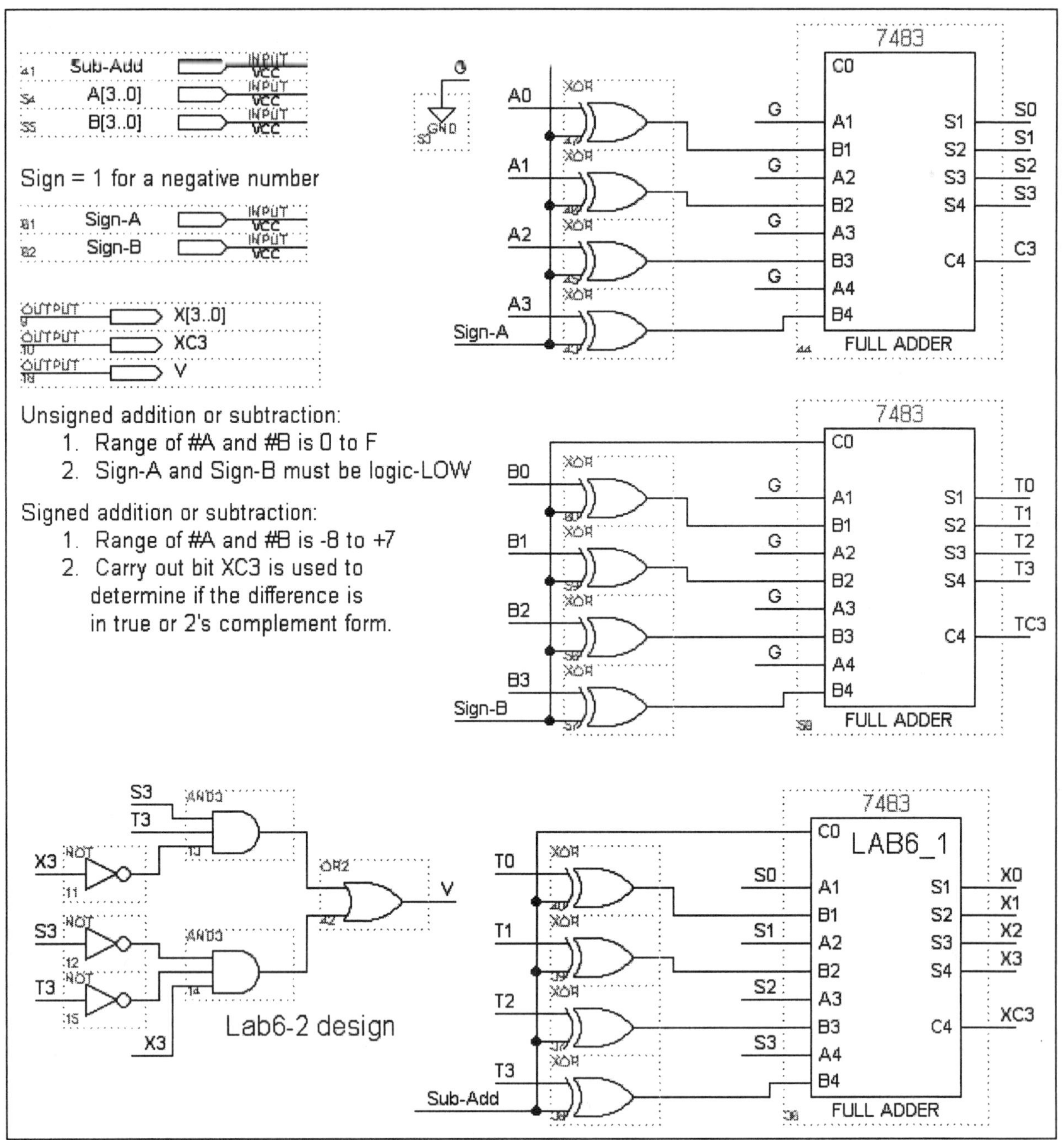

Figure 7

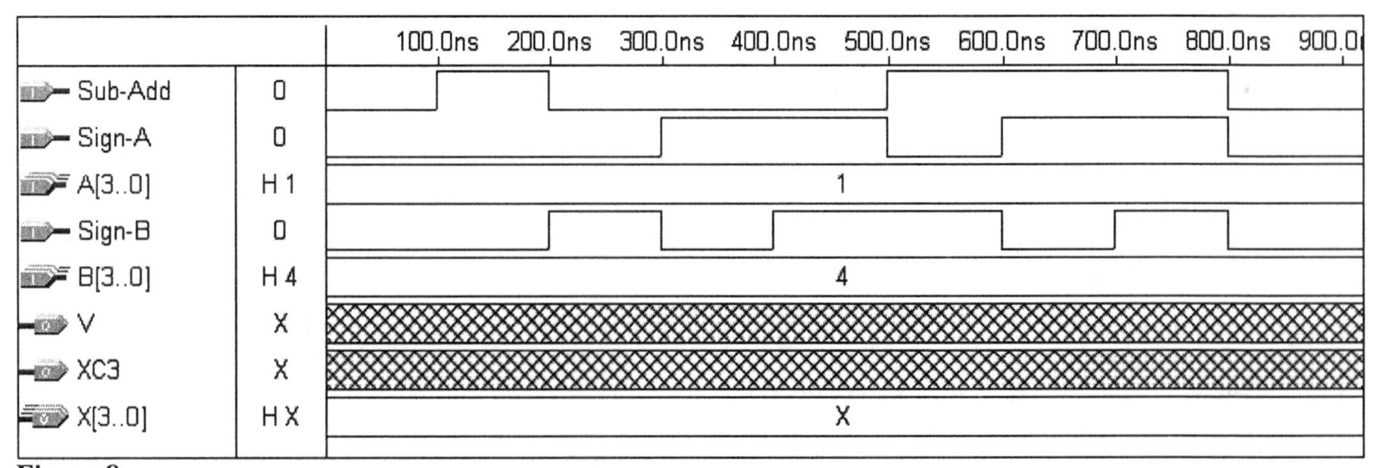

Figure 8

8. Demonstrate the Graphic and Waveform Editors files to the instructor. Obtain the signature of approval directly on the answer page for this lab.

9. Save your files as **Lab6_3,** then exit the Graphic and Waveform Editors.

10. Write a technical summary based on the results obtained for this lab. Your summary should include an embedded graphic to supplement your discussion.

11. Staple and submit the following in the sequence listed to your instructor for grading.

    - Cover page
    - Typed summary
    - Hard copy of the Graphic Editor file, **Part 3, Step 7A**
    - Hard copy of the Waveform Editor file, **Part 3, Step 7B**
    - The completed answer page for this lab.

Lab 6: Adding and Subtracting Answer Pages

Name: ______________________

Part 1

5. (Adds/Subtracts)

6.

7		________$_2$
+5	+	________$_2$
	=	________$_2$
	=	________$_H$

A		________$_2$
+C	+	________$_2$
	=	________$_2$
	=	________$_H$

D		________$_2$
+3	+	________$_2$
	=	________$_2$
	=	________$_H$

0 ________$_2$ ________$_2$

−A − ________$_2$ + ________$_2$

\+ ___ (C_{IN})

= ________$_2$

The carry out is a logic-(LOW/HIGH).
The answer is (positive/negative).
The answer is in (true/2's complement) form.

3 ________$_2$ ________$_2$

−1 − ________$_2$ + ________$_2$

\+ ___

= ________$_2$

The carry out is a logic-(LOW/HIGH).
The answer is (positive/negative).
The answer is in (true/2's complement) form.

7. (Yes/No)

8. (Yes/No)

Part 2

1.

A_3	B_3	S_3	$A_3 + B_3 = S_3$	V	Product term
0	0	0	(+) + (+) = (+)	___	________________
0	0	1	(+) + (+) = (−)	___	________________
0	1	0	(+) + (−) = (+)	___	________________
0	1	1	(+) + (−) = (−)	___	________________
1	0	0	(−) + (+) = (+)	___	________________
1	0	1	(−) + (+) = (−)	___	________________
1	1	0	(−) + (−) = (+)	___	________________
1	1	1	(−) + (−) = (−)	___	________________

Table 4

SOP: V = ____

7. 0 to 400 ns: (Adding/Subtracting) 400 ns to 1 μs: (Adding/Subtracting).

8. From 0 to 200 ns

F		________$_2$
+C	+	________$_2$
	=	________$_2$
		________$_H$

From 200 to 400 ns

6		________$_2$
+3	+	________$_2$
	=	________$_2$
		________$_H$

8. (continued)

From 400 – 600 ns

	Convert to Binary	Add Using 2's Complement
D	___________$_2$	___________$_2$
– A	– ___________$_2$	+ ___________$_2$
		= ___________$_2$

When adding, S3 = (0/1), T3 = (0/1), and X3 = (0/1)

This addition process was (valid/invalid) so V = (0/1)

From 600 – 800 ns

	Convert to Binary	Add Using 2's Complement
4	___________$_2$	___________$_2$
– 1	– ___________$_2$	+ ___________$_2$
		= ___________$_2$

When adding, S3 = (0/1), T3 = (0/1), and X3 = (0/1)

This addition process was (valid/invalid) so V = (0/1)

	200.0ns	400.0ns	600.0ns	800.0ns
V				
XC3				
X[3..0]				

Figure 6

Part 3

5.

1 + 4	1 – 4	1 +(–) 4	(–) 1 + 4	(–) 1 + (–) 4	1 – (–) 4	(–) 1 – 4	(–) 1 – (–) 4
V = ___	V = ___	V = ___	V = ___	V = ___	V = ___	V = ___	V = ___

8. Demonstrated to: ________________________________ Date: ________________

Grade: ______________

Lab 7: Comparators

Objectives:

1. Use the Exclusive-OR gate as a comparator
2. Build a 4-bit magnitude comparator using basic logic gates
3. Use the 7485 4-bit magnitude comparator for 4-bit and 8-bit binary comparisons

Materials List:

- Max+plus II software by Altera Corporation
- University board with CPLD (optional)
- Computer requirements:
  - Minimum 486/66 with 8 MB RAM
- Floppy disk

Discussion:

A comparator is a circuit that determines the relationship of two numbers. Typical outputs of the comparator are greater than (A > B), less than (A < B), and equal to (A = B). Either the exclusive-OR or the exclusive-NOR gates are used for equality comparisons. AND, OR, and NOT gates are combined to determine inequality.

Examine the circuit and waveforms shown in Figure 1. Output AequalB is a logic-HIGH when inputs A and B are the same logic level. This circuit produces an "active-HIGH" output based on equality.

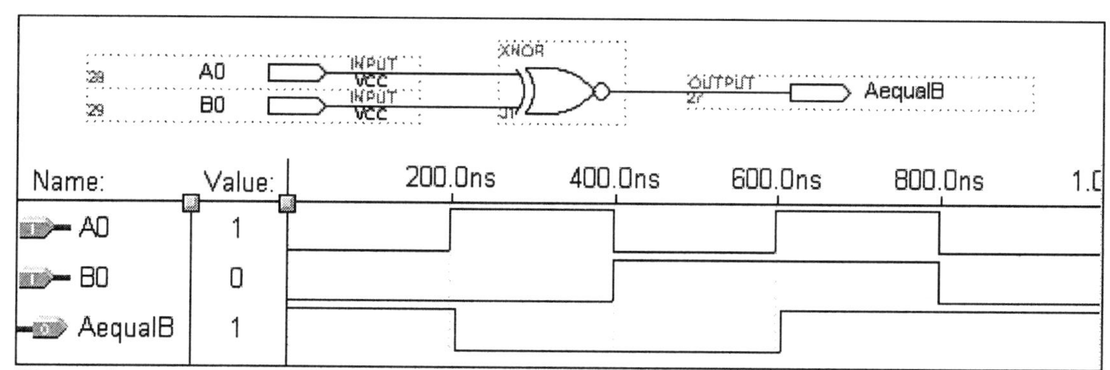

Figure 1

When comparing two 4-bit numbers, $A_3A_2A_1A_0$ to $B_3B_2B_1B_0$, equality can be determined if each of the corresponding bits are identical. Intermediate signals, $i_3i_2i_1i_0$, may be outputs of XOR gates or XNOR gates used to determine equality. For instance, examine the following statements.

$$\text{XOR:} \quad i_3 = A_3 \oplus B_3 = 0 \text{ for equality}$$

$$\text{XNOR:} \quad i_3 = \overline{A_3 \oplus B_3} = 1 \text{ for equality}$$

If using XOR gates to determine equality, the intermediate signals may be ORed to produce an active-LOW output or NORed to produce an active-HIGH output when A is equal to B. Either the AND or NAND gate may be used with the XNOR gates to produce the desired active level on the output for equality. (Figure 2)

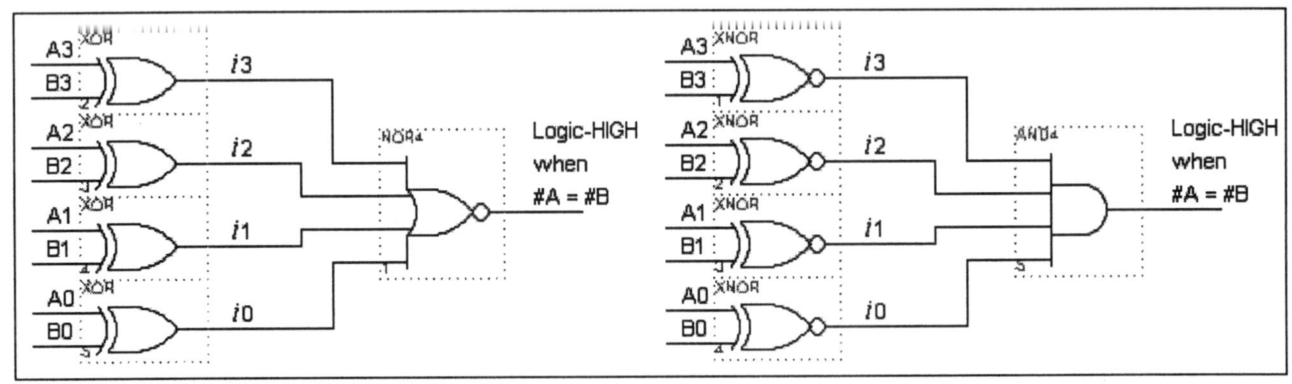

Figure 2

The equation for equality, based on XNOR gates, is: $(\text{A} = \text{B}) = \overline{\text{A}_3 \oplus \text{B}_3} \bullet \overline{\text{A}_2 \oplus \text{B}_2} \bullet \overline{\text{A}_1 \oplus \text{B}_1} \bullet \overline{\text{A}_0 \oplus \text{B}_0}$ (1)

When A is greater than B, use equation 2. $(\text{A} > \text{B}) = \text{A}_3\overline{\text{B}_3} + i_3\text{A}_2\overline{\text{B}_2} + i_3 i_2\text{A}_1\overline{\text{B}_1} + i_3 i_2 i_1\text{A}_0\overline{\text{B}_0}$ (2)

When A is less than B, use equation 3. $(\text{A} < \text{B}) = \text{B}_3\overline{\text{A}_3} + i_3 B_2\overline{\text{A}_2} + i_3 i_2 B_1\overline{\text{A}_1} + i_3 i_2 i_1 B_0\overline{\text{A}_0}$ (3)

Lucky for us, the 7485 integrated circuit is a 4-bit magnitude comparator with cascadable input and outputs for bigger number comparisons. The 7485 will be used for 4- and 8-bit comparisons.

Part 1 Procedure

1. Open the Max+plus II software. Assign the project name **compare2**, MAX7000S as the device family, and EPM7128LC84 as the device.

2. Open a new Graphic Editor and construct the circuit of Figure 3.

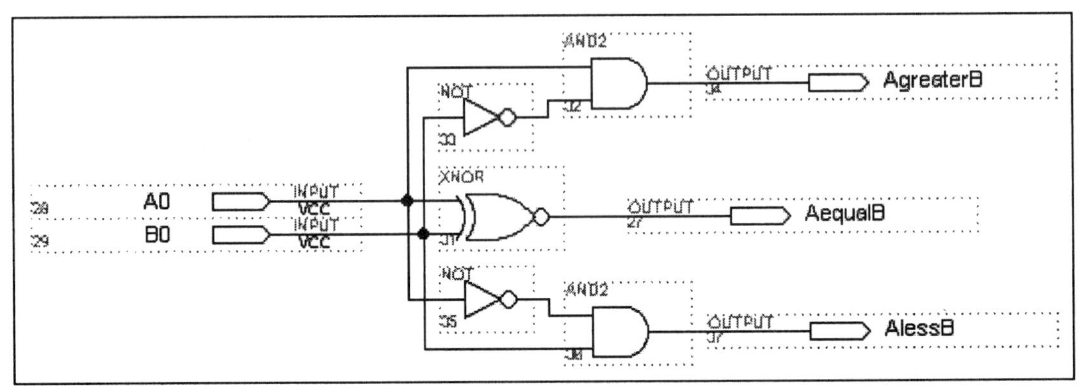

Figure 3

3. Open a new Waveform Editor, set Grid Size to 200 ns, then create waveforms A0 and B0 as shown in Figure 1. Create the three output waveforms identified in Figure 3.

4. Run the Compiler and Simulator. Correct all errors before continuing.

5. Analyze the output waveforms with respect to the inputs, then complete Table 1.

6. Save all files to Drive A as **compare2**, then exit the Graphic and Waveform Editors.

Inputs		Outputs		
B_0	A_0	A>B	A<B	A=B
0	0	___	___	___
0	1	___	___	___
1	0	___	___	___
1	1	___	___	___

Table 1

Part 2 Procedure

1. Open the Max+plus II software. Assign the project name **compare3**.

2. Open a new Graphic Editor and construct the circuit shown in Figure 4.

3. Open a new Waveform Editor, set Grid Size to 100 ns, and create the waveforms shown in Figure 5.

4. Run the Compiler and Simulator. Correct all errors before continuing.

5. Neatly and accurately draw the output waveforms in the space provided in Figure 5.

6. Analyze the output waveforms with respect to inputs, then complete Table 2.

7. Is input A1 dominant over input A0? (Yes/No)

8. Name the output(s) that take on an active-HIGH characteristic. ______________________

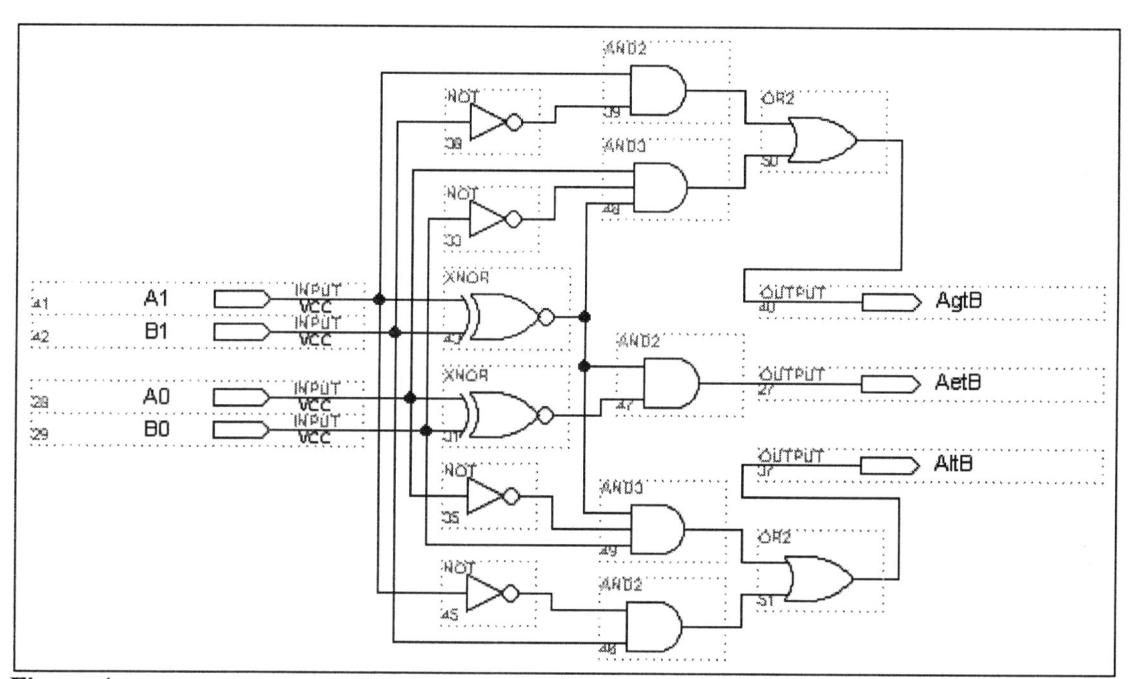

Figure 4

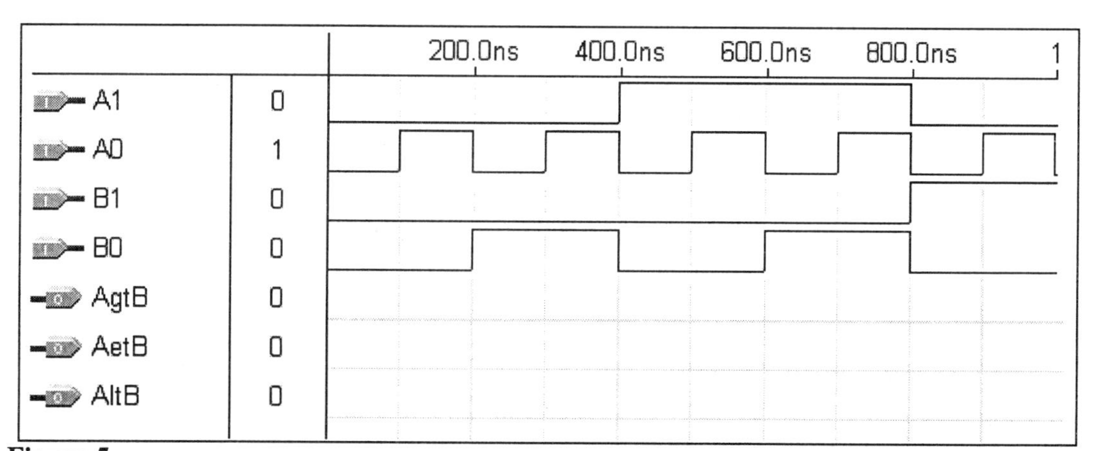

Figure 5

Time	Inputs		Outputs		
	B_1B_0	A_1A_0	A>B	A<B	A=B
50 ns	0 0	0 0			
150 ns	0 0	0 1			
250 ns	0 1	0 0			
350 ns	0 1	0 1			
450 ns	0 0	1 0			

Time	Inputs		Outputs		
	B_1B_0	A_1A_0	A>B	A<B	A=B
550 ns	0 0	1 0			
650 ns	0 1	1 1			
750 ns	0 1	1 0			
850 ns	1 0	0 1			
950 ns	1 0	0 0			

Table 2

9. Name the output(s) that take on an active-HIGH characteristic. ________________________

10. Save all files to Drive A as **compare3**, then exit the Graphic and Waveform Editors.

Part 3 Procedure

1. Open the Max+plus II software. Assign the project name **compare4**.

2. Open a new Graphic Editor and construct the circuit shown in Figure 7.

3. Open a new Waveform Editor, set the Grid Size to 100 ns, and construct the waveforms shown in Figure 8.

4. Highlight waveform A[3..0] and click the Over Write Group button in the Draw tool bar (see Figure 6). Enter "C" for the Group Value. Assign the Group Value "7" to waveform B[3..0].

Figure 6

5. Compile and simulate the circuit. Draw the output waveforms in the space provided in Figure 8.

6. Did the logic level of the cascade inputs affect the outputs of the 7485 when A[3..0] = "C" and B[3..0] = "7"? (Yes/No) Explain why or why not.

____

____

7. Change B[3..0] to a value of "D" from 0 ns to 1 μs.

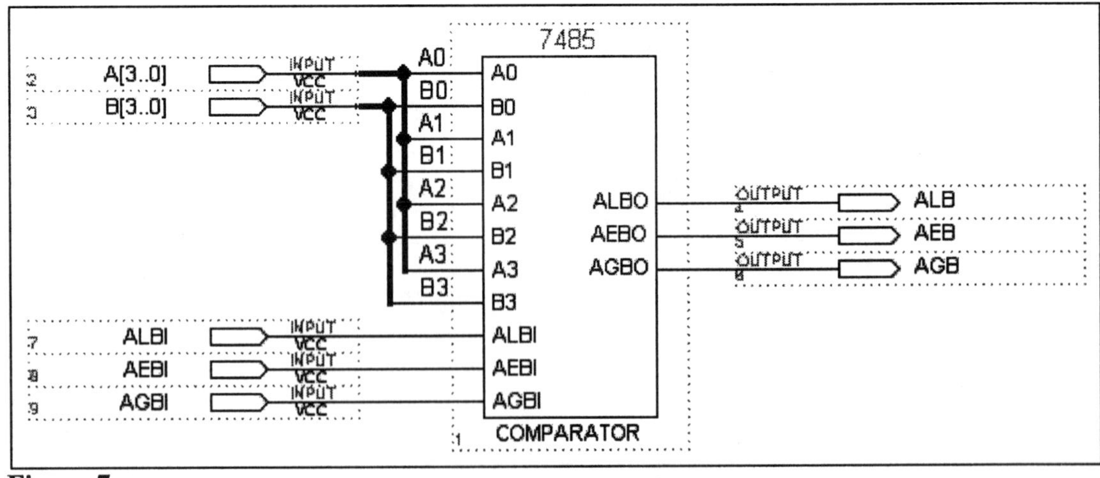

Figure 7

Name:	Value:	200.0ns	400.0ns	600.0ns	800.0ns	1.0
A[3..0]	H X	C				
B[3..0]	H X	7				
ALBI	X					
AGBI	X					
AEBI	X					
ALB	X					
AGB	X					
AEB	X					

Figure 8

8. Explain how the circuit operation (after simulation) was altered when B[3..0] is "D."

   ____

9. Change B[3..0] to a value of "C" from 0 ns to 1 μs, then click the Simulate button.

10. Complete Table 3 based on the results when A[3..0] and B[3..0] contain "C_H."

Time Segment when	Cascade Inputs A>B	Cascade Inputs A<B	Cascade Inputs A=B	Outputs A>B	Outputs A<B	Outputs A=B
0 ns to 100 ns	___	___	___	___	___	___
100 to 200 ns	___	___	___	___	___	___
200 to 300 ns	___	___	___	___	___	___
300 to 400 ns	___	___	___	___	___	___
400 to 500 ns	___	___	___	___	___	___
500 to 600 ns	___	___	___	___	___	___
600 to 700 ns	___	___	___	___	___	___
700 to 800 ns	___	___	___	___	___	___
800 to 900 ns	___	___	___	___	___	___
900 ns to 1 μs	___	___	___	___	___	___

Table 3

11. Based on the results recorded for Table 3, which cascade input is dominant? ____________

12. Save all files to Drive A as **compare4**, then exit the Graphic and Waveform Editors.

Part 4 Procedure

1. Open the Max+plus II software and assign the project name **compare5**.

2. Create an 8-bit magnitude comparator using two 7485s with data inputs A[7..0] and B[7..0]. The cascade inputs must be hard-wired to either Vcc or ground as required by the IC specifications.

3. Demonstrate that the 8-bit magnitude comparator produces the desired output comparisons, A[7..0] > B[7..0], A[7..0] = B[7..0], and A[7..0] < B[7..0], for various values of numbers A and B.

4. Demonstrate your 8-bit magnitude comparator to your instructor. Obtain the signature of approval on the answer page for this lab.

5. Obtain hard copies of the Graphic and Waveform Editors showing the 8-bit magnitude comparator. Label these hard copies **Part 4, Step 5A** and **Part 4, Step 5B**.

6. Exit the Max+plus II software, saving all files to Drive A as **compare5** upon exit.

7. Write a summary pertaining to the results obtained from this lab. Include tables and figures to supplement your discussion.

8. Place all papers for this lab in the following sequence, then submit the materials to your instructor for grading.

    - Cover page
    - Typed summary
    - The completed answer page
    - Hard copy of the Graphic Editor, **Part 4, Step 5**
    - Hard copy of the Waveform Editor, **Part 4, Step 5**

Lab 7: Comparators Answer Pages

Name: ______________________________

Part 1

Inputs		Outputs		
B_0	A_0	A>B	A<B	A=B
0	0	___	___	___
0	1	___	___	___
1	0	___	___	___
1	1	___	___	___

Table 1

Part 2

7. (Yes/No)

8. ______________________________

200.0ns 400.0ns 600.0ns 800.0ns 1

AgtB

AetB

AltB

Figure 5

Time	Inputs		Outputs		
	B_1B_0	A_1A_0	A>B	A<B	A=B
50 ns	0 0	0 0			
150 ns	0 0	0 1			
250 ns	0 1	0 0			
350 ns	0 1	0 1			
450 ns	0 0	1 0			
550 ns	0 0	1 0			
650 ns	0 1	1 1			
750 ns	0 1	1 0			
850 ns	1 0	0 1			
950 ns	1 0	0 0			

Table 2

9. ______________________________

Part 3

6. (Yes/No) Explain why or why not.

____

____

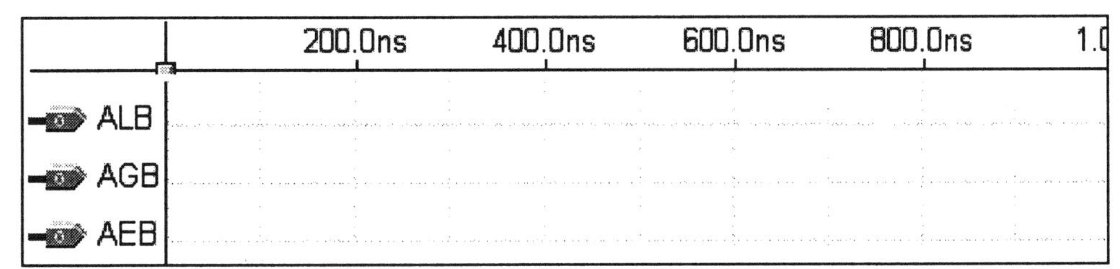

Figure 6

8. ____

Time Segment when	Cascade Inputs			Outputs		
	A>B	A<B	A=B	A>B	A<B	A=B
0 ns to 100 ns	___	___	___	___	___	___
100 to 200 ns	___	___	___	___	___	___
200 to 300 ns	___	___	___	___	___	___
300 to 400 ns	___	___	___	___	___	___
400 to 500 ns	___	___	___	___	___	___
500 to 600 ns	___	___	___	___	___	___
600 to 700 ns	___	___	___	___	___	___
700 to 800 ns	___	___	___	___	___	___
800 to 900 ns	___	___	___	___	___	___
900 ns to 1 μs	___	___	___	___	___	___

Table 3

11. ________________

Part 4

4. Demonstrated to: ________________________________ Date: _______________

Grade: _________

LAB 8: Parity

Objectives:

1. Implement a basic parity circuit using XOR gates
2. Use the 74180 as a parity generator
3. Use the 74180 as a parity checker
4. Analyze a parity generator/checker circuit and modify the circuit to eliminate undesirable glitches in the output signal

Materials List:

- Max+plus II software by Altera Corporation
- University Board by Altera Corporation (optional)
- Computer requirement:
  Minimum 486/66 with 8 MB RAM
- Floppy disk

Discussion:

Transmitted data in a digital system may be altered due to external electrical interference or an internal defective part. If the system generates and transmits a parity bit with the data, the parity checker at the receiver end will generate an error signal. A parity bit is generated during the memory write cycle time and is checked during the memory read cycle time. If an error is detected, a memory error message is displayed on the monitor and the system is halted. An error detected in data received by a modem may send an interrupt signal to the transmitter requesting that the data be retransmitted. The operator may not know that transmission errors occurred.

This lab shows how to use the XOR gate to create a parity generator, then illustrates how to wire a 74280 integrated circuit as a parity generator or checker. The parity generator, parity checker, and control signals will be combined to create a parity error detection circuit typical in a computer or modem system.

Part 1 Procedure

1. Open the Max+plus II software and assign the project name **parity1** and MAX7000S as the device family.

2. Open a new Graphic Editor file, then construct the circuit shown in Figure 1.

3. Open a new Waveform Editor file, set Grid Size to 100 ns, and construct the waveforms shown in Figure 2.

4. Sketch the output waveform in the space provided in Figure 2 after simulating the circuit.

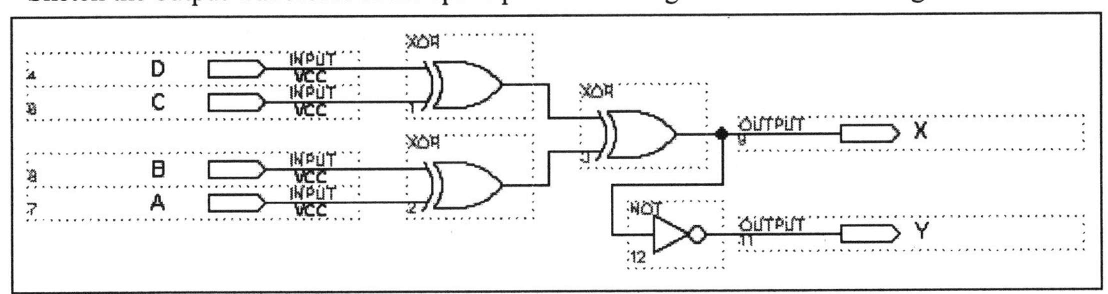

Figure 1

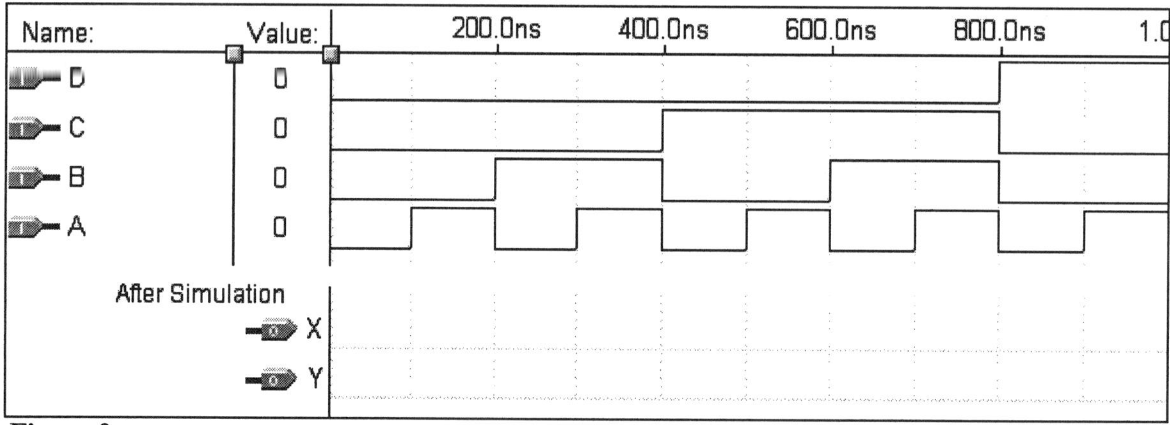

Figure 2

Time Segment	X D C B A (in binary)	Number of 1s in X D C B A	Y D C B A (in binary)	Number of 1s in Y D C B A
0 ns-100 ns				
100 ns-200 ns				
200 ns-300 ns				
300 ns-400 ns				
400 ns-500 ns				
500 ns-600 ns				
600 ns-700 ns				
700 ns-800 ns				
800 ns-900 ns				
900 ns-1 μs				

Table 1

5. Complete Table 1.

6. Based on the results obtained in Figure 1, the "X" output is an (ODD/EVEN) parity generator and the "Y" output is an (ODD/EVEN) parity generator.

7. Save both editor files to Drive A as **parity1**, then exit the Graphic and Waveform Editors.

Part 2 Procedure

Rarely will you find a parity circuit constructed out of XORs. Designers will use integrated circuits such as the 74280, a 9-Bit Odd/Even Parity Generator/Checker used in this lab. This IC can be used for either Even or Odd parity systems and has a cascadable input for *n*-bits.

1. Open the Max+plus II software and assign the project name **parity2**.

2. Open a new Graphic Editor file, then construct the circuit shown in Figure 3.

3. Open a new Waveform Editor file, set Grid Size to 100 ns, and construct the waveforms shown in Figure 4. Set **Starting Value:** to 17 and **Increment By:** to 23 in the Overwrite Count Value dialog box.

4. Sketch the output waveform in the space provided in Figure 4 after simulating the circuit.

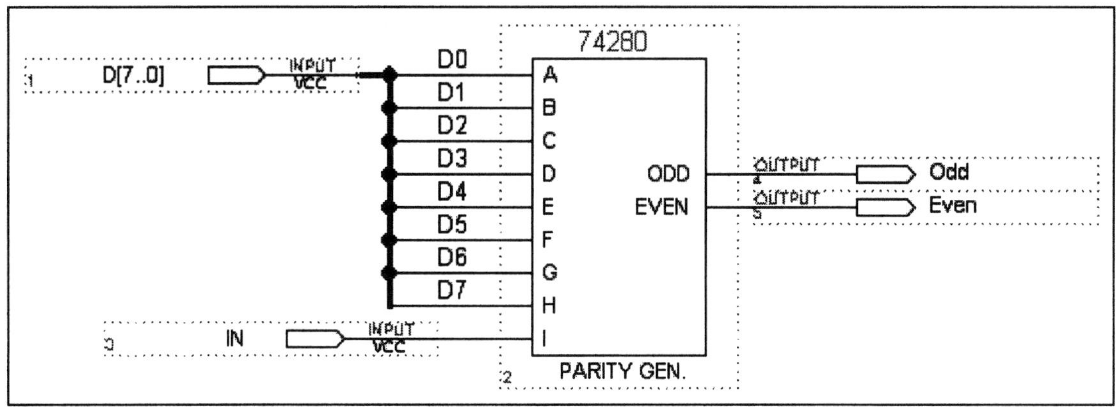

Figure 3

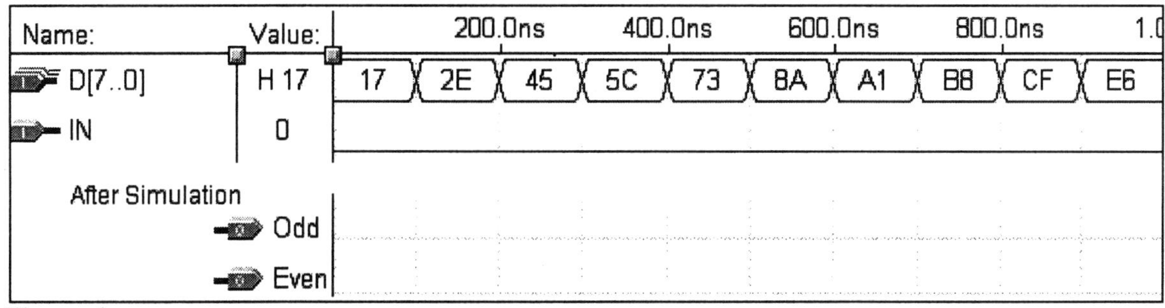

Figure 4

5. Complete Table 2 showing the odd and even output bits for each data applied to the 74280 in Figure 3.

D[7..0]	D[7..0] (in binary)	Odd Output Bit	Even Output Bit
17	____________$_2$	_____	_____
2E	____________$_2$	_____	_____
45	____________$_2$	_____	_____
5C	____________$_2$	_____	_____
73	____________$_2$	_____	_____
8A	____________$_2$	_____	_____
A1	____________$_2$	_____	_____
B8	____________$_2$	_____	_____
CF	____________$_2$	_____	_____
E6	____________$_2$	_____	_____

Table 2

6. Write a paragraph explaining the results obtained in Figure 4 with respect to the circuit in Figure 3.

____

____

____

____

7. Save both editor files to Drive A as **parity2**, then exit the Graphic and Waveform Editors.

Part 3 Procedure

A system using parity error detection will be designed for either EVEN or ODD parity. If used in a computer for reading or writing RAM memory, the parity choice is determined by the designer. Usually the communications software used for transmitting data (over the Internet) will auto detect the parity being transmitted; however, for some software, the operator must set his or her parity to match the parity of the system to which it is connected. The rest of this lab is based on a parity generator/checker system that may be used inside a computer for reading and writing RAM memory.

1. Open the Max+plus II software and assign the project name **parity3**.

2. Open a new Graphic Editor file, then construct the circuit shown in Figure 5. Insert, but do not wire, the GND symbol at this time.

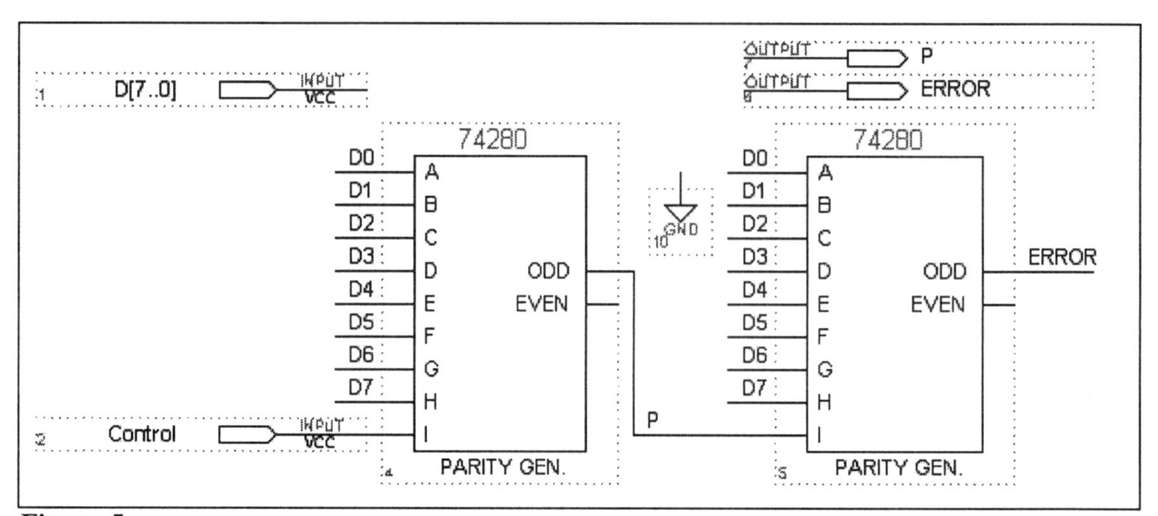

Figure 5

3. Open a new Waveform Editor file, set Grid Size to 100 ns, and construct the waveforms shown in Figure 6. For the D[7..0] waveform, set **Starting Value:** to 35 and **Increment By:** to 27 in the Overwrite Count Value dialog box.

4. Click the Simulate button, then sketch the outputs displayed by the Simulator in the area provided in Figure 7. The Error waveform has numerous spikes, called glitches, that occur when the data is changing states. Glitches are common to asynchronous circuits. Additional circuitry would be necessary to synchronize the timing of when the error bit is to be read.

5. Complete Table 3 for the time segment 0 ns to 500 ns.

6. The parity output bit, P, represents (EVEN/ODD) parity for the time segment 0 ns to 500 ns when the input control line is a logic-LOW.

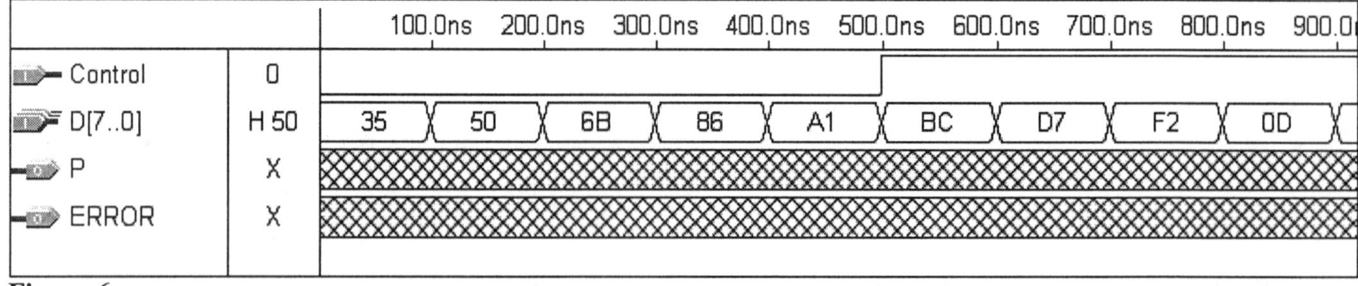

Figure 6

	100.0ns	200.0ns	300.0ns	400.0ns	500.0ns	600.0ns	700.0ns	800.0ns	900.0ns
P									
ERROR									

Figure 7

Control	D[7..0]	Data[7..0] (in binary)	ODD (1 or 0)	ODDC (1 or 0)
0	35	________$_2$	____	____
0	50	________$_2$	____	____
0	6B	________$_2$	____	____
0	86	________$_2$	____	____
0	A1	________$_2$	____	____

Table 3

7. The Error output from 0 ns to 500 ns represents active-(LOW/HIGH) levels when an error is detected.

8. Complete Table 4 for the time segment 500 ns to 1 μs.

Control	D[7..0]	Data[7..0] (in binary)	P (1 or 0)	Error (1 or 0)
1	BC	________$_2$	____	____
1	D7	________$_2$	____	____
1	F2	________$_2$	____	____
1	0D	________$_2$	____	____
1	28	________$_2$	____	____

Table 4

9. The parity output, P, represents (EVEN/ODD) parity for the time segment 500 ns to 1 μs when the input control line is a logic-HIGH.

10. The Error output from 500 ns to 1 μs represents active- (LOW/HIGH) levels when an error is detected.

11. Based on your observation, which 74280 IC in Figure 5 is the Parity Generator? (Left/Right)

12. You should notice that Error is a constant value, if you ignore the undesirable glitches when the circuit does not detect an error (either from 0 to 500 ns when Control = 0 or from 500 ns to 1 μs when Control = 1). If an error occurs, Error will become "active" to represent the error.

13. To simulate an error, connect the GND symbol to the D0 input of the 74280 symbol on the right. Delete the D0 label on this input that has been grounded.

14. Click the Compile and Simulate buttons.

15. The Error output is now "active" whenever bit 0 of the data bus is a logic-HIGH. No errors occur when bit 0 of the data bus is a logic-LOW since GND is also a logic–LOW.

16. Demonstrate the Graphic and Waveform Editors files to the instructor. Obtain the signature of approval on the answer page for this lab.

17. Obtain a hard copy of the Graphic and Waveform Editors files. Label these hard copies **Part 3, Step 17A** and **Part 3, Step 17B**, respectively.

18. Save both editor files to drive A as **parity3**, then exit the Graphic and Waveform Editors.

19. Write a 1- to 2-page summary pertaining to the results obtained from this lab. Compare and contrast the results obtained for Figure 5 when D0 of the parity checker was grounded. Include embedded figures and waveforms in your summary.

20. Place all papers for this lab in the following sequence, then submit the lab to your instructor for grading.

- Cover page
- Typed summary
- The completed answer page for this lab
- Hard copy of the Graphics Editor, **Part 3, Step 17A**
- Hard copy of the Waveform Editor, **Part 3, Step 17B**

Lab 8: Parity Answer Pages

Name:______________________________

Part 1

	200.0ns	400.0ns	600.0ns	800.0ns	1.0
X					
Y					

Figure 2

Time Segment	X D C B A (in binary)	Number of 1s in X D C B A	Y D C B A (in binary)	Number of 1s in Y D C B A
0 ns–100 ns	________	____	________	____
100 ns–200 ns	________	____	________	____
200 ns–300 ns	________	____	________	____
300 ns–400 ns	________	____	________	____
400 ns–500 ns	________	____	________	____
500 ns–600 ns	________	____	________	____
600 ns–700 ns	________	____	________	____
700 ns–800 ns	________	____	________	____
800 ns–900 ns	________	____	________	____
900 ns–1 µs	________	____	________	____

Table 1

6. X: (ODD/EVEN) Y: (ODD/EVEN)

Part 2

	200.0ns	400.0ns	600.0ns	800.0ns	1.0
Odd					
Even					

Figure 4

6. ______________________________

____

____

____

____

____

____

____

____

D[7..0]	D[7..0] (in binary)	Odd Output Bit	Even Output Bit
17	________$_2$	____	____
2E	________$_2$	____	____
45	________$_2$	____	____
5C	________$_2$	____	____
73	________$_2$	____	____
8A	________$_2$	____	____
A1	________$_2$	____	____
B8	________$_2$	____	____
CF	________$_2$	____	____
E6	________$_2$	____	____

Table 2

Part 3

	100.0ns	200.0ns	300.0ns	400.0ns	500.0ns	600.0ns	700.0ns	800.0ns	900.0ns
P									
ERROR									

Figure 7

6. EVEN ODD

7. LOW HIGH

9. EVEN ODD

10. LOW HIGH

11. LEFT RIGHT

Control	D[7..0]	Data[7..0] (in binary)	ODD (1 or 0)	ODDC (1 or 0)
0	35	________$_2$	____	____
0	50	________$_2$	____	____
0	6B	________$_2$	____	____
0	86	________$_2$	____	____
0	A1	________$_2$	____	____

Table 3

Control	D[7..0]	Data[7..0] (in binary)	P (1 or 0)	Error (1 or 0)
1	BC	________$_2$	____	____
1	D7	________$_2$	____	____
1	F2	________$_2$	____	____
1	0D	________$_2$	____	____
1	28	________$_2$	____	____

Table 4

16. Demonstrated to: ______________________________ Date: ____________

Grade: ________

Lab 9: Encoders

Objectives:

1. Construct a basic encoder using logic gates
2. Verify that the 74147 and 74148 ICs operate according to their respective function tables in the data specification sheets
3. Cascade multiple 74148s to create a 16-line to 4-line encoder

Materials List:

- Max+plus II software by Altera Corporation
- Computer requirements:
  Minimum 486/66 with 8 MB RAM
- University Board
- Floppy disk

Discussion:

The encoder is a circuit that converts intelligent information like the keys on a keyboard to a binary-based code of 1s and 0s. Two encoders in chip form are the 74147, 10-line to 4-line BCD encoder, and the 74148 8-line to 3-line octal encoder. Both are priority encoders in that the highest numbered active input dictates the output logic levels. The logic symbols (Figure 1) show all inputs and outputs with bubbles, identifying these pins as active-LOW. The bit pattern on the outputs will represent the 1's complement of the input selected.

Even though the 74147 is a 10-line to 4-line encoder, there are only 9 physical inputs. By not selecting any input on the 74147, none of the outputs will be active, thus representing zero.

The 74148 enable input, E_{IN}, and enable output, E_{ON}, allow multiple 74148s to be cascaded for bigger encoder systems.

Table 1 represents the DCBA output logic levels for ten inputs of a 10-line to 4-line encoder. The table shows active-HIGH inputs and active-HIGH outputs. A designer would write output sum of product expressions for outputs A, B, C, and D, then simplify these complex expressions using Boolean identities and rules or Karnaugh maps. By observation, you may notice that output A is a logic-HIGH when input 1 or input 3 or input 5 or input 7 or input 9 are active. This may be expressed by a 5-input OR gate. Output B is a logic-HIGH when inputs 2, 3, 6, or 7 are a logic-HIGH. This is a 4-input OR gate. By similar observation, outputs C and D may be determined, hence resulting in a circuit that looks like Figure 2.

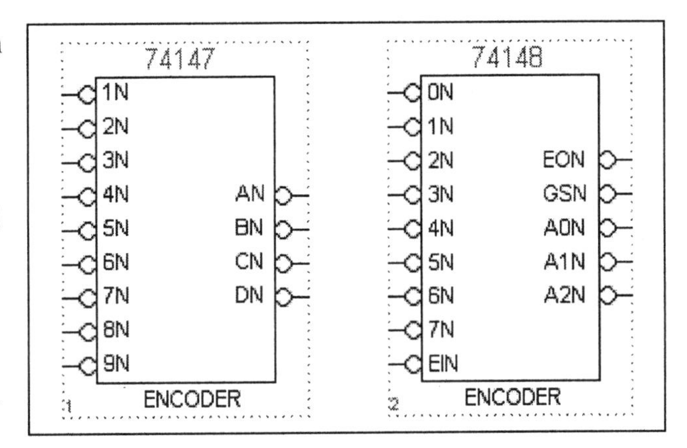

Figure 1

Will this circuit work as an encoder? Yes, as long as only one input is active at a time. However, the 74147 and 74148 ICs are priority encoders, in that only the highest numeric active input will be encoded. The 74148 also has an active-LOW enable input and enable output for octal expansion or to turn the IC on or off.

It is advisable to use a Web browser to search, download, and read the data specification sheets for the 74147 and 74148 ICs before completing this lab.

No.	Inputs										Outputs			
	0	1	2	3	4	5	6	7	8	9	D	C	B	A
0	0	0	0	0	0	0	0	0	0	0	0	0	0	0
1	0	1	0	0	0	0	0	0	0	0	0	0	0	1
2	0	0	1	0	0	0	0	0	0	0	0	0	1	0
3	0	0	0	1	0	0	0	0	0	0	0	0	1	1
4	0	0	0	0	1	0	0	0	0	0	0	1	0	0
5	0	0	0	0	0	1	0	0	0	0	0	1	0	1
6	0	0	0	0	0	0	1	0	0	0	0	1	1	0
7	0	0	0	0	0	0	0	1	0	0	0	1	1	1
8	0	0	0	0	0	0	0	0	1	0	1	0	0	0
9	0	0	0	0	0	0	0	0	0	1	1	0	0	1

Table 1

Figure 2

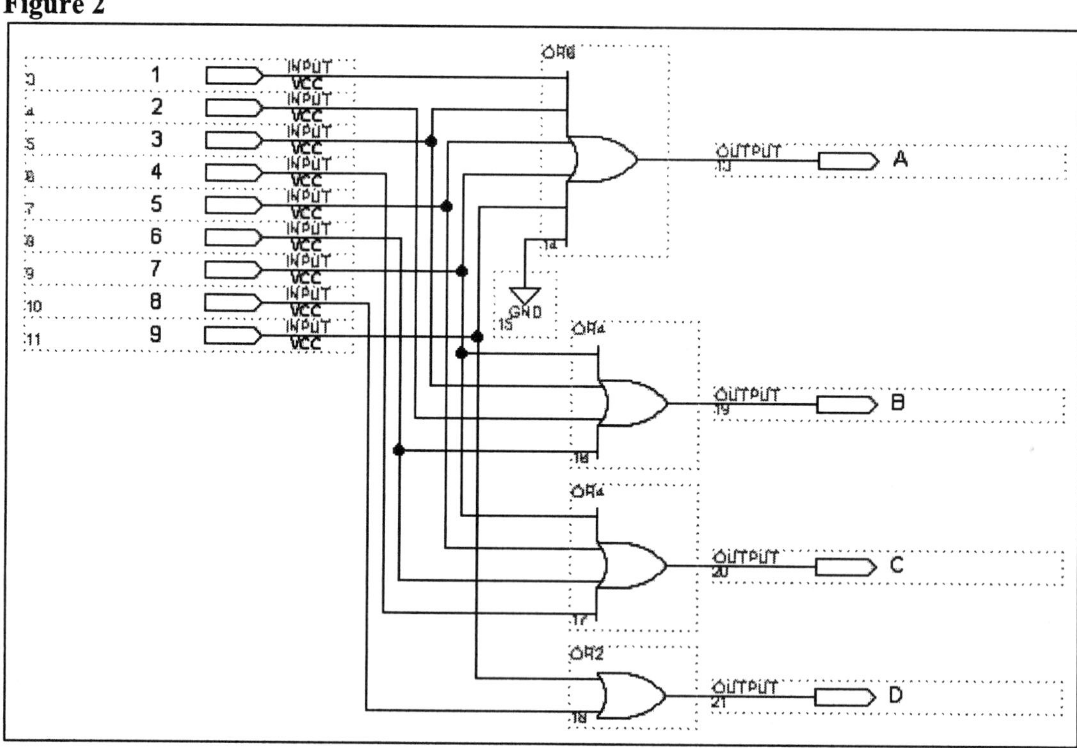

Part 1 Procedure

1. Open the Max+plus II software. Assign the project name **encode1** and MAX7000S as the device family.

2. Open a new Graphic Editor and construct the circuit shown in Figure 2.

3. Open a new Waveform Editor, set Grid Size to 100 ns, and create the waveforms shown in Figure 3.

4. Complete Table 2 identifying which input and output are active during each time segment listed.

5. During the time segments 0 to 900 ns when only one input was active at a time, did the output BCD code represent that active input? (Yes/No)

6. Click the Compile and Simulate buttons. Correct all errors before continuing, then sketch the output waveforms in the area provided at the bottom of Figure 3.

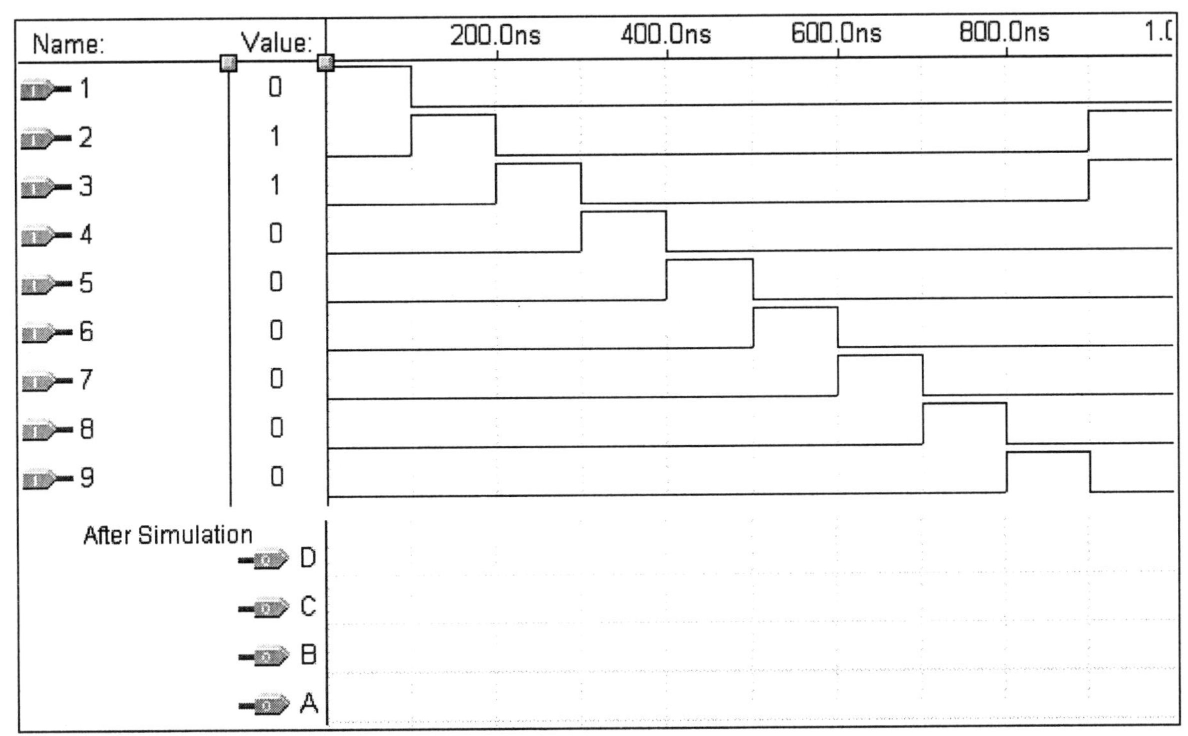

Figure 3

Time Segment	Active Input	DCBA Output	Time Segment	Active Input	DCBA Output
0-100 ns 100-200 ns 200-300 ns 300-400 ns 400-500 ns			500-600 ns 600-700 ns 700-800 ns 800-900 ns 900 ns-1 μs		

Table 2

7. Explain what happened during the 900 ns to 1 μs interval when multiple inputs are active simultaneously.

___

8. Would the circuit of Figure 2 be used in an application to convert numbers typed on a keypad to their BCD equivalents? (Yes/No) Explain why or why not.

___

9. Save all files to Drive A as **encode1**, then exit the Graphic and Waveform Editors.

Part 2 Procedure

The circuit of Figure 2 may work just fine for some applications; however, when multiple inputs are active, it may be necessary to only recognize the larger numbered input. The designer may need to extract the SOP based on all input bits, highs and lows, and simplify the output expressions for each D, C, B, and A output. If the application requires a 10-line to 4-line encoder, it might be easier to select the 74147 IC.

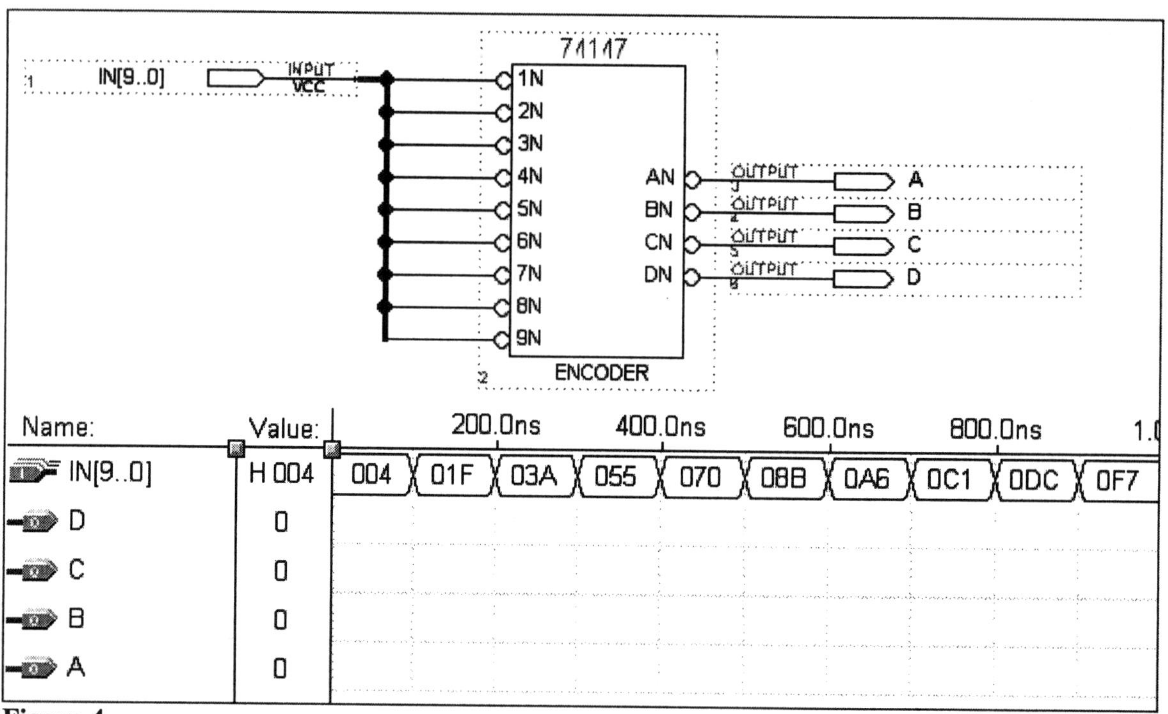

Figure 4

1. Open the Max+plus II software. Assign the project name **encode2.**

2. Open a new Graphic Editor and construct the circuit shown in Figure 4.

3. Open a new Waveform Editor and create the waveforms shown in Figure 4. Enter these values, incrementing by 27, using the Count Overwrite Value dialog box.

4. The Compiler shows 9 or more errors in your circuit. What were the errors in the circuit and what was necessary to correct these errors?

 ____

 ____

5. Run the Simulator. Correct all errors before continuing.

6. Which input is reflected by the logic levels on the DCBA outputs? ____________

7. Why are the DCBA outputs constant even though the inputs are varying?

 ____

8. Highlight the IN[9..0] waveform and overwrite this waveform with **3FD** by entering this number into the Overwrite Group Value dialog box.

9. Simulate the circuit. Record the DCBA bit pattern on the outputs:
 D = __ C = __ B = __ A = __

10. Convert 3FD to its 12-bit binary equivalent.

 $3FD_H$ = $B_{11}B_{10}B_9B_8$ $B_7B_6B_5B_4$ $B_3B_2B_1B_0$
 ____

11. Bits B_{11}, B_{10}, and B_0 are *not* used in Figure 4 so the DCBA outputs are determined by B_9 through B_1.

12. Convert the hex numbers in Table 3 to binary, then predict the DCBA output of the 74147 with these hex numbers applied.

13. Apply the bit pattern of Table 3 as IN[9..0] to the circuit of Figure 4, re-simulate the circuit, and verify your DCBA predictions in Table 3.

14. Save all files to Drive A as **encode2**, then exit the Graphic and Waveform Editors.

Hex	$B_{11}B_{10}B_9B_8\ B_7B_6B_5B_4\ B_3B_2B_1B_0$	D C B A
3FD	______________	______
3FB	______________	______
3F7	______________	______
3EF	______________	______
3DF	______________	______
3BF	______________	______
37F	______________	______
3FF	______________	______

Table 3

Part 3 Procedure

The 74148 has active-LOW inputs and active-LOW outputs similar to the 74147, except the 74148 has an enable input, E_{IN}, and enable output, E_{ON}, used for cascading multiple chips. The G_{SN} output may be used as a flag (or interrupt) signal identifying when a conversion takes place.

1. Open the Max+plus II software. Assign the project name **encode3**.

2. Open a new Graphic Editor and construct the circuit and waveforms shown in Figure 5.

3. Open a new Waveform Editor and create the waveforms shown in Figure 6.

4. Sketch the output waveforms after simulation in the area provided in Figure 6.

5. The inputs to the 74148 are active-(HIGH/LOW).

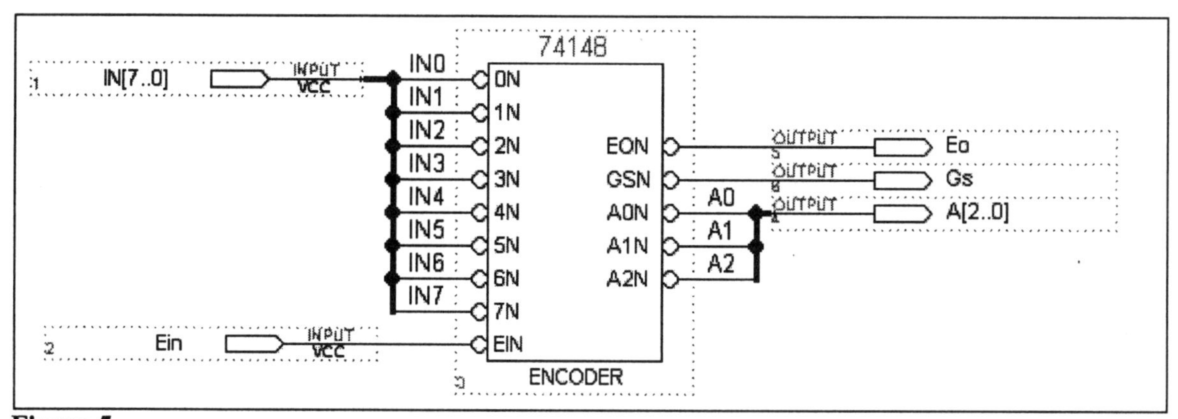

Figure 5

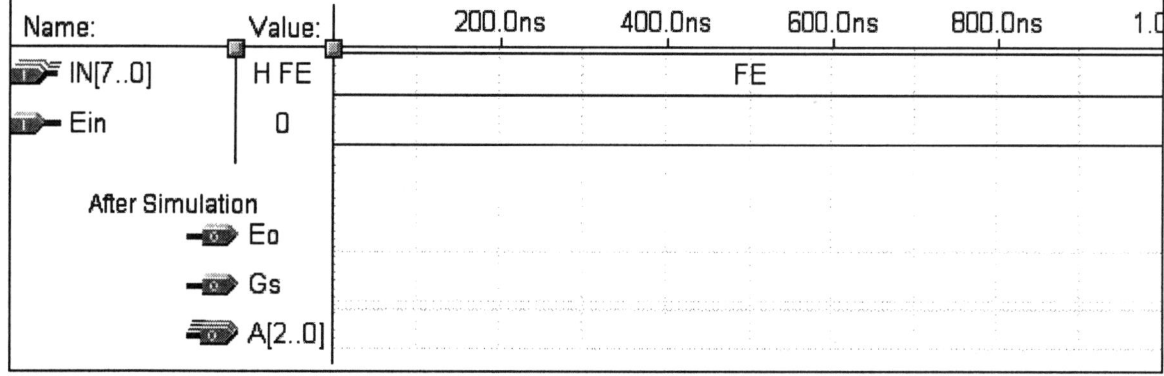

Figure 6

6. Based on the waveforms for Figure 6, which input(s) in Figure 5 are "active"? ________ and ________

7. The logic levels on A2, A1, and A0 in Figure 6 represent which active input? 0 1 2 3 4 5 6 7

8. Complete Table 4 by identifying what hexadecimal code is necessary for IN[7..0] that will cause only the corresponding input to become active. Record the logic levels on the outputs for each row.

E_{IN}	Active Input	IN[7..0]	Outputs Eo Gs A2 A1 A0
0	0	FE	___ ___ ___ ___ ___
0	1	______	___ ___ ___ ___ ___
0	2	______	___ ___ ___ ___ ___
0	3	______	___ ___ ___ ___ ___
0	4	______	___ ___ ___ ___ ___
0	5	______	___ ___ ___ ___ ___
0	6	______	___ ___ ___ ___ ___
0	7	______	___ ___ ___ ___ ___
1	X	X	___ ___ ___ ___ ___

Table 4

9. Save all files to Drive A as **encode3**, then exit the Graphic and Waveform Editors.

Part 4 Procedure

1. Open Max+plus II software. Assign the project name **encode4**.

2 Open a new Graphic Editor. Cascade two 74148 ICs to make a 16-line, IN[15..0], to 4-line, DCBA, encoder network. Include an output flag that will become active only when one or more IN[15..0] inputs and E_{IN} are active.

3. Open a new Waveform Editor and create a set of waveforms for IN[15..0] and E_{IN} that will produce the desired outputs to demonstrate that the circuit operates properly.

4. Demonstrate your 16-line to 4-line encoder to your instructor. Obtain the signature of approval on the answer page for this lab.

5. Obtain hard copies of your 16-line to 4-line encoder. Label these hard copies **Part 4, Step 5A** and **Part 4, Step 5B**.

6. Save all files to Drive A as **encode5**, then exit the Graphic and Waveform Editors.

7. Write a 1-to 2-page summary with embedded graphics pertaining to the results obtained from this lab.

8. Staple all papers in the following sequence, then submit the materials to your instructor for grading.

    - Cover page
    - Typed summary
    - The completed answer page
    - Hard copy of the Graphics Editor, **Part 4, Step 5A**
    - Hard copy of the Waveform Editor, **Part 4, Step 5B**

Lab 9: Encoders Answer Pages

Name: ______________________________

Part 1

200.0ns 400.0ns 600.0ns 800.0ns 1.(

D
C
B
A

Figure 3

Time Segment	Active Input	DCBA Output	Time Segment	Active Input	DCBA Output
0-100 ns			500-600 ns		
100-200 ns			600-700 ns		
200-300 ns			700-800 ns		
300-400 ns			800-900 ns		
400-500 ns			900 ns-1 μs		

Table 2

5. (Yes/No)

7. ____

8. (Yes/No) Explain why or why not.

____

Part 2

4. ____

____

6. ________________

7. ____

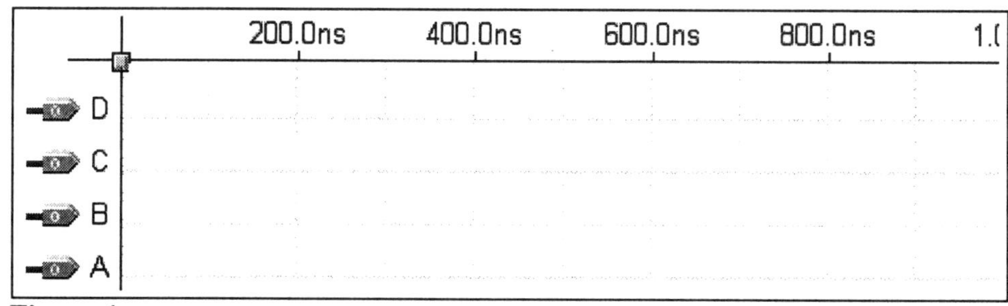

Figure 4

9. D = __ C = __ B = __ A = __

10.

$3FD_H$ = $B_{11}B_{10}B_9B_8\ B_7B_6B_5B_4\ B_3B_2B_1B_0$ ______________________________

Hex	$B_{11}B_{10}B_9B_8\ B_7B_6B_5B_4\ B_3B_2B_1B_0$	D C B A
3FD	____________________	______
3FB	____________________	______
3F7	____________________	______
3EF	____________________	______
3DF	____________________	______
3BF	____________________	______
37F	____________________	______
3FF	____________________	______

Table 3

Part 3

5. (HIGH/LOW)

	200.0ns	400.0ns	600.0ns	800.0ns	1.0
Eo					
Gs					
A[2..0]					

Figure 6

6. ________ and ________

7. 0 1 2 3 4 5 6 7

Ein	Active Input	IN[7..0]	Outputs Eo Gs A2 A1 A0
0	0	FE	___ ___ ___ ___ ___
0	1	______	___ ___ ___ ___ ___
0	2	______	___ ___ ___ ___ ___
0	3	______	___ ___ ___ ___ ___
0	4	______	___ ___ ___ ___ ___
0	5	______	___ ___ ___ ___ ___
0	6	______	___ ___ ___ ___ ___
0	7	______	___ ___ ___ ___ ___
1	X	X	___ ___ ___ ___ ___

Table 4

10. ____

Part 4

4. Demonstrated to: ________________________________ Date: ___________

Grade: _________

Lab 10: Decoders

Objectives:

1. Use the multiple input AND or NAND gate as a decoder
2. Examine the effects of the enable inputs of the 74138 3-line to 8-line decoder
3. Cascade multiple 74138 ICs to create an *N*-line to 2^N-line decoder

Materials list:

- Max+plus II software by Altera Corporation
- Computer requirements:
  Minimum 486/66 with 8 MB RAM
- University Board
- Floppy disk

Discussion:

The decoder is a circuit that converts binary information into intelligent information. The binary code may be data outputted by a computer, the result of an Arithmetic Logic Unit (ALU), or the calculator. The intelligence may be lit light emitting diodes (LEDs), a numeric display, or monitor.

Basic decoders are the AND or NAND gates that provide active-HIGH or active-LOW outputs when a certain input condition exists. Consider the examples in Figure 1. Gate 1 produces a logic-HIGH output when inputs D **and** A are a logic-HIGH, while inputs C **and** B are a logic-LOW. Gate 2 produces a logic-LOW output when inputs D **and** C **and** B **and** A are a logic-HIGH. Normally, several decoder gates will be included in an integrated circuit, such as the 7442, 4-line BCD to 10-line decimal decoder.

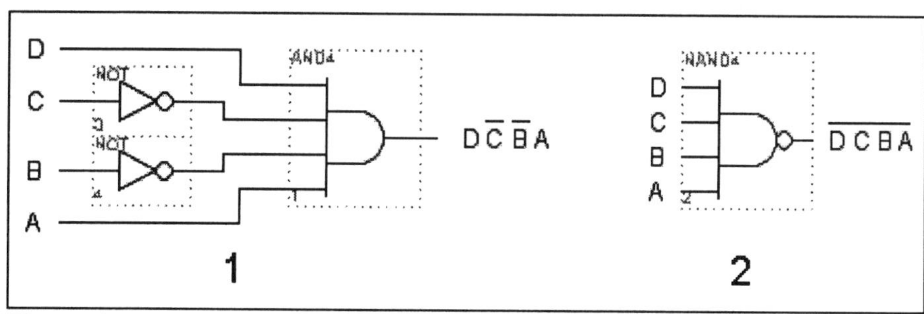

Figure 1

The 74138, 3-line to 8-line decoder, and the 74154, 4-line to 16-line decoder, include separate gate enable inputs for cascading. Quite often the 74138 or 74154 are used for memory or input/output address decoding in a microprocessor-based system. Search the web for and download and read the data specification sheets for the 74138 IC before completing this lab.

Part 1 Procedure

1. Open the Max+plus II software. Assign the project name **decode1,** MAX7000S as the device family, and EPM7128SLC84 as the device.

2. Open a new Graphic Editor and create the circuit shown in Figure 2.

3. Open a new Waveform Editor. Create input A as a 5 MHz square wave and input B as a 2.5 MHz square wave. Compile and simulate your circuit and sketch the input and output, I/O, waveforms in Figure 3.

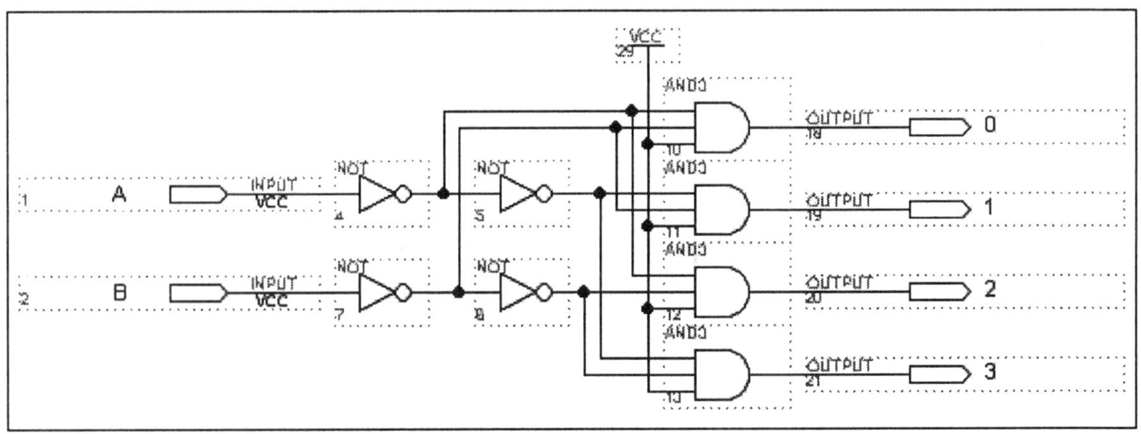

Figure 2

Name:	Value:	250.0ns	500.0ns	750.0ns	1.0
A	0				
B	0				
0	0				
1	0				
2	0				
3	0				

Figure 3

4. Examine the waveforms in Figure 3. Identify which output is active based on the BA code applied.
   When BA = 00, output ____ is active;
   When BA = 01, output ____ is active;
   When BA = 10, output ____ is active;
   When BA = 11, output ____ is active.

5. The outputs of the circuit in Figure 2 are active- (LOW/HIGH).

6. The circuit demonstrated in Figure 2 is a _____-line to _____-line decoder.

7. What modifications to the circuit are necessary to make the outputs active-LOW?

   ____

8. The third input to each AND gate were connected to Vcc. How would the circuit respond if these inputs were grounded instead?

   ____

9. Save all files to Drive A as **decode1**, then exit the Graphic and Waveform Editors.

Part 2 Procedure

1. Open the Max+plus II software. Assign the project name **decode2,** MAX7000S as the device family, and EPM7128SLC84 as the device.

2. Open the **decodel.scf** file that was created in Part 1 of this lab. Save this file as **decode2.scf.**

3. Open the **decodel.gdf** file that was created in Part 1 of this lab. Save this file as **decode2.gdf.**

4. Delete the Vcc symbol in the **decode.gdf** file, replacing it with an input symbol labeled Enable.

5. What must be the logic level of the Enable input to cause the decoder to decode inputs A and B? (logic-LOW/logic-HIGH)

6. Delete the Enable input symbol, replacing it with the input circuit shown in Figure 4.

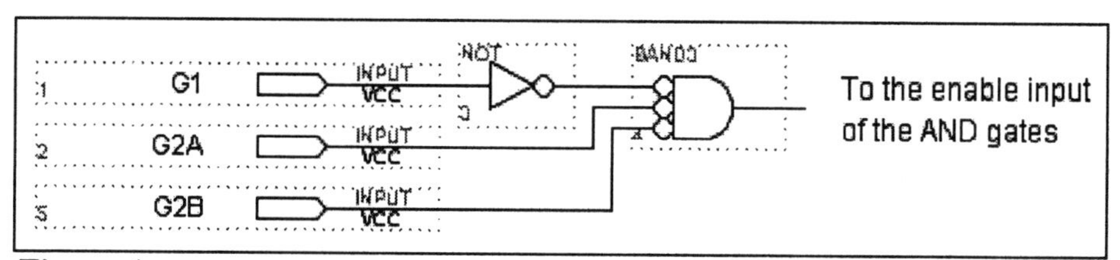

Figure 4

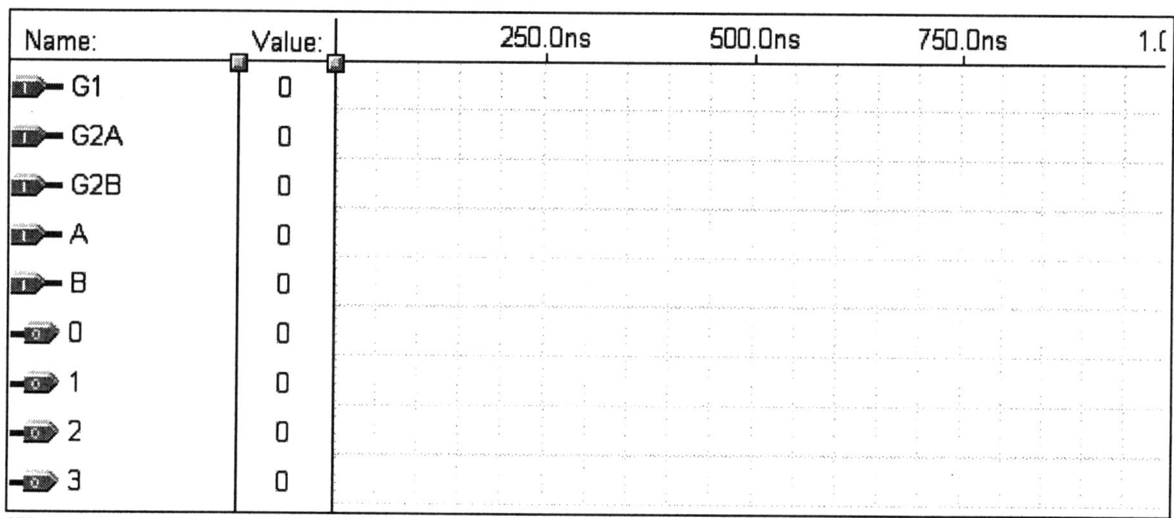

Figure 5

7. Add input waveforms G1, G2A, and G2B to your Waveform Editor file as shown in Figure 5.

8. Select proper logic levels for G1, G2A, and G2B so the encoder is enabled, allowing the outputs to respond to the BA inputs.

9. Be sure that the BA inputs are as described in Part 1, Step 3 of this lab before continuing.

10. Run the Compiler and Simulator. Sketch all I/O waveforms in the area provided in Figure 5.

11. To enable the 2-line to 4-line decoder, G1 must be a logic- (LOW/HIGH), G2A must be a logic- (LOW/HIGH), and G2B must be a logic- (LOW/HIGH).

12. Save all files to Drive A as **decode2**, then exit the Graphic and Waveform Editors.

Part 3 Procedure

1. Open the Max+plus II software. Assign the project name **decode3,** MAX7000S as the device family, and EPM7128SLC84 as the device.

2. Open a new Graphic Editor and construct the circuit of Figure 6.

3. Open a new Waveform Editor and create the waveforms shown in Figure 7.

4. Run the Compiler and Simulator. Sketch the resulting output waveforms in the area provided in Figure 7.

5. Complete Table 1 by recording the logic levels of the inputs and outputs for each time segment listed based on the results recorded in Figure 7.

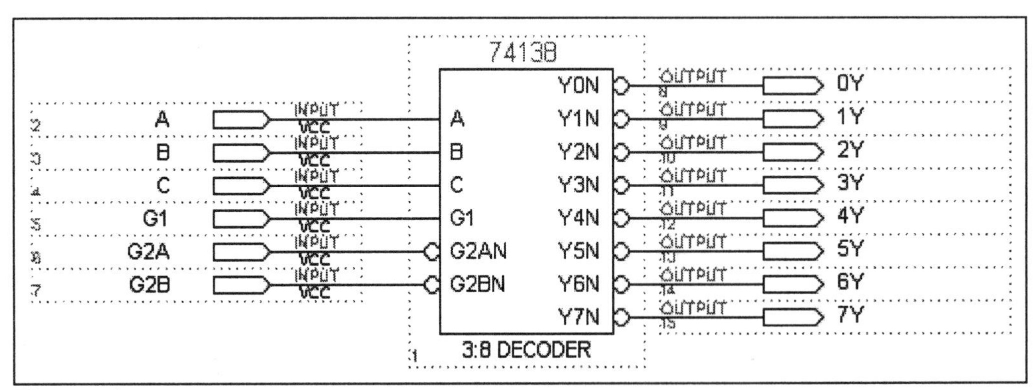

Figure 6

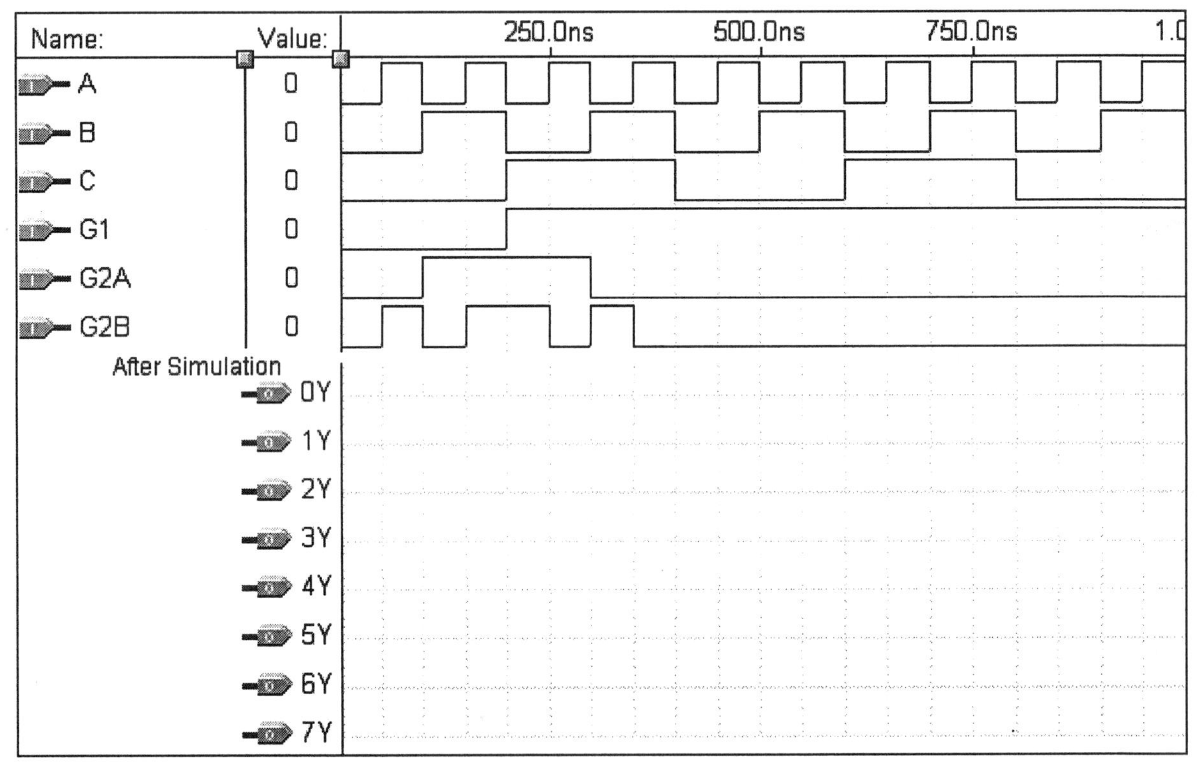

Figure 7

Time	Inputs G1 G2A G2B C B A	Outputs 0Y 1Y 2Y 3Y 4Y 5Y 6Y 7Y
0 to 200 ns	___ X X X X X	__ __ __ __ __ __ __ __
200 to 300 ns	X ___ X X X X	__ __ __ __ __ __ __ __
300 to 350 ns	X X ___ X X X	__ __ __ __ __ __ __ __
350 to 400 ns	___ ___ ___ ___ ___ ___	__ __ __ __ __ __ __ __

Table 1 X = irrelevant

Time	Inputs G1 G2A G2B C B A	Outputs 0Y 1Y 2Y 3Y 4Y 5Y 6Y 7Y
400 to 450 ns	___ ___ ___ ___ ___ ___	___ ___ ___ ___ ___ ___ ___ ___
450 to 500 ns	___ ___ ___ ___ ___ ___	___ ___ ___ ___ ___ ___ ___ ___
500 to 550 ns	___ ___ ___ ___ ___ ___	___ ___ ___ ___ ___ ___ ___ ___
550 to 600 ns	___ ___ ___ ___ ___ ___	___ ___ ___ ___ ___ ___ ___ ___
600 to 650 ns	___ ___ ___ ___ ___ ___	___ ___ ___ ___ ___ ___ ___ ___
650 to 700 ns	___ ___ ___ ___ ___ ___	___ ___ ___ ___ ___ ___ ___ ___
700 to 750 ns	___ ___ ___ ___ ___ ___	___ ___ ___ ___ ___ ___ ___ ___
750 to 800 ns	___ ___ ___ ___ ___ ___	___ ___ ___ ___ ___ ___ ___ ___
800 to 850 ns	___ ___ ___ ___ ___ ___	___ ___ ___ ___ ___ ___ ___ ___
850 to 900 ns	___ ___ ___ ___ ___ ___	___ ___ ___ ___ ___ ___ ___ ___
900 to 950 ns	___ ___ ___ ___ ___ ___	___ ___ ___ ___ ___ ___ ___ ___
950 ns to 1 μs	___ ___ ___ ___ ___ ___	___ ___ ___ ___ ___ ___ ___ ___

Table 2

6. Complete Table 2 by recording the logic levels of the inputs and outputs for each time segment listed based on the results recorded in Figure 7.

7. Based on the results recorded for G1 in Table 1 and Table 2, G1 is active- (LOW/HIGH).

8. Based on the results recorded for G2A in Table 1 and Table 2, G2A is active- (LOW/HIGH).

9. Based on the results recorded for G2B in Table 1 and Table 2, G2B is active- (LOW/HIGH).

10. Based on the results recorded for the outputs in Table 1 and Table 2, the outputs are active- (LOW/HIGH).

11. Based on the results recorded for Table 1 and Table 2, which inputs are dominant?

    G1 G2A G2B C B A

12. Assuming the 74138 is turned on, what is the relationship of the output that is active and the CBA code applied?

    ____

13. Save all files to Drive A as **decode3**, then exit the Graphic and Waveform Editors.

Part 4 Procedure

1. Open the Graphic and Waveform Editors. Assign the project name **decode4**.

2. Examine Table 3.

3. Table 3 represents 32 rows, listing all possible combinations of EDCBA. Well not really, notice that the CBA code for Y0 to Y7 is identical to the CBA code for Y8 to Y15, Y16 to Y32, and Y24 to Y31. For the first 8 rows of the function table, Y0 to Y7, ED is 00. From Y8 to Y15, ED is 01, from Y16 to Y23, ED = 10, and from Y24 to Y31, ED = 11.

4. Cascade four 74138 ICs with five inputs, E, D, C, B, and A, to create a 5-line to 32-line decoder network.

Inputs E D C B A	Active Output
0 0 0 0 0	Y0
. . .	. . .
0 0 1 1 1	Y7
0 1 0 0 0	Y8
. . .	. . .
0 1 1 1 1	Y15
1 0 0 0 0	Y16
. . .	. . .
1 0 1 1 1	Y23
1 1 0 0 0	Y24
. . .	. . .
1 1 1 1 1	Y31

Table 3

5. Set Grid Size to 30 ns and create Waveforms A, B, C, D, and E using the following multipliers in the Overwrite Count Value dialog box.

   Waveform A: Multiplied by: 1
   Waveform B: Multiplied by: 2
   Waveform C: Multiplied by: 4
   Waveform D: Multiplied by: 8
   Waveform E: Multiplied by: 16

6. Run the Compiler and Simulator. Correct all errors before continuing.

7. Obtain a hard copy of the Graphic and Waveform Editor files. Label these hard copies **Part 4, Step 7A** and **Part 4, Step 7B**.

8. Demonstrate the 5-line to 32-line decoder to your instructor. Obtain the signature on the answer page for this lab.

9. Write a 1-to 2-page summary pertaining to the results obtained from this lab. The summary must contain and make reference to an embedded figure and a table, both related to this lab contents.

10. Place all papers for this lab in the following sequence, then submit the lab to your instructor for grading.

    - Cover page
    - Typed summary
    - The completed answer pages for this lab
    - Hard copy of the Graphic Editor, **Part 4, Step 7A**
    - Hard copy of the Waveform Editor, **Part 4, Step 7B**

Lab 10: Decoders Answer Pages

Name: ______________________

Part 1

Name:	Value:	250.0ns	500.0ns	750.0ns	1.0
A	0				
B	0				
0	0				
1	0				
2	0				
3	0				

Figure 3

4. 00: Output ______ 01: Output ______ 10: Output ______ 11: Output ______

5. (LOW/HIGH) 6. ______line to ______

7. ____

8. ____

Part 2

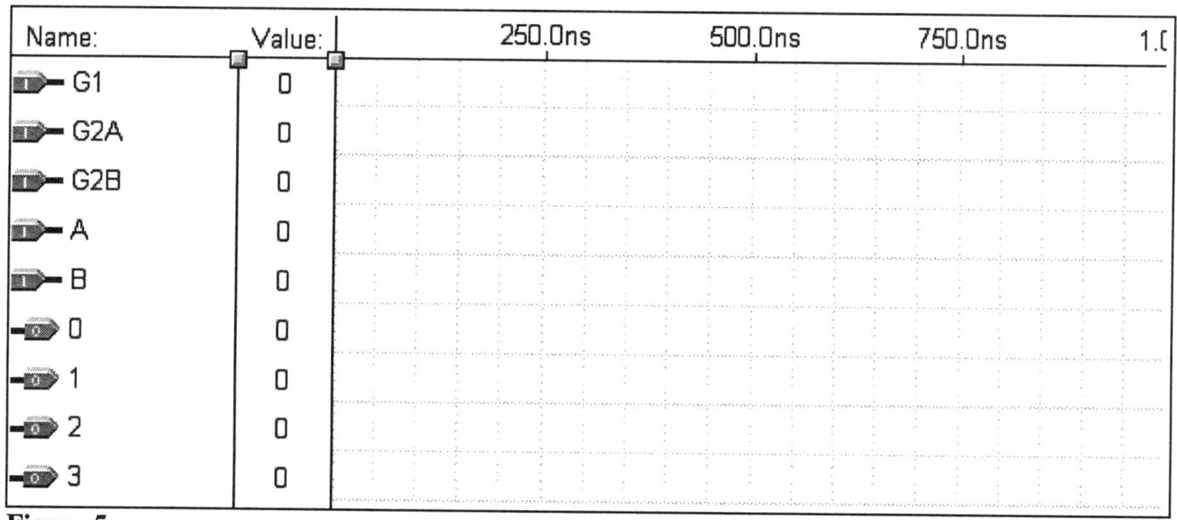

Figure 5

5. (logic-LOW/logic-HIGH)

11. G1: (LOW/HIGH) G2A: (LOW/HIGH) G2B: (LOW/HIGH)

Part 3

Name:	Value:	100.0ns	200.0ns	300.0ns	400.0ns	500.0ns	600.0ns	700.0ns	800.0ns	900.0ns
0Y	0									
1Y	0									

Figure 7a

Name:	Value:	100.0ns 200.0ns 300.0ns 400.0ns 500.0ns 600.0ns 700.0ns 800.0ns 900.0ns
2Y	0	
3Y	0	
4Y	0	
5Y	0	
6Y	0	
7Y	0	

Figure 7

Time	Inputs G1 G2A G2B C B A	Outputs 0Y 1Y 2Y 3Y 4Y 5Y 6Y 7Y
0 to 200 ns	___ X X X X X	__ __ __ __ __ __ __ __
200 to 300 ns	X ___ X X X X	__ __ __ __ __ __ __ __
300 to 350 ns	X X ___ X X X	__ __ __ __ __ __ __ __
350 to 400 ns	__ ___ ___ __ __ __	__ __ __ __ __ __ __ __

Table 1 X = irrelevant

Time	Inputs G1 G2A G2B C B A	Outputs 0Y 1Y 2Y 3Y 4Y 5Y 6Y 7Y
400 to 450 ns	__ ___ ___ __ __ __	__ __ __ __ __ __ __ __
450 to 500 ns	__ ___ ___ __ __ __	__ __ __ __ __ __ __ __
500 to 550 ns	__ ___ ___ __ __ __	__ __ __ __ __ __ __ __
550 to 600 ns	__ ___ ___ __ __ __	__ __ __ __ __ __ __ __
600 to 650 ns	__ ___ ___ __ __ __	__ __ __ __ __ __ __ __
650 to 700 ns	__ ___ ___ __ __ __	__ __ __ __ __ __ __ __
700 to 750 ns	__ ___ ___ __ __ __	__ __ __ __ __ __ __ __
750 to 800 ns	__ ___ ___ __ __ __	__ __ __ __ __ __ __ __
800 to 850 ns	__ ___ ___ __ __ __	__ __ __ __ __ __ __ __
850 to 900 ns	__ ___ ___ __ __ __	__ __ __ __ __ __ __ __
900 to 950 ns	__ ___ ___ __ __ __	__ __ __ __ __ __ __ __
950 ns to 1 μs	__ ___ ___ __ __ __	__ __ __ __ __ __ __ __

Table 2

7. (LOW/HIGH) 8. (LOW/HIGH) 9. (LOW/HIGH)

10. (LOW/HIGH) 11. G1 G2A G2B C B A

12. ____

Part 4

8. Demonstrated to: ______________________ Date: ____________

Grade: ___________

Lab 11: Multiplexers

Objectives:

1. Analyze a 2-line to 1-line multiplexer
2. Evaluate the 74157 quad 2-line to 1-line multiplexer
3. Analyze the output of a 74153 based on the enable and select input
4. Analyze the effects of the control inputs of a 74151 8-line to 1-line multiplexer
5. Create an octal 2-line to 1-line multiplexer using the 74157 quad 2-line to 1-line multiplexer

Materials list:

- ♦ Max+plus II software by Altera Corporation
- ♦ Computer requirements:
  Minimum 486/66 with 8 MB RAM
- ♦ University Board
- ♦ Floppy disk

Discussion:

The basic 2-line to 1-line multiplexer can be illustrated by a toggle switch with two inputs; the common lug of the switch is the output (see Figure 1). Which data input appears on the output depends on the switch setting. The multiplexer has many data inputs and one data output.

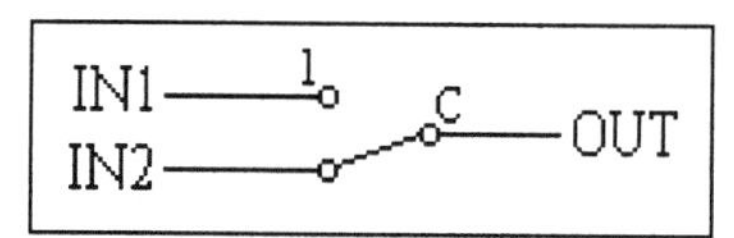

Figure 1

Figure 2 illustrates a 2-line to 1-line multiplexer using basic logic gates. The select input determines which input data appears on the output only if the Enable input is logic-HIGH. The 74157 contains four 2-line to 1-line multiplexers that share the same select and enable inputs.

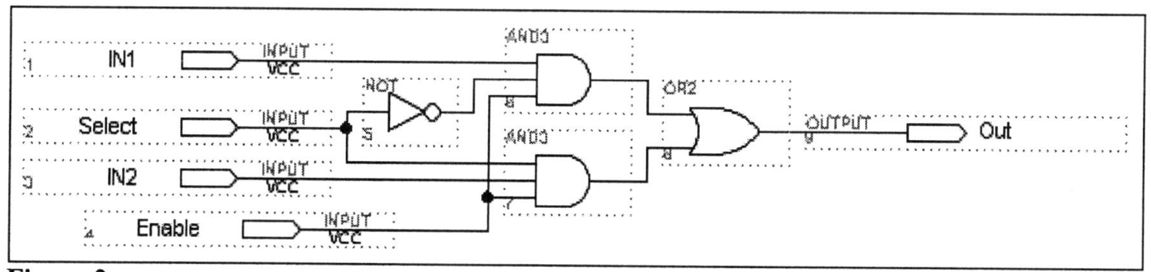

Figure 2

Multiple input multiplexers illustrated in this lab are the 74157 quad 2-line to 1-line, 74153 dual 4-line to 1-line, and 74151 8-line to 1-line multiplexers. The 74157 will be used in Parts 2 and 6 of this lab.

The 74153 is a dual 4-line to 1-line multiplexer. Each multiplexer has independent enable and data inputs but share the same BA selects. The 74153 will be used in Part 3 of this lab.

The 74151 is an 8-line to 1-line multiplexer that provides either an inverting or noninverting output. The 74151 has a single enable input and three data selects. The 74151 will be used in Part 4 of this lab.

Use a Web browser to search for, download, and read the data specification sheets for the 74151, 74153, and 74157 multiplexer ICs before completing this lab.

Part 1 Procedure

1. Open the Max+plus II software. Assign the project name **mux1** and MAX7000S as the device family.

2. Open the Graphic Editor and construct the circuit shown in Figure 2.

Name:	Value:	200.0ns	400.0ns	600.0ns	800.0ns	1.0
IN1	0					
IN2	0					
Select	0					
Enable	0					
Out	0					

Figure 3

3. Open a new Waveform Editor and create the waveforms shown in Figure 3.

4. Run the Compiler and Simulator. Sketch the resulting output signal in the area provided in Figure 3.

5. Invert the enable signal, then re-simulate the circuit.

6. Based on the results obtained, the enable input to Figure 2 is active- (LOW/HIGH).

7. When the multiplexer is enabled and the select input is a logic-LOW, which input data appears on the output of the circuit? (Data IN1/Data IN2)

8. When the multiplexer is enabled and the select input is a logic-HIGH, which input data appears on the output of the circuit? (Data IN1/Data IN2)

9. Which input to the 2-line to 1-line multiplexer of Figure 2 is dominant?
   IN1 IN2 Select Enable

10. Save all files to Drive A as **mux1**, then exit the Graphic and Waveform Editors.

Part 2 Procedure

1. Open the Max+plus II software. Assign the project name **mux2**.

2. Open a new Graphic Editor and construct the circuit shown in Figure 4.

3. Open a new Waveform Editor and create the waveforms shown in Figure 4.

4. Sketch output Y[3..0] in the space provided in Figure 4, after simulation.

5. Note the Y[3..0] output was constant LOW during the first 200 ns. Why? ____________________________

   ____

6. From your observations of the output waveform in Figure 4, GN would be active- (LOW/HIGH).

7. Which data input appeared on output Y when A/B was a logic-LOW? (A[3..0]/B[3..0])

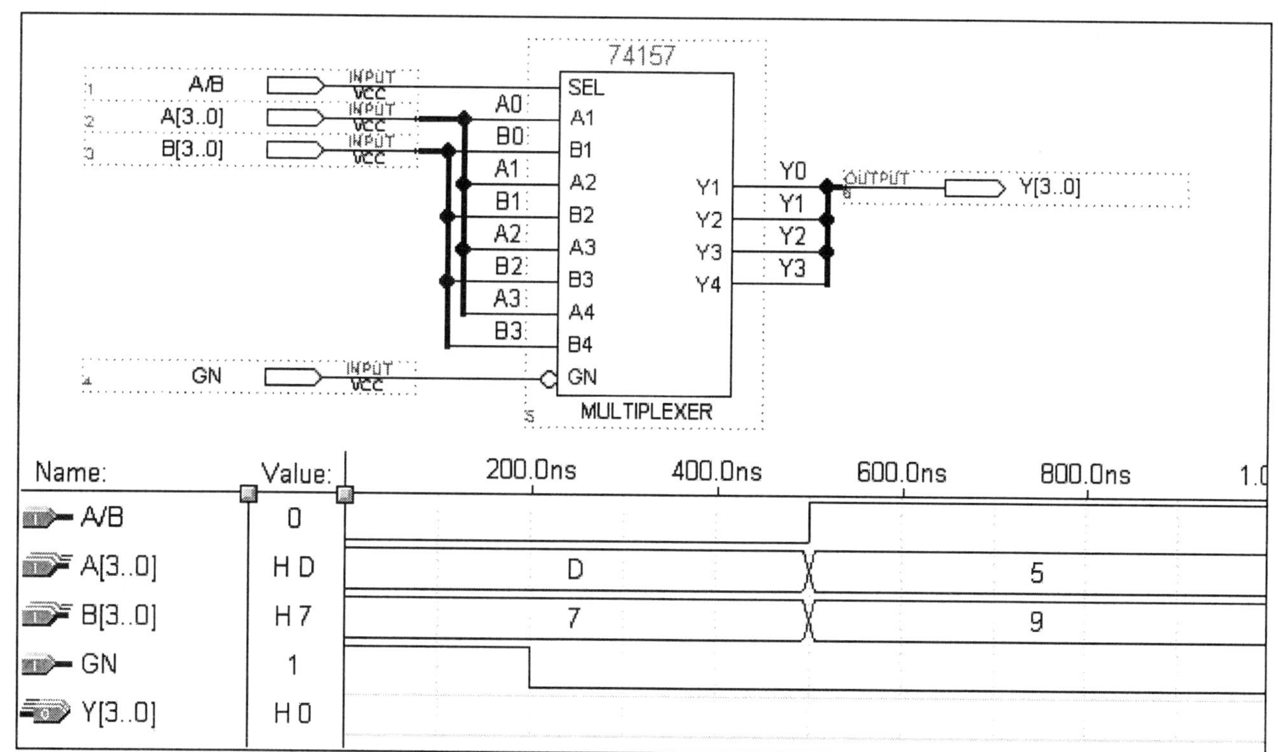

Figure 4

8. Which data input appeared on output Y when A/B was a logic-HIGH? (A[3..0]/B[3..0])

9. Save all files to Drive A as **mux2**, then exit the Graphic and Waveform Editors.

Part 3 Procedure

1. Open the Max+plus II software. Assign the project name **mux3**.

2. Open a new Graphic Editor and construct the circuit shown in Figure 5.

3. Open a new Waveform Editor, set Grid Size to 500 ns, and create the waveforms shown in Figure 5.

4. Click the Compiler and Simulator buttons. Correct all errors before continuing.

5. Sketch the output waveform in the area provided in Figure 5 after simulating the circuit.

6. During the time segment 0 to 500 ns, the BA code is (00, 01, 10, 11).

7. From 0 to 500 ns, 1Y takes on the value of (A0, A1, A2, A3, B0, B1, B2, B3).

8. From 0 to 500 ns, 2Y takes on the value of (A0, A1, A2, A3, B0, B1, B2, B3).

9. During the time segment 500 ns to 1 μs, the BA code is (00, 01, 10, 11).

10. From 500 ns to 1 μs, 1Y takes on the value of (A0, A1, A2, A3, B0, B1, B2, B3).

11. From 500 ns to 1 μs, 2Y takes on the value of (A0, A1, A2, A3, B0, B1, B2, B3).

12. Invert Waveform B, then re-simulate the circuit.

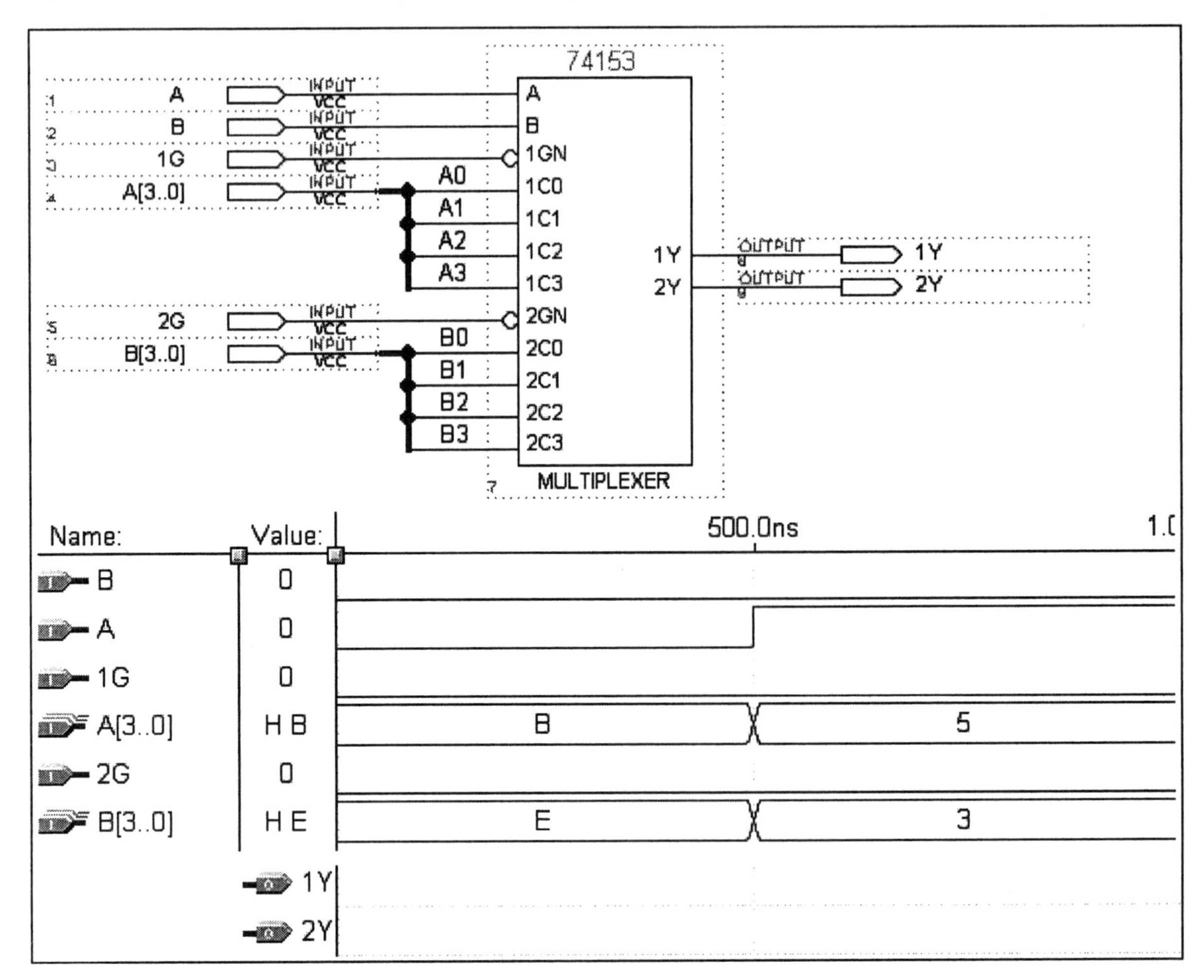

Figure 5

13. During the time segment 0 to 500 ns, the BA code is (00, 01, 10, 11).

14. From 0 to 500 ns, 1Y takes on the value of (A0, A1, A2, A3, B0, B1, B2, B3).

15. From 0 to 500 ns, 2Y takes on the value of (A0, A1, A2, A3, B0, B1, B2, B3).

16. During the time segment 500 ns to 1 μs, the BA code is (00, 01, 10, 11).

17. From 500 ns to 1 μs, 1Y takes on the value of (A0, A1, A2, A3, B0, B1, B2, B3).

18. From 500 ns to 1 μs, 2Y takes on the value of (A0, A1, A2, A3, B0, B1, B2, B3).

19. Invert waveform 1G, then re-simulate the circuit.

20. Based on the results, 1G is active- (LOW/HIGH).

21. A logic-LOW on 2G (enables/inhibits) (1Y/2Y).

22. Save all files to Drive A as **mux3**, then exit the Graphic and Waveform Editors.

Part 4 Procedure

1. Open the Max+plus II software. Assign the project name **mux4**.

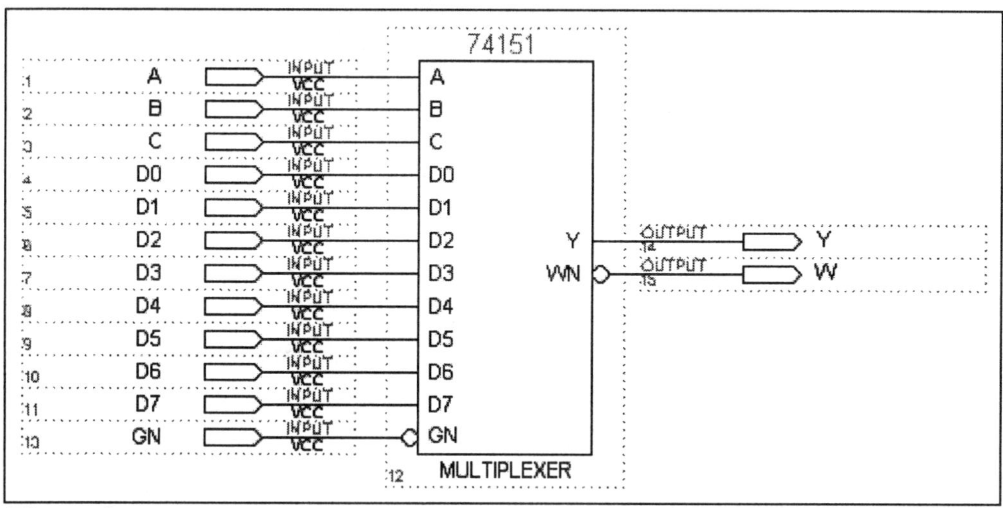

Figure 6

2. Open a new Graphic Editor. Construct the circuit shown in Figure 6.

3. Open the Waveform Editor, set Grid Size to 50 ns, and construct the waveforms shown in Figure 7.

4. Compile and simulate the circuit. Correct all errors before continuing.

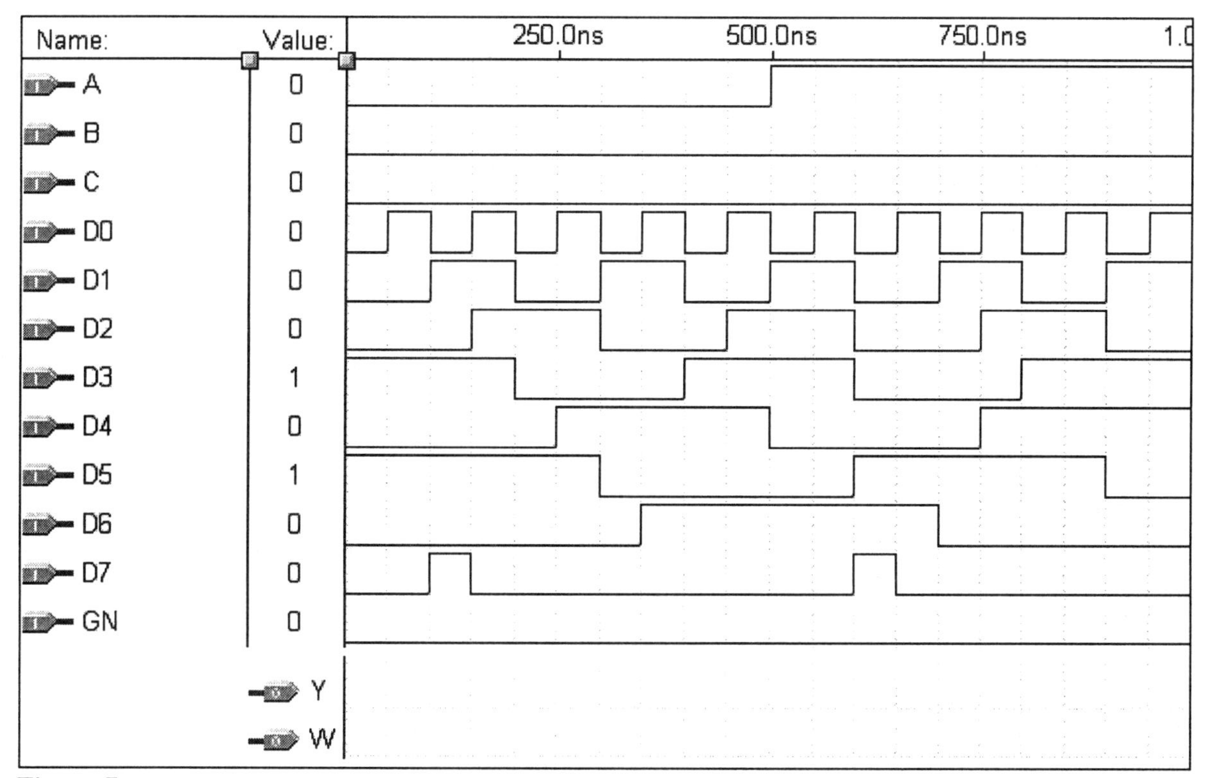

Figure 7

Time Segment	C B A	"Y" takes on the value of (Select one)
0-500 ns	__ __ __	(D_7 D_6 D_5 D_4 D_3 D_2 D_1 D_0)
500 ns-1 μs	__ __ __	(D_7 D_6 D_5 D_4 D_3 D_2 D_1 D_0)

Table 1

5. Sketch the output waveforms in the space provided in Figure 7.

6. Complete Table 1 based on the waveforms in Figure 7.

7. Invert Waveform B. Re-simulate the circuit.

8. Complete Table 2 based on the new set of waveforms after inverting Waveform B.

Time Segment	C B A	"Y" takes on the value of (Select one)
0-500 ns	__ __ __	(D_7 D_6 D_5 D_4 D_3 D_2 D_1 D_0)
500 ns-1 μs	__ __ __	(D_7 D_6 D_5 D_4 D_3 D_2 D_1 D_0)

Table 2

9. Invert Waveform C. Re-simulate the circuit.

10. Complete Table 3 based on the new set of waveforms after inverting Waveform C.

Time Segment	C B A	"Y" takes on the value of (Select one)
0-500 ns	__ __ __	(D_7 D_6 D_5 D_4 D_3 D_2 D_1 D_0)
500 ns-1 μs	__ __ __	(D_7 D_6 D_5 D_4 D_3 D_2 D_1 D_0)

Table 3

11. When C = 1, B = 0, and A = 0, which data input will appear on output Y? ______________

12. When C = 1, B = 0, and A = 1, which data input will appear on output Y? ______________

13. Which output in Figure 10 inverts the data? ______________

14. Save all files to Drive A as **mux4**, then exit the Graphic and Waveform Editors.

Part 5 Procedure

1. Open new Graphic and Waveform Editors. Assign the project name **mux5**.

2. Use a 74151 multiplexer to implement the function shown in the table to the right.

3. Create a set of input waveforms showing the CBA pattern of Table 5 and apply the proper data bits that will produce output Y.

4. Once you verify that the circuit operation matches the data shown in Table 4, obtain a hard copy of the Graphic and Waveform Editor files. Label these hard copies **Part 5, Step 4A** and **Part 5, Step 4B**, respectively.

C B A	Output Y
0 0 0	1
0 0 1	0
0 1 0	0
0 1 1	1
1 0 0	0
1 0 1	0
1 1 0	1
1 1 1	1

Table 4

5. Write the Boolean expression for output Y with respect to the data and select inputs.

   Y = ____

6. Save all files to Drive A as **mux5**, then exit the Graphic and Waveform Editors.

Part 6 Procedure

1. Open the Max+plus II software. Assign the project name **mux6**.

2. Open a new Graphic Editor.

3. Cascade 74157 ICs to multiplex A[7..0] with A[15..8] to obtain Y[7..0]. The low byte lines of A[15..0] will appear on Y[7..0] when the select line is a logic-LOW. The high byte lines of A[15..0] will appear on Y[7..0] when the select line is a logic-HIGH.

4. Open a new Waveform Editor.

5. Assign waveform A[15..0] to be $7b95_{16}$. Create the necessary control waveforms to pass A[7..0], then A[15..8] to Y[7..0].

6. Compile and simulate your circuit. Correct all errors before continuing.

7. Obtain a hard copy of the Graphic and Waveform Editor files. Label these hard copies **Part 6, Step 7A** and **Part 6, Step 7B**, respectively.

8. Demonstrate the octal 2-line to 1-line multiplexer to the instructor. Obtain the signature of approval on the answer page for this lab.

9. Save all files to Drive A as **mux6**, then exit the Graphic and Waveform Editors.

10. Write a 1- to 2-page summary containing and making reference to at least one embedded figure and a table obtained from this lab.

11. Place all papers for this lab in the following sequence, then submit the lab to your instructor for grading.

    - Cover page
    - Typed summary
    - The completed lab
    - Printout of the Graphic Editor, **Part 5, Step 4A**
    - Printout of the Waveform Editor, **Part 5, Step 4B**
    - Printout of the Graphic Editor, **Part 6, Step 7A**
    - Printout of the Waveform Editor, **Part 6, Step 7B**

Lab 11: Multiplexers Answer Pages

Name: ______________________________

Part 1

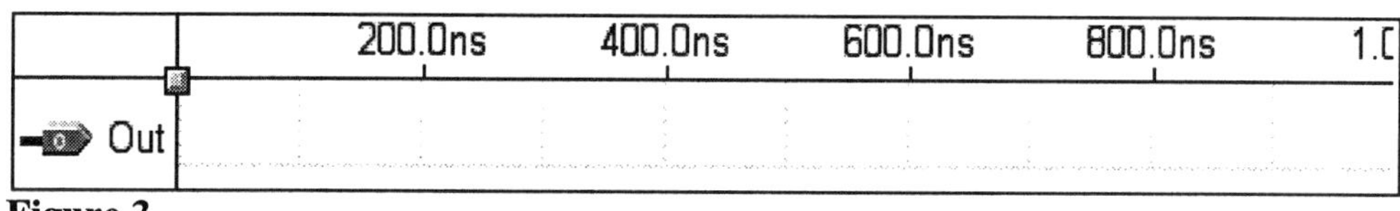

Figure 3

6. (LOW/HIGH) 7. (Data IN/Data IN2)

8. (Data IN1/Data IN2) 9. IN1 IN2 Select Enable

Part 2

5. ____

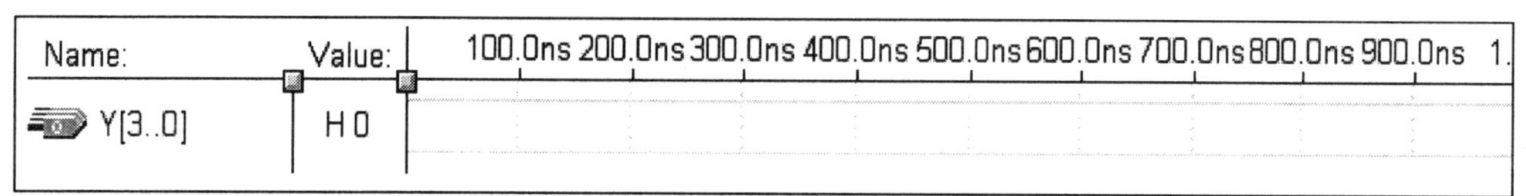

Figure 4

8. (LOW/HIGH) 9. (A[3..0]/B[3..0]) 10. (A[3..0]/B[3..0])

Part 3

6. (00, 01, 10, 11) 7. (A0, A1, A2, A3, B0, B1, B2, B3)

8. (A0, A1, A2, A3, B0, B1, B2, B3) 9. (00, 01, 10, 11)

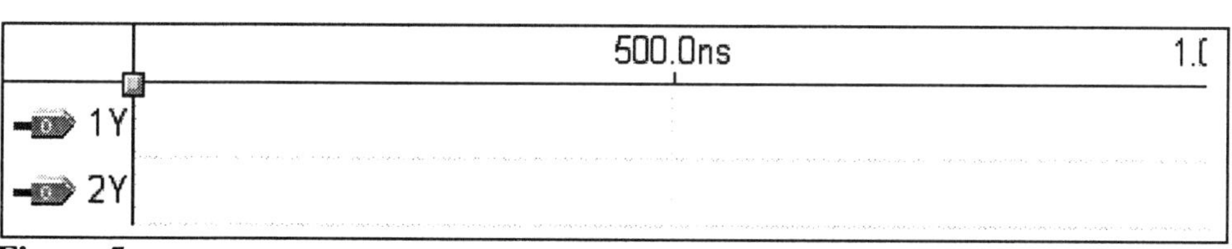

Figure 5

10. (A0, A1, A2, A3, B0, B1, B2, B3) 11. (A0, A1, A2, A3, B0, B1, B2, B3)

13. (00, 01, 10, 11) 14. (A0, A1, A2, A3, B0, B1, B2, B3)

15. (A0, A1, A2, A3, B0, B1, B2, B3) 16. (00, 01, 10, 11)

17. (A0, A1, A2, A3, B0, B1, B2, B3) 18. (A0, A1, A2, A3, B0, B1, B2, B3)

20. (LOW/HIGH) 21. (enables/inhibits) (1Y / 2Y)

Part 4

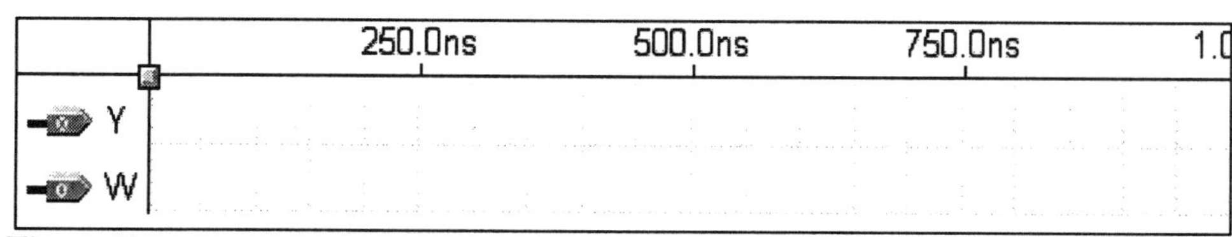

Figure 7

Time Segment	C B A	"Y" takes on the value of (Select one)
0-500 ns	__ __ __	(D_7 D_6 D_5 D_4 D_3 D_2 D_1 D_0)
500 ns-1 μs	__ __ __	(D_7 D_6 D_5 D_4 D_3 D_2 D_1 D_0)

Table 1

Time Segment	C B A	"Y" takes on the value of (Select one)
0-500 ns	__ __ __	(D_7 D_6 D_5 D_4 D_3 D_2 D_1 D_0)
500 ns-1 μs	__ __ __	(D_7 D_6 D_5 D_4 D_3 D_2 D_1 D_0)

Table 2

Time Segment	C B A	"Y" takes on the value of (Select one)
0-500 ns	__ __ __	(D_7 D_6 D_5 D_4 D_3 D_2 D_1 D_0)
500 ns-1 μs	__ __ __	(D_7 D_6 D_5 D_4 D_3 D_2 D_1 D_0)

Table 3

11. ________________ 12. ________________ 13. ________________

Part 5

5. ____

Part 6

8. Demonstrated to: ______________________________ Date: ______________

Grade: ____________

Lab 12: Demultiplexers

Objectives:

1. Evaluate a basic 1-line to 2-line demultiplexer
2. Analyze a 2-line to 4-line multiplexer
3. Use the 74138 as a demultiplexer with no data inversion
4. Modify the 73138 demultiplexer to invert data
5. Create a circuit using an 8-line to 1-line multiplexer and a 1-line to 8-line demultiplexer

Materials List:

- Max+plus II software by Altera Corporation
- Computer requirements:
  Minimum 486/66 with 8 MB RAM
- University Board
- Floppy disk

Discussion:

A demultiplexer has a single data input and will send that data to one of many outputs. The number of outputs is determined by total possible combinations of the number of select inputs.

The basic 1-line to 2-line demultiplexer is shown in Figure 1. Data applied to the input will appear on one of the outputs, depending on the setting of the switch. Other names for the demultiplexer are data selector or data distributor.

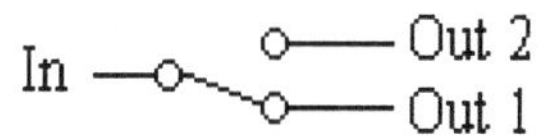

Figure 1

Figure 2 shows a demultiplexer using logic gates. The logic level on the select input determines on which output the data will appear.

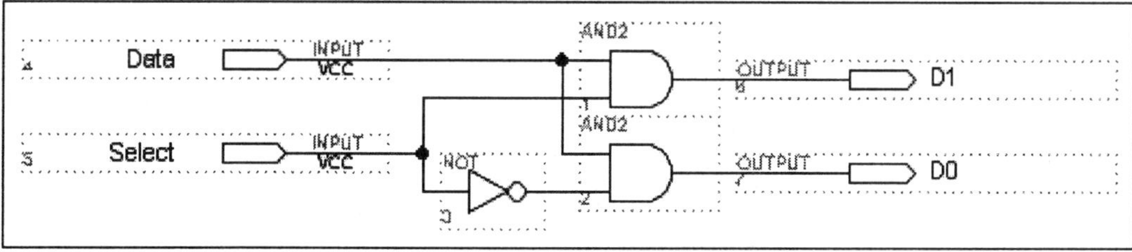

Figure 2

Any decoder that has an enable input may be used as a demultiplexer. Circuits that are designed to pass data from a single input to one of many outputs is a demultiplexer, where as a circuit whose outputs either turn on or turn off another circuit is a decoder.

This lab will illustrate the operation of several demultiplexers using logic gates and integrated circuits.

Part 1 Procedure

1. Open the Max+plus II software. Assign the project name **demux1** and MAX7000S as the device family.

2. Open a new Graphic Editor and construct the circuit shown in Figure 2.

3. Open a new Waveform Editor, set Grid Size to 50 ns, and then create the waveforms shown in Figure 3.

4. Sketch the output waveforms in the area provided in Figure 3, after simulation.

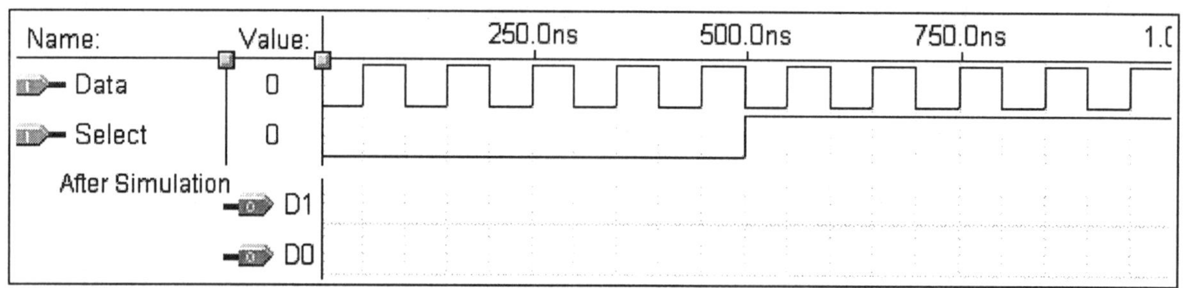

Figure 3

5. What output did the input data appear on when the select line was a logic-LOW? (D1/D0)

6. What output did the input data appear on when the select line was a logic-HIGH? (D1/D0)

7. The data input is active- (LOW/HIGH).

8. The data outputs are active- (LOW/HIGH).

9. Was the data inverted as data was passed through the circuit in Figure 2? (Yes/No)

10 Save all files to Drive A as **demux1**, then exit the Graphic and Waveform Editors.

Part 2 Procedure

1. Open the Max+plus II software. Assign the project name **demux2**.

2. Open a new Graphic Editor file and construct the circuit shown in Figure 4.

3. Open the Waveform Editor, set Grid Size to 50 ns, and create the waveforms shown in Figure 5. The casual-minded observer may recognize that the circuit in Figure 5 is very similar to Figure 2 in Lab 10.

4. Sketch the output waveforms after simulation in the area provided in Figure 5.

5. What output did the data appear on when the BA code was 00_2? (Y0 , Y1 , Y2 , Y3)

6. What output did the data appear on when the BA code was 01_2? (Y0 , Y1 , Y2 , Y3)

7. What output did the data appear on when the BA code was 10_2? (Y0 , Y1 , Y2 , Y3)

8. What output did the data appear on when the BA code was 11_2? (Y0 , Y1 , Y2 , Y3)

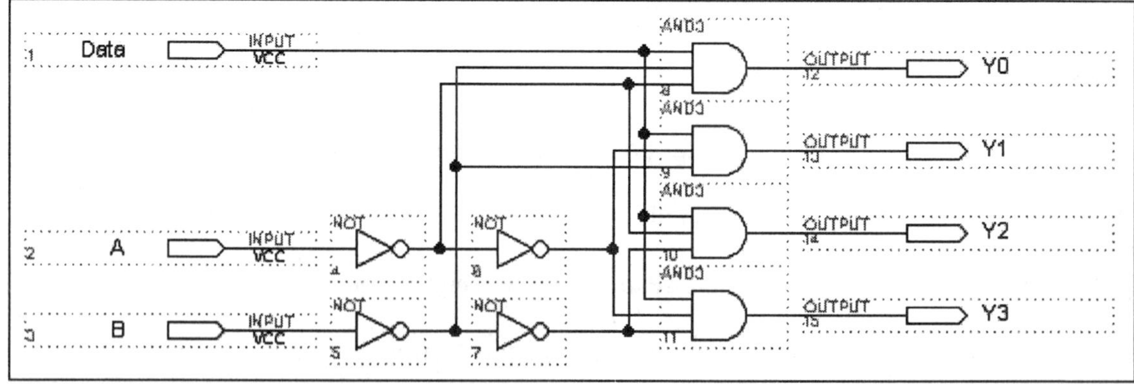

Figure 4

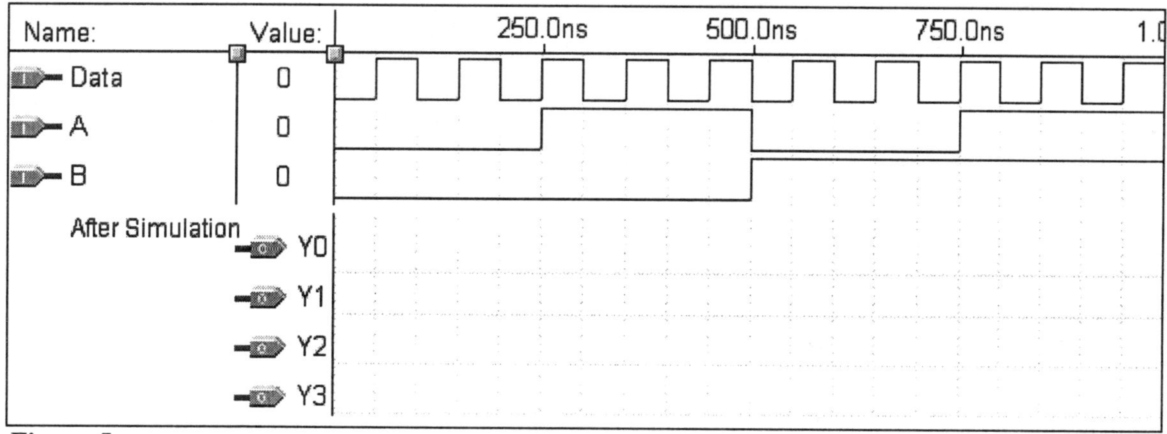

Figure 5

9. Was the data inverted as data was passed through the circuit in Figure 4? (Yes/No)

10. Save all files to Drive A as **demux2**, then exit the Graphic and Waveform Editors.

Part 3 Procedure

1. Open the Max+plus II software. Assign the project name **demux3**.

2. Open a new Graphic Editor and create the circuit and waveforms shown in Figure 6.

3. Open a new Waveform Editor, set Grid Size to 25 ns, and then create the waveforms shown in Figure 7.

4. Sketch the output waveforms after simulation in the area provided in Figure 7.

5. Complete Table 1 for the circuit of Figure 6 based on the results shown in Figure 7.

6. What pins in Figure 6 are the select lines? (CBA , G1, G2, Y)

7. What pins in Figure 6 are the enable lines? (CBA , G1, G2, Y)

8. The outputs in Figure 6 are active- (LOW/HIGH).

9. Was the data in Figure 6 inverted? (Yes/No)

10. The enable inputs in Figure 6 are active- (LOW/HIGH).

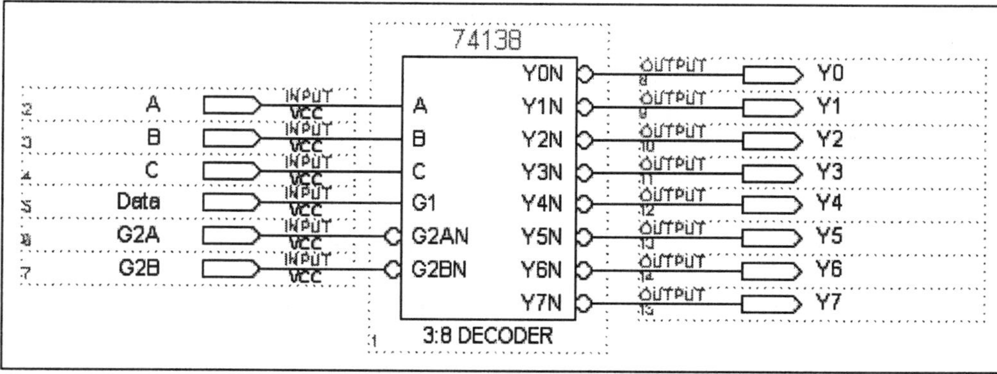

Figure 6

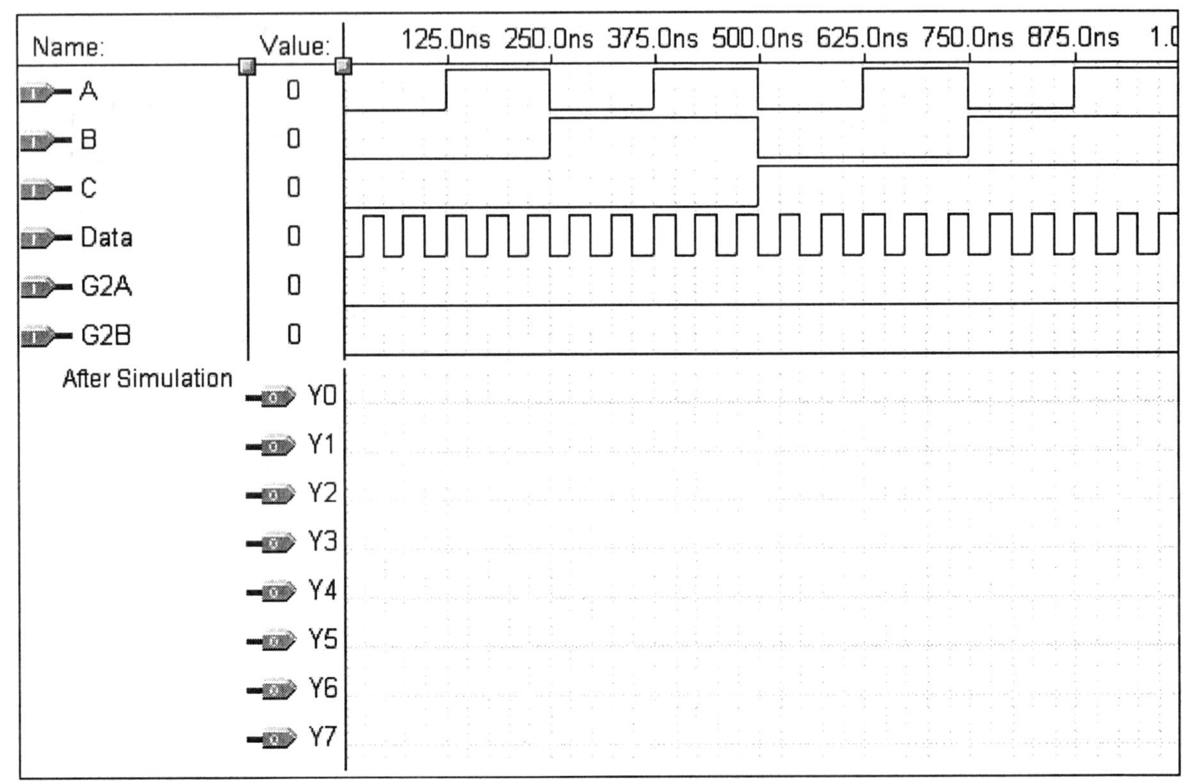

Figure 7

Time	G2B	G2A	C B A	Active Output
0 - 125 ns	___	___	___ ___ ___	_____
125 - 250 ns	___	___	___ ___ ___	_____
250 - 375 ns	___	___	___ ___ ___	_____
375 - 500 ns	___	___	___ ___ ___	_____
500 - 625 ns	___	___	___ ___ ___	_____
625 - 750 ns	___	___	___ ___ ___	_____
750 - 875 ns	___	___	___ ___ ___	_____
875 ns - 1 μs	___	___	___ ___ ___	_____

Table 1

11. What inputs in Figure 6 are dominant? (CBA , G1, G2, Y)

12. Explain in appropriate detail how the circuit in Figure 6 would operate if either G2 inputs were at a logic-HIGH level.

____

____

____

____

13. Save all files to Drive A as **demux3**, then exit the Graphic and Waveform Editors.

Part 4 Procedure

1. Open the **demux3.gdf** and **demux3.scf** files that were created in Part 3 of this lab. Save these files as **demux4.gdf** and **demux4.scf**, respectively.

2. Bring the Graphic Editor to the foreground.

3. Disconnect the Data input symbol from the G1 input and delete the G2A input symbol.

4. Connect the Data input symbol to the G2A input of the 74138.

5. Connect an input symbol to the G1 input. Label this symbol **G1**, then save the Graphic Editor file.

6. Bring the Waveform Editor to the foreground.

7. Change the G2A waveform label to G1, invert the G1 waveform, then save the Waveform Editor file.

8. Recompile and simulate the **demux4** circuit.

9. Does the **demux4** circuit invert the data applied? (Yes/No)

10. To enable the **demux4** circuit, G1 must be a logic- (LOW/HIGH).

11. To enable the **demux4** circuit, G2B must be a logic- (LOW/HIGH).

12. Obtain a hard copy of the Graphic and Waveform files. Label these hard copies **Part 4, Step 12A** and **Part 4, Step 12B**, respectively.

13. Save all files to Drive A as **demux4**, then exit the Graphic and Waveform Editors.

Part 5 Procedure

1. Open the Graphic and Waveform Editors in the Max+plus II software. Assign the project name **demux5**.

2. Figure 8 illustrates a circuit containing an 8-line to 1-line multiplexer and a 1-line to 8-line demultiplexer that share common select inputs. Select proper components from the Max+plus II parts bins and create the circuit so that the Data Out is in phase with the Data In. Use a bus line for the select inputs but *not* for the data inputs or data outputs. Hard-wire all enable inputs to turn on both ICs.

3. Create eight data input signals to the multiplexer, each with a distinct bit pattern.

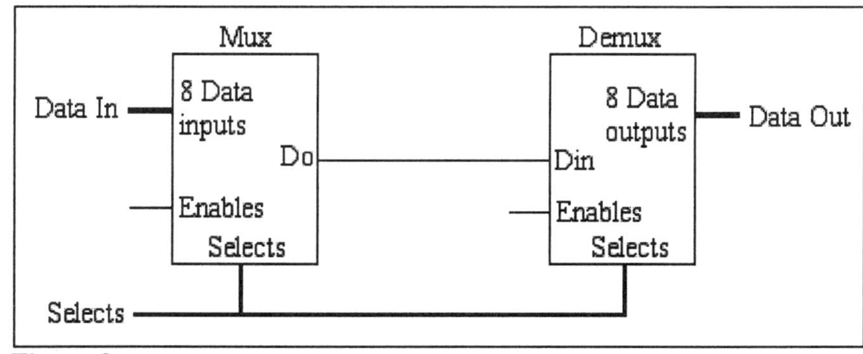

Figure 8

4. Create and set the CBA bus to 000_2.

5. Create eight data output signals to the demultiplexer.

6. Compile, and then simulate the circuit. Correct all errors before continuing.

7. If successful, the waveform on output 0 of the demultiplexer should be identical to the waveform on input 0 of the multiplexer.

8. Change the CBA value from 0 to 7, sequentially, to verify that each demultiplexer output matches the respective multiplexer input.

9. Demonstrate the circuit you created to your instructor. Obtain the signature of approval on the answer page for this lab.

10. Obtain a hard copy of the Graphic and Waveform Editors files. Label these hard copies **Part 5, Step 10A** and **Part 5, Step 10B**, respectively.

11. Save all files to Drive A as **demux5**, then exit the Graphic and Waveform Editors.

12. Write a 1- to 2-page summary containing at least one embedded graphic pertaining to the results obtained from this lab.

13. Place all papers for this lab in the following sequence, then submit the lab to your instructor for grading.

    - Cover page
    - Typed summary
    - The completed answer page for this lab
    - Hard copy of the Graphic Editor, **Part 4, Step 12A**
    - Hard copy of the Waveform Editor, **Part 4, Step 12B**
    - Hard copy of the Graphic Editor, **Part 5, Step 10A**
    - Hard copy of the Waveform Editor, **Part 5, Step 10B**

Lab 12: Demultiplexers Answer Pages

Name: ______________________________

Part 1

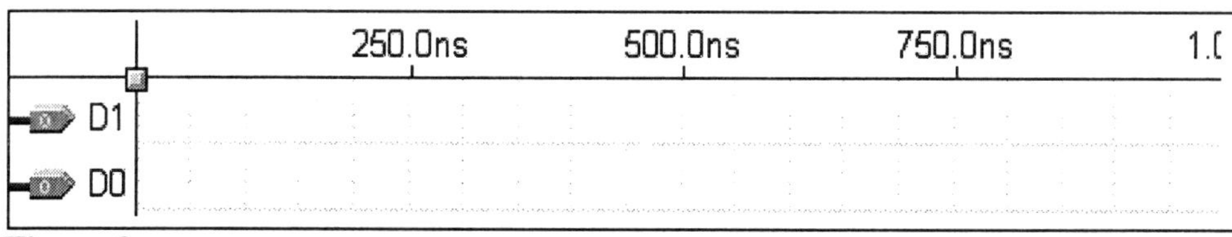

Figure 3

5. (D0/D1) 6. (D0/D1)

7. (LOW/HIGH) 8. (LOW/HIGH)

9. Yes No

Part 2

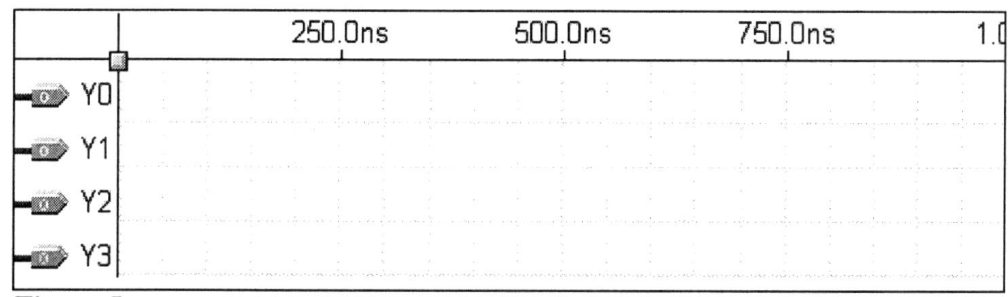

Figure 5

5. (Y0 , Y1 , Y2 , Y3) 6. (Y0 , Y1 , Y2 , Y3)

7. (Y0 , Y1 , Y2 , Y3) 8. (Y0 , Y1 , Y2 , Y3)

8. (Y0 , Y1 , Y2 , Y3) 9. (Yes/No)

Part 3

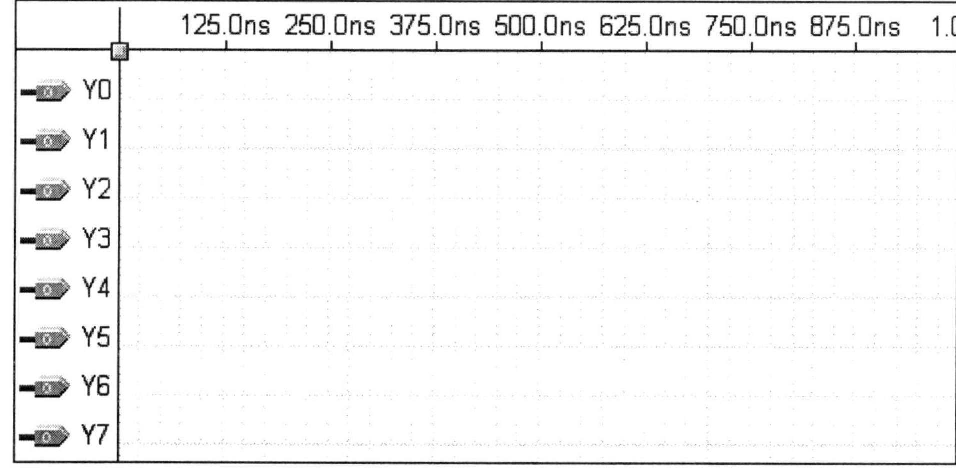

Figure 7

Time	G2B	G2A	C B A	Active Output
0-125 ns	___	___	___ ___ ___	_____
125-250 ns	___	___	___ ___ ___	_____
250-375 ns	___	___	___ ___ ___	_____
375-500 ns	___	___	___ ___ ___	_____
500-625 ns	___	___	___ ___ ___	_____
625-750 ns	___	___	___ ___ ___	_____
750-875 ns	___	___	___ ___ ___	_____
875 ns - 1 μs	___	___	___ ___ ___	_____

Table 1

6. (CBA , G1 , G2 , Y)

7. (CBA , G1 , G2 , Y)

8. (LOW/HIGH)

9. (Yes/No)

10. (LOW/HIGH)

11. (CBA , G1 , G2 , Y)

12. ____

____

____

Part 4

9. (Yes/No)

10. (LOW/HIGH)

11. (LOW/HIGH)

Part 5

9. Demonstrated to: ______________________ Date: __________

Grade: ______________

Lab 13: Latches

Objectives:

1. Construct and test the S-R latch
2. Construct and test the gated S-R latch
3. Construct and test the gated D-latch
4. Examine the 74373 latch for address latching

Material Required:

Max+plus II software by Altera Corporation

Discussion: The SR and $\overline{S}\,\overline{R}$ Latch

A latch is a bistable multivibrator device. This means that the latch is an electronic circuit designed to exist in one of two possible states and is stable in either state. These two states are called SET and RESET (or CLEAR). This device typically has two outputs that are labeled Q and $\overline{Q}$. This labeling convention indicates that the two outputs have opposite logic-levels. By definition, a latch is SET (SET = 1, or more accurately, is active) when Q = 1 and $\overline{Q} = 0$, and a latch is RESET (CLEAR, RESET = 0) when Q = 0 and $\overline{Q} = 1$. Conversely, for "Active Low" inputs, the latch is set when $\overline{SET} =$ 0 and $\overline{RESET} =$ 1.

The function tables for the SR latch (Table 1) and $\overline{S}\,\overline{R}$ latch (Table 2) are shown below.

S	R	Q	$\overline{Q}$	Mode
0	0	Q	$\overline{Q}$	No change
0	1	0	1	Reset
1	0	1	0	Set
1	1	0	0	Avoid

Table 1

$\overline{S}$	$\overline{R}$	Q	$\overline{Q}$	Mode
0	0	1	1	Avoid
0	1	1	0	Set
1	0	0	1	Reset
1	1	Q	$\overline{Q}$	No change

Table 2

Part 1 Procedure

1. Open the Max+plus II software and assign the project name **sr-not**.

2. Open a new Graphic Editor and construct the circuit shown in Figure 1.

3. Open a new Waveform Editor and create the waveforms shown in Figure 2, and then sketch the corresponding output waveforms after simulation.

4. Referring to the output waveforms, complete Table 1 for the $\overline{S}\,\overline{R}$ latch.

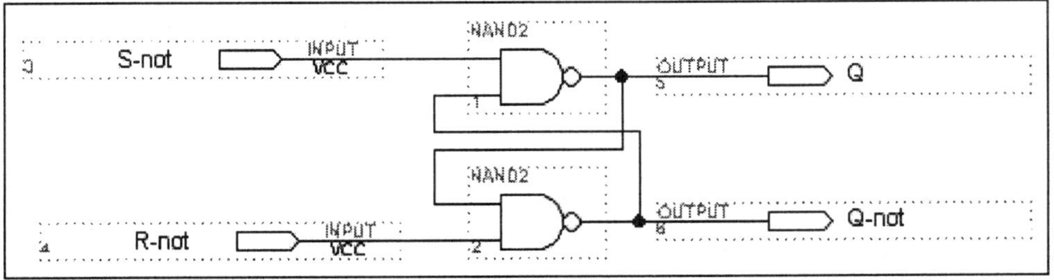

Figure 1

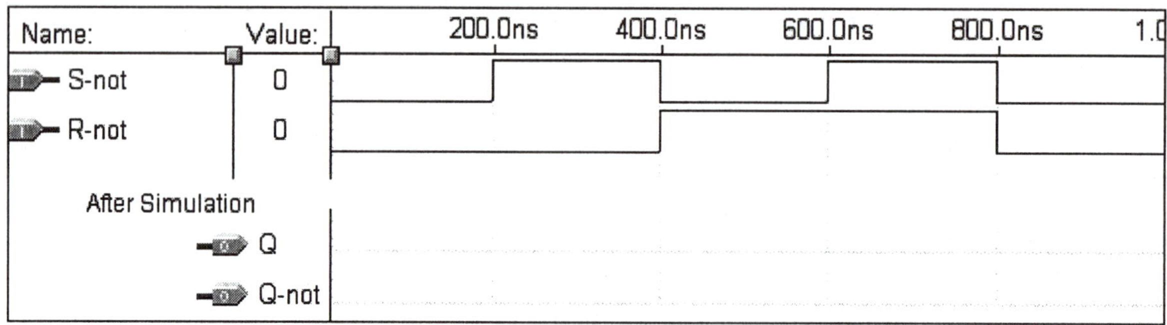

Figure 2

Time	$\overline{S}$	$\overline{R}$	Q	$\overline{Q}$	Mode
0 – 200 ns	0	0	1	0	AVOID
200 – 400 ns	1	0	0	1	RESET
400 – 600 ns	0	1	1	0	SET
600 – 800 ns	1	1	1	0	NO CHANGE
800 ns – 1 μs	X	X	X	X	

Table 3

5. Do the results recorded in Table 3 match the function table for the $\overline{S}\,\overline{R}$ latch in Table 2? (Yes/No) [Yes circled]

6. If the current state of inputs is set, reset, or no change, then the output condition for the next state is predictable according to the table. However, if the current state is the avoid condition going to the next state (that is, a no change condition), then an unpredictable result will occur. This condition is called __________.

7. Save all files to Drive A as **s-r-latch**, then exit the Graphic and Waveform Editors.

Part 2 Procedure

1. Open the Max+plus II software. Assign the project name **sr-latch**.

2. Open a new Graphic Editor and construct the circuit shown in Figure 3.

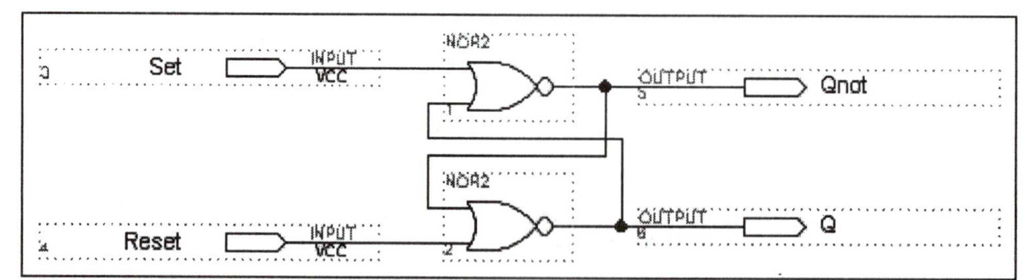

Figure 3

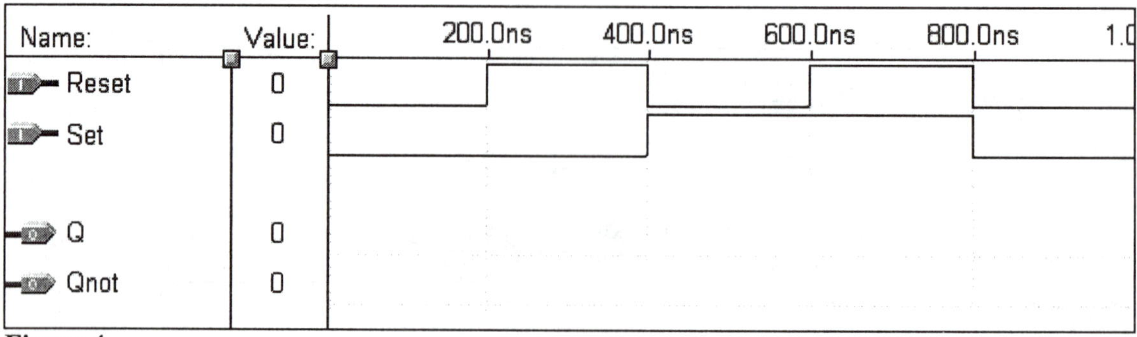

Figure 4

3. Open a new Waveform Editor and create the waveforms shown in Figure 4 and sketch the output waveforms after simulation.

4. Referring to the waveforms in Figure 4, complete the following truth table (Table 4) for the S-R latch.

Time	S	R	Q	$\overline{Q}$	Mode
0 – 200 ns	0	0	X	X	NO CHANGE
200 – 400 ns	1	0	0	1	RESET
400 – 600 ns	0	1	1	0	SET
600 – 800 ns	1	1	0	0	AVOID
800 ns – 1 μs	X	X	X	X	

Table 4

5. Do the results recorded in Table 4 match the function table for the SR latch in Table 1? (Yes/No)

6. Save all files to Drive A as **s-r-latch**, then exit the Graphic and Waveform Editors.

Part 3 Procedure

Discussion: The Gated S-R and Gated D Latch

Steering gates added to the inputs of the SR latch (Figure 3) produce a gated SR latch shown in Figure 5. The gate must be a logic-HIGH to pass the S and R signals through the AND gates in order to affect the outputs of the SR latch. If the gate is a logic-LOW, the steering gates prevent the S and R inputs from affecting the outputs of the latch.

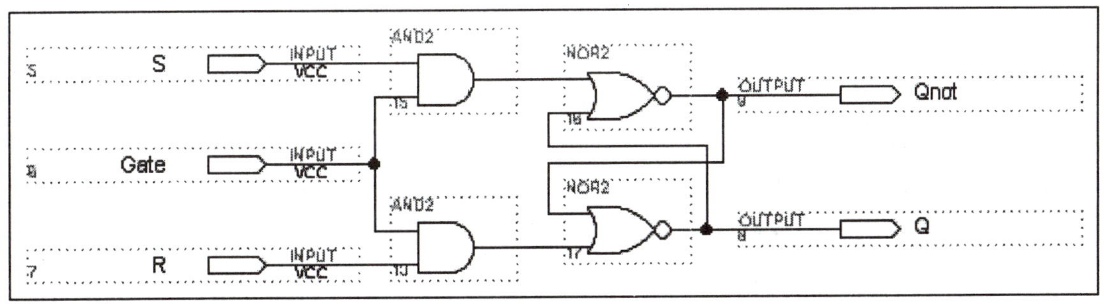

Figure 5

1. Open the Max+plus II software and assign the project name **gated-sr**.

2. Open a new Graphic Editor and construct the circuit shown in Figure 5.

3. Open a new Waveform Editor and create the waveforms shown in Figure 6.

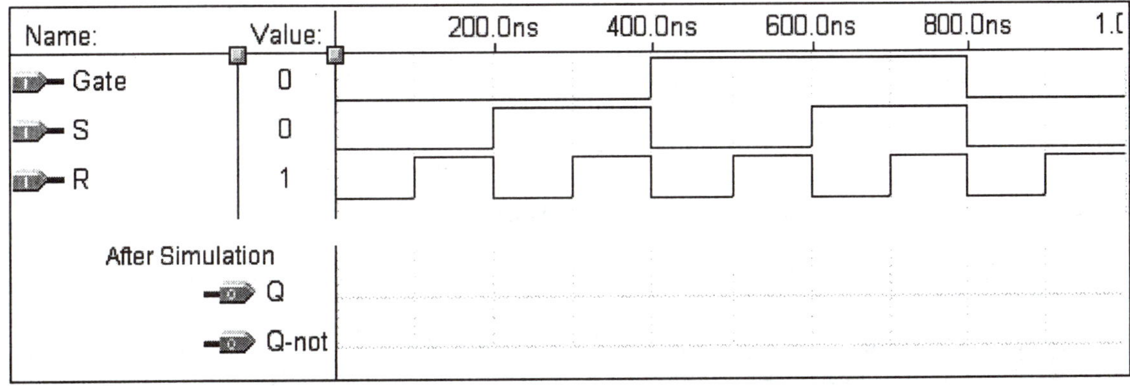

Figure 6

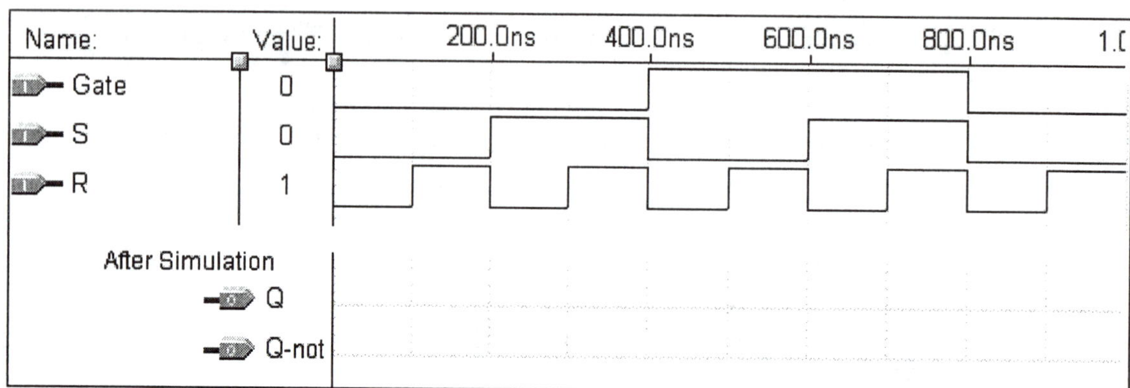

Figure 6

4. Sketch the output waveforms at the bottom of Figure 6 after simulation.

5. Since the gate signal starts at a logic-LOW state, the gated SR latch initially is turned (On/Off). Therefore the mode of operation from 0 to 400 ns is (No change/Set/Reset/Avoid). From 400 ns to 500 ns, the mode of operation is (No change/Set/Reset/Avoid). At 500 ns, the mode of operation is (No change/Set/Reset/Avoid) and the outputs are now in a predictable state.

6. Referring to the waveforms in Figure 6, complete the truth table (Table 5) for the gated S-R latch.

	Gate	S	R	Q	$\overline{Q}$	Mode
ON	1	0	0	X	X	
	1	0	1	0	1	
	1	1	0	1	0	
	1	1	1	0	0	
OFF	0	X	X	X	X	

Table 5 X is irrelevant

7. If all three inputs were in the HIGH state and changed to a LOW state at the same time, the latch would go through a condition called the______________ condition. Would the same condition occur if only the GATE input had gone LOW? ________ (Yes/No)

8. To avoid the condition referred to in Step 7, the gated S-R latch will be modified by connecting an inverter between the S and R inputs. This ensures that the inputs would never be the same, eliminating the no change and avoid conditions. The gated-D latch will be examined in Part 4.

9. Save all files to Drive A as **gated-sr**, then exit the Graphic and Waveform Editors.

Part 4 Procedure

1. Open the Max+plus II software and assign the project name **gated-d**.

2. Open a new Graphic Editor and construct the circuit shown in Figure 7.

3. Open a new Waveform Editor and create the waveforms shown in Figure 7.

4. Click the Compile and Simulate buttons. Correct all errors before continuing.

5. Referring to the waveforms in Figure 8, complete the truth table (Table 6) for the gated-D latch.

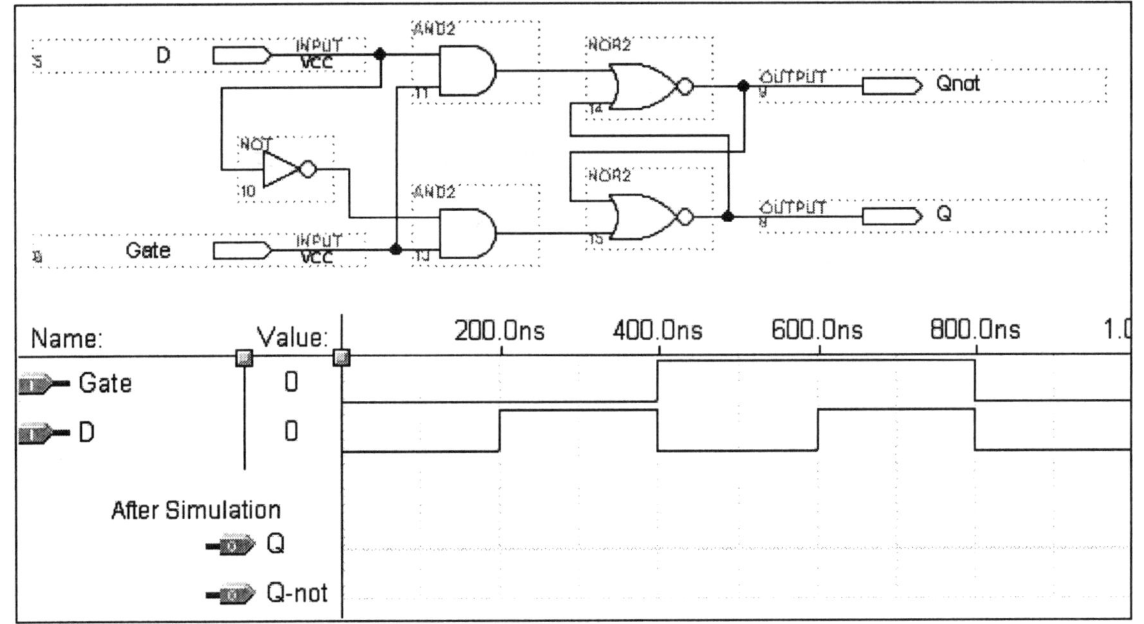

Figure 7

Gate	D	Q	$\overline{Q}$
____	____	____	____
____	____	____	____
____	____	____	____
____	____	____	____

Table 6

6. How many inputs does your table show? _______________. How many input combinations are possible with this latch? _______________.

7. There is a saying that "Q follows D when the gate is 'active' in a gated-D latch." Does your truth table agree with this saying? (Yes/No) Why or why not?

____

8. Does the invalid condition of S-R latches exist for the gated-D latch? (Yes/No)

9. Write a paragraph explaining how the gated-D latch operates, based on the function table and your waveforms in Figure 7.

____

____

____

10. Save all files to Drive A as **gated-d**, then exit the Graphic and Waveform Editors.

Part 5 Procedure

Discussion:

The latch is ideally suited for temporary storage of information between processing units and input/output devices. When the enable is "active," data on the input of the latch will pass to the Q output (transparent). When the enable becomes "inactive," the data on the Q outputs is "latched" and will be retained until the latch is again enabled or power is lost (volatile memory).

The 74373 has eight latches that share the same enable input. Microprocessors (uP) or micro-controllers (uC) often will have multiplexed address and data lines, AD[7..0], sharing common pins. It is necessary to isolate the data D[7..0] and address A[7..0] lines external to the microprocessor. As illustrated in Figure 8, the 74373 will be used to "latch" onto the address lines, freeing up the address/data pins of the microprocessor to be used by data. Proper timing is determined by control signals that enable the latch to "memorize" the address and control the tristate outputs of the latch.

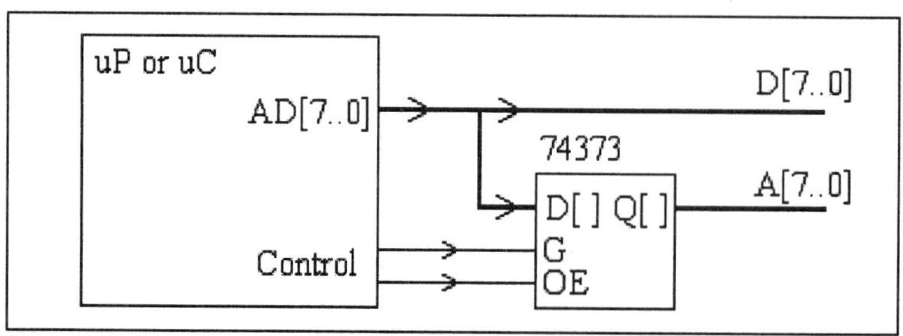

Figure 8

1. Open the Max+plus II software. Assign the project name **latch373**.

2. Open a new Graphic Editor and create the waveforms shown in Figure 9.

3. Open a new Waveform Editor, set Grid size to 100 ns, and create the waveforms shown in Figure 10. D[7..0] starts at 75 and increments by 174.

4. Run the Compiler and Simulator. Correct all errors before continuing.

5. Sketch the resulting output waveforms in the area provided at the bottom of Figure 10.

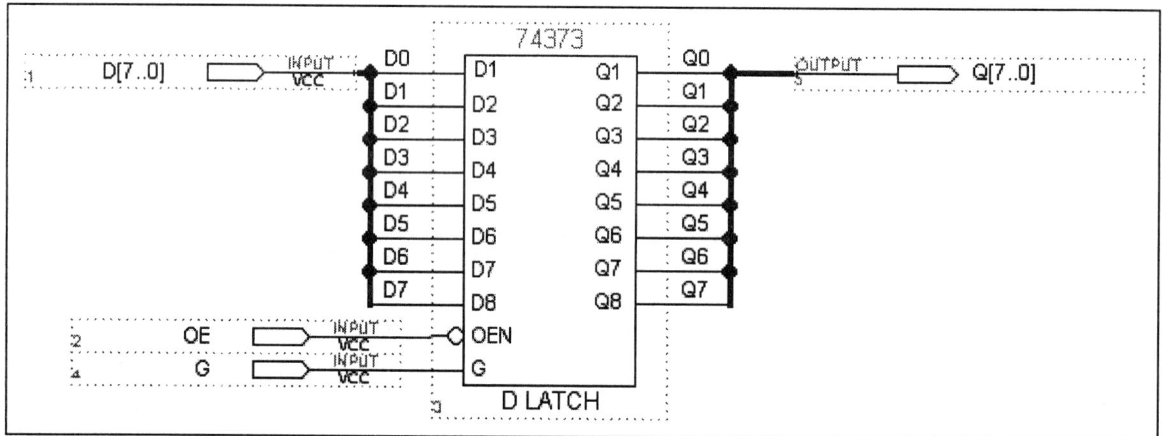

Figure 9

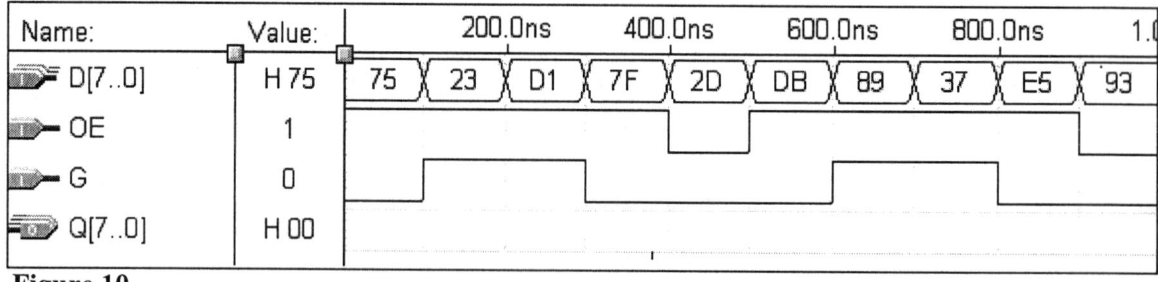

Figure 10

6. What time segment from 0 to 500 ns is the latch "ON" ? ____________
(When the Q outputs are transparent)

7. What data was "latched" in the first 500 ns of time? ____________

8. In what time segment did this data appear on the outputs? ____________

9. What data was "latched" in the second 500 ns time frame? (From 500 ns to 1 μs) ____________

10. What time segment did the data appear on the outputs? ____________

11. What do the ZZs in the output waveform represent? ______________________________

12. Save all files to Drive A as **latch373**, then exit the Graphic and Waveform Editors.

13. Demonstrate the circuit and waveforms for Part 5 of this lab to the instructor. Obtain the signature of approval directly on the answer page for this lab.

14. Obtain hard copies of the Graphic and Waveform Editors. Label these hard copies **Part 5, Step 14A** and **Part 5, Step 14B**, respectively.

15. Write a 1- to 2-page summary containing and making reference to at least one embedded figure from this lab.

16. Place all papers for this lab in the following sequence, then submit the lab to your instructor for grading.

    - Cover page
    - Typed summary
    - The completed answer pages for this lab
    - Printout of the Graphic Editor, **Part 5, Step 14A**
    - Printout of the Waveform Editor, **Part 5, Step 14B**

Lab 14: The 555 Timer

Objectives:

1. Construct an oscillator using the 555 timer
2. Design an oscillator with a specific duty cycle

Materials List:

- Power Supply
- Oscilloscope
- 555 Timer IC
- 4.7 KΩ resistor
- 47 KΩ resistor
- 0.001 µF capacitor
- 0.01 µF capacitor

Discussion:

There are three types of multivibrators: the astable, bistable, and monostable. The bistable multivibrator, or flip-flop, has two stable states. When set, the flip-flop will remain set until a condition on the input forces the output to change. The same is true if the flip-flop is in reset mode. A monostable (one-shot) multivibrator has a single stable state. An input condition forces the multivibrator to become unstable for a period of time as determined by an external resistor and capacitor, then automatically reverts back to its stable state. The astable multivibrator doesn't have a stable state so it oscillates.

The TTL compatible 555 timer may be used as an astable multivibrator (oscillator) or as a monostable multivibrator (one-shot). Illustrated in Figure 1, the 555 timer has three resistors in series to provide two reference voltages, V_{REF1} and V_{REF2}. The output of Comparator 1 becomes +Vcc when the threshold voltage at Pin 6 becomes more positive than V_{REF1}, causing the latch to reset. When the negated Q output is a logic-HIGH, the transistor is saturated causing the collector-to-emitter voltage to be approximately 0.2 V. When the Trigger voltage at Pin 2 becomes less positive than V_{REF2}, set occurs causing the negated Q output to go to a logic-LOW. This in turn shuts off the transistor.

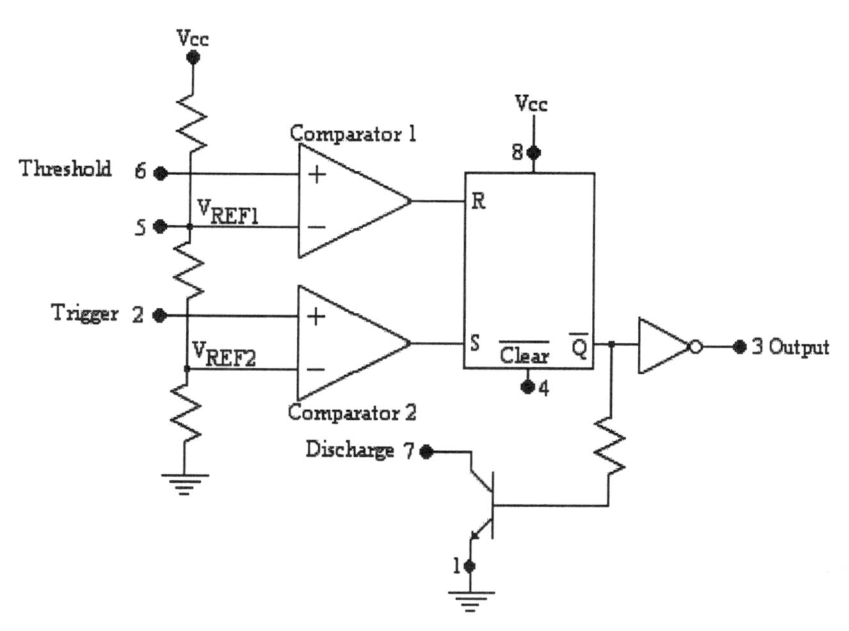

Figure 1

External components needed to cause the 555 timer IC to become a *free running* oscillator are shown in Figure 2. Capacitor C1 charges through R2 and R1 towards Vcc. When E_{C1} reaches the threshold, V_{REF1}, voltage required on Pin 6, the internal transistor saturates, effectively shorting Pin 7, allowing capacitor C1 to discharge. Once E_{C1} drops below the trigger voltage, V_{REF2}, on Pin 2, the internal transistor turns off, allowing C1 to charge again.

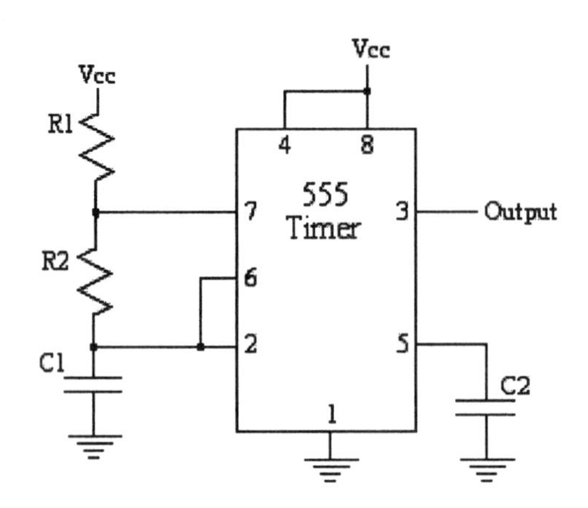

Figure 2

The charge time for C1 is determined by: $t_{CHARGE} = 0.7(R_1 + R_2)C_1$ (1)

The discharge time for C1 is determined by: $t_{DISCHARGE} = 0.7R_2C_1$ (2)

The period of a complete cycle of the output frequency is the sum of the charge and discharge rates of capacitor C1. The period, T, may be calculated by : $T = 0.7(R_1 + 2R_2)C_1$ (3)

The output frequency, f, may be determined by taking the reciprocal of the period, T.

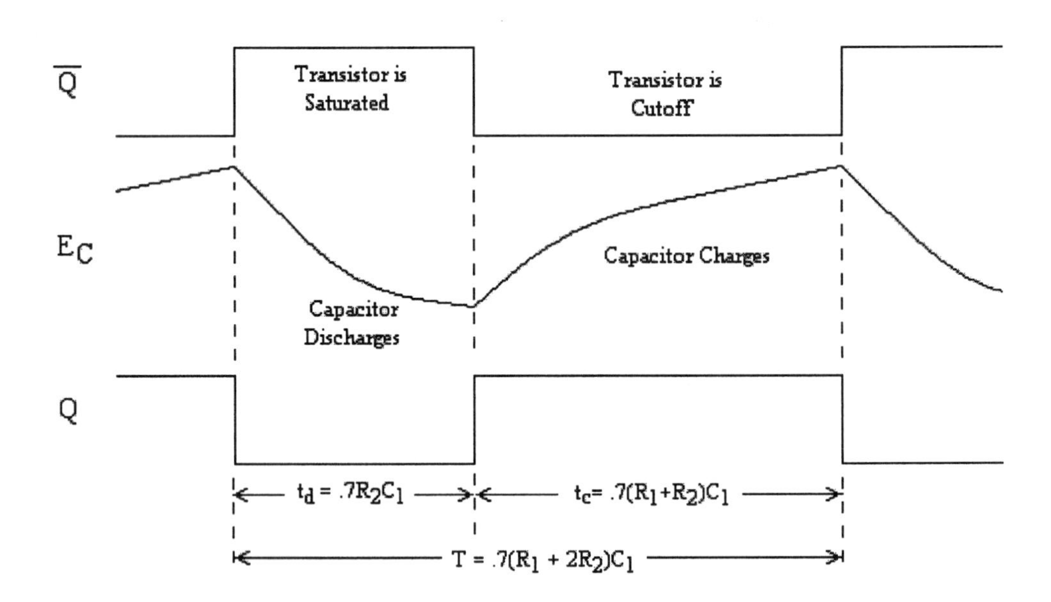

Figure 3

The waveforms in Figure 3 show the relationship of output Q (pin 3) with respect to the charge/discharge of capacitor C1 (pin 2). The pulse and space width are controlled by selecting different values of R1 and R2. As per equations 1 and 2, changing R1 only affects the charge time but changing R2 affects both charge and discharge times.

To ensure proper operation as an oscillator, R1 should always be greater than 1 KΩ, R1 + R2 should be less than 6.6 MΩ, and C1 should be greater than 50 pFd.

Part 1 Procedure

1. Construct the circuit shown in Figure 2 using the values listed below.

   - Vcc = 5 V DC
   - 1 = 4.7 KΩ
   - R2 = 47 KΩ
   - C1 = 0.001 μFd
   - C2 = 0.01 μFd

2. Based on the component values given and Equations 1 through 4, calculate the following: Show all work!
   - Charge time for C1

   - Discharge time for C1

   - Frequency of the output signal

   - Duty cycle (pulse width/period) of the output signal

3. Use an oscilloscope to measure the frequency at the output (Pin 3) of the circuit constructed.

   $f_{MEASURED}$ = ______________

4. Accurately read the time for each segment of the output waveform displayed on the oscilloscope and record the results for the designated areas on the waveform shown in Figure 4.

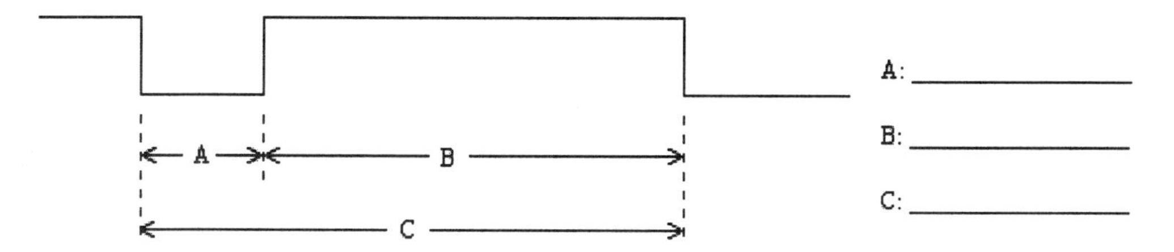

Figure 4

5. Based on the recorded time measurements for the pulse width and period, calculate the duty cycle for the waveform displayed on the oscilloscope.

   DC = PW/T = ______ / ______ = __________

   %DC = DC × 100 = ____________%

Part 2 Procedure

1. Assume you want to create a 25-KHz pulse waveform with a 67% duty cycle. Assuming C1 is 0.001 μFd and C2 is 0.01 μFd, calculate values of R_1 and R_2 that will produce the desired waveform. Remember that R_1 must be greater than 1 KΩ.

A. Given:

$$f = \frac{1.44}{(R_1 + 2R_2)C_1}$$

Solve for $\mathbf{R_1+2R_2}$: ____________

B. Given: Duty cycle $= \left(\frac{\mathbf{R_1 + R_2}}{\mathbf{R_1 + 2R_2}}\right)$

Based on 67% duty cycle, calculate R_2 in terms of R_1
Solution:

$$.67 = \left(\frac{R_1 + R_2}{R_1 + 2R_2}\right)$$

$$.67(R_1 + 2R_2) = R_1 + R_2$$

$$.67R_1 + 1.34R_2 = R_1 + R_2$$

$$.34R_2 = .33R_1$$

$$R_2 = \frac{.33R_1}{.34}$$

$$\text{so } R_2 \cong R_1$$

C. Based on your answer for Step 1A and relationship of R_1 and R_2 as determined in Step 1B, select a value for $\mathbf{R_1}$ and $\mathbf{R_2}$. $\mathbf{R_1}$ = ____________ $\mathbf{R_2}$ = ____________

2. Construct the circuit shown in Figure 2 based on your calculated values from Part 2, Step 1. Use potentiometers for $\mathbf{R_1}$ and $\mathbf{R_2}$, set to the calculated resistances.

3. Use an oscilloscope to measure the frequency at the output (Pin 3) of the circuit constructed.

$f_{MEASURED}$ = ________________

4. Accurately read the time for each segment of the output waveform displayed on the oscilloscope and record the results for the designated areas on the waveform shown in Figure 5.

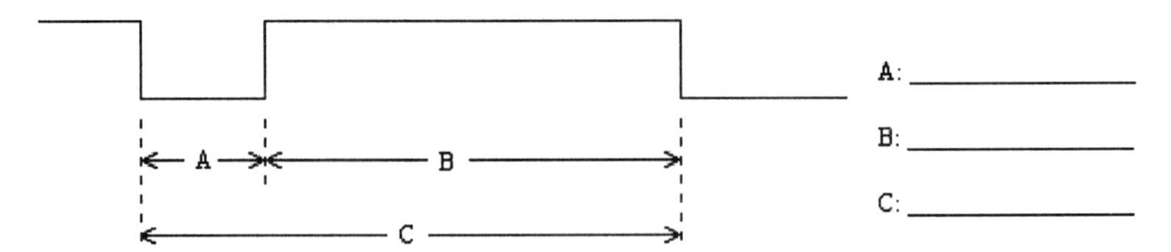

A: ____________

B: ____________

C: ____________

Figure 5

5. Based on the recorded time measurements for the pulse width and period, calculate the duty cycle for the waveform displayed on the oscilloscope.

DC = PW/T = ______ / ______ = __________

%DC = DC × 100 = ____________ %

6. Demonstrate the circuit and displayed waveforms confirming the 25 KHz with 67% duty cycle to your instructor. Obtain the signature of approval directly on the answer page for this lab.

7. Write a 1- to 2-page summary pertaining to the results obtained for this lab. Include embedded graphics of the circuit constructed for Part 2, and a computer generated output waveform showing the pulse width, space width, and period similar to Figure 5.

8. Submit the following for grading:
   - Cover page
   - Typed summary
   - Completed answer page for this lab

Lab 14: The 555 Timer Answer Pages

Name: ____________________

Part 1

2. (Show work)

- Charge time for C1

- Discharge time for C1

- Frequency of the output signal

- Duty cycle (pulse width/period) of the output signal

3. $f_{MEASURED}$ = ________________

4. A. ______________

   B. ______________

   C. ______________

5. DC = PW/T = ______ / _______ = __________

   %DC = DC × 100 = ____________%

Part 2

1A. $\mathbf{R}_1+2\mathbf{R}_2$: __________

1C. $\mathbf{R}_1$ = __________, $\mathbf{R}_2$ = __________

3. $f_{MEASURED}$ = ________________

4. A. ______________

   B. ______________

   C. ______________

5. DC = PW/T = ______ / _______ = __________

   %DC = DC × 100 = ____________%

6. Demonstrated to: ____________________________ Date: __________

Lab 15: Flip-Flops

Objectives:

1. Become familiar with the operation of the D flip-flop
2. Become familiar with the operation of the J-K flip-flop
3. Compare the 7476 and 74LS76 Dual J-K flip-flops
4. Evaluate a computer output port for latched and unlatched data transfer

Materials List:

- Max+plus II software by Altera Corporation
- University Board by Altera Corporation (optional)
- Computer requirements:
  Minimum 486/66 with 8 MB RAM
- Floppy disk

Discussion:

Flip-flops may also be used as temporary storage devices and are most appropriate for clock-dependent circuits such as counters and registers. The flip-flop differs from the gated latch in that the clock input of the flip-flop (FF) has an edge-detector circuit. This means that while the latch will change states to correspond to the logic levels on the input(s) whenever the gate (or enable) input is at the level corresponding to its active state, the flip-flop will only recognize the logic levels on the inputs when the clock input logic level is changed from a LOW to HIGH (leading edge or positive going transition), or the clock is changing from HIGH to LOW (trailing edge or negative going transition). Even though numerous flip-flops are available, this lab focuses on the D-type, Toggle, and J-K flip-flops. Like the latch, the flip-flop is a bistable multivibrator.

The D-type flip-flop

The D and J-K flip-flops are probably the most common flip-flops in use today. Both versions were developed as improvements on the S-R version. The D flip-flop did away with the "illegal" state by ensuring that both inputs would never be the same. This, however, left the D flip-flop with one input and the capability to monitor only one signal at a time. This fact makes the D flip-flop ideal for register use.

Edge-triggered devices have the advantage of rejuvenating an input for only a few nanoseconds during the edge of the clock pulse, thus excluding "noise" inputs at all other times. The disadvantage is that the clock pulse must be a "clean" or "sharp" pulse wave. If the edge is jagged, the device may "see" several "edges" and trigger several times.

Part 1 Procedure

1. Open the Max+plus II software. Assign the project name **d-fflop**.

2. Open a new Graphic Editor and construct the circuit shown in Figure 1. Use the DFF symbol in the primitives directory.

3. Open a new Waveform Editor, set Grid Size to 50 ns, and then create the waveforms shown in Figure 1. Sketch the output waveforms in the space provided in Figure 2 after simulation.

4. List the times that the Q output goes to a logic-HIGH. ________________________________

5. List the times that the Q output goes to a logic-LOW. ________________________________

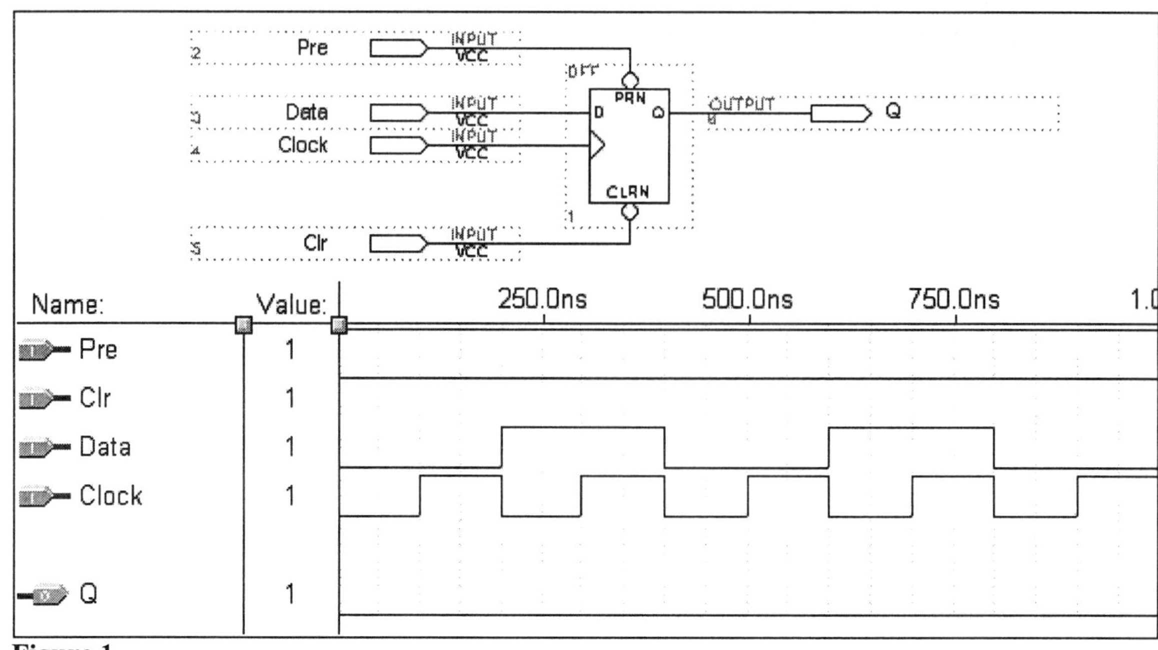

Figure 1

Name:	Value:	250.0ns	500.0ns	750.0ns	1.0
Q	1				

Figure 2

6. When output Q changes, Q follows the _____________ when the ______________ goes to a logic-(HIGH/LOW).

7. Be sure the Snap to Grid in the Options menu is turned ON. Modify the Pre and Clr waveforms as shown in Figure 3. Sketch output Q after simulation in the space provided in Figure 3.

8. What input transition caused output Q to go to a logic-HIGH in the first 400 ns? ______________________

9. At what time did output Q go to a logic-HIGH in the first 400 ns? ____________ ns

10. Was this transition in output Q (during the first 400 ns) clock dependent? (Yes/No)

11. Before 600 ns, Q went to a logic-LOW. This transition occurred at what time? ___________ ns

12. Was the first transition of Q from a logic-HIGH to a logic-LOW clock dependant? (Yes/No)

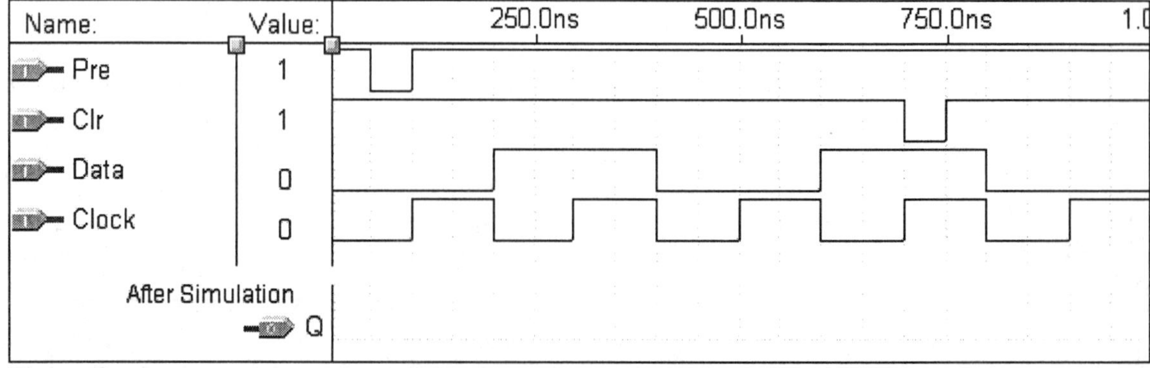

Figure 3

13. Why did Q go to a logic-LOW? ______________________________

14. A glitch occurred in the output waveform near 750 ns. Explain what caused this glitch.

____

15. At what time did the glitch in the output waveform occur? __________ ns

16. The D flip-flop used in Figure 10 is (positive/negative) (edge/level) triggered.

17. The D and clock inputs are (synchronous/asynchronous).

18. Does the clock input have the same effect on the preset and clear inputs that it has on the D input? (Yes/No)

19. The preset and clear inputs are (synchronous/asynchronous) inputs.

20. From the timing diagram, the preset and clear inputs are active- (HIGH/LOW) inputs.

21. Save all files to Drive A as **d-fflop**, then exit the Graphic and Waveform Editors.

Part 2 Procedure

The Toggle flip-flop

Discussion:

The toggle flip-flop is derived from the D-type flip-flop by wiring the $\overline{Q}$ output back to the D input. Assuming Q is a logic-LOW, the D input would be a logic-HIGH. When a clock pulse is applied, Q takes on the value of D so Q is now a logic-HIGH. This now makes input D a logic-LOW. On the next clock pulse, Q then goes low. As a result, the Q output changes states, or toggles, every time a clock pulse is applied.

1. Open the Max+plus II software. Assign the project name **t-fflop**.

2. Open a new Graphic Editor and construct the circuit shown in Figure 4. Use the DFF symbol in the primitives directory.

3. Open a new Waveform Editor, set Grid Size to 50 ns, and then create the waveforms shown in Figure 4.

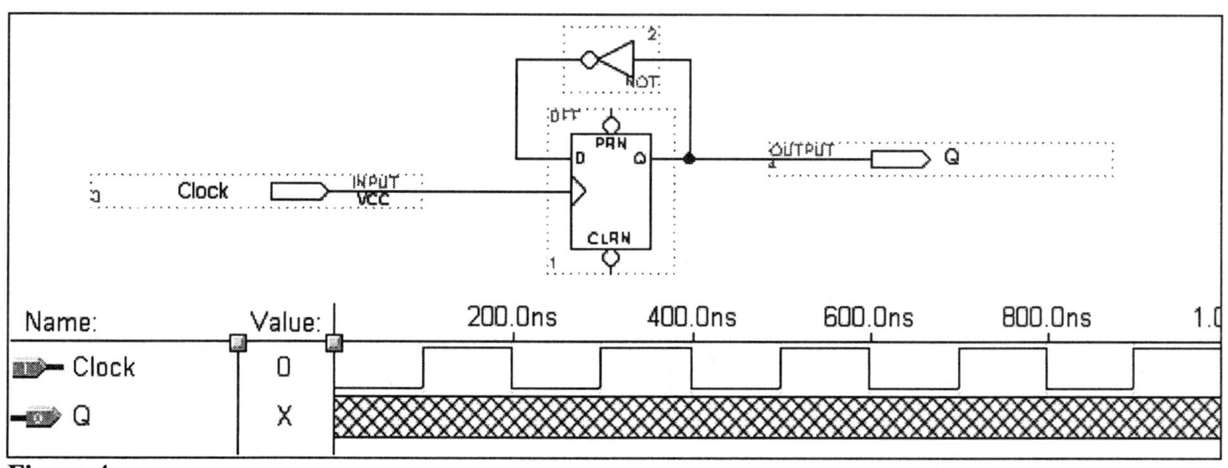

Figure 4

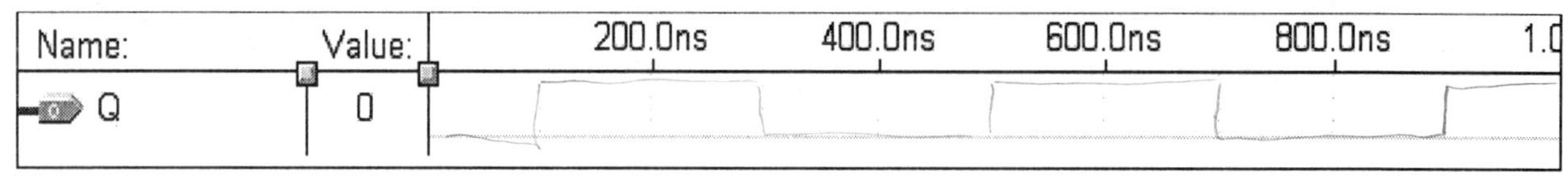

Figure 5

4. Compile and simulate the circuit. Correct all errors before continuing.

5. Sketch the output waveform in the space provided in Figure 5.

6. What is the period of one complete clock cycle? ___________ (Provide units)

7. What is the period of one complete cycle on the Q output of the T flip-flop? ___________

8. What is the relationship between the period of the Q output with respect to the period of the clock input?

   ____

9. What is the frequency of the clock? ___________

10. What is the frequency of the Q output? ___________

11. Calculate the duty cycle of the Q output waveform. DC = ___________

12. Based on the duty cycle, the output of the toggle flip-flop is a ___________ wave.

13. Save all files to Drive A as **t-fflop**, then exit the Graphic and Waveform Editors.

14. Altera has included a toggle flip-flop in the primitive parts list labeled TFF. This flip-flop has an active-HIGH enable input, T. Open a new Graphic Editor file and assign the project name **pulse**. Construct the circuit shown in Figure 6 using the TFF symbol.

15. Open a new Waveform Editor and create the waveforms shown in Figure 6.

16. Compile and simulate the circuit. Correct all errors before continuing.

17. Sketch the output waveform in the space provided in Figure 7.

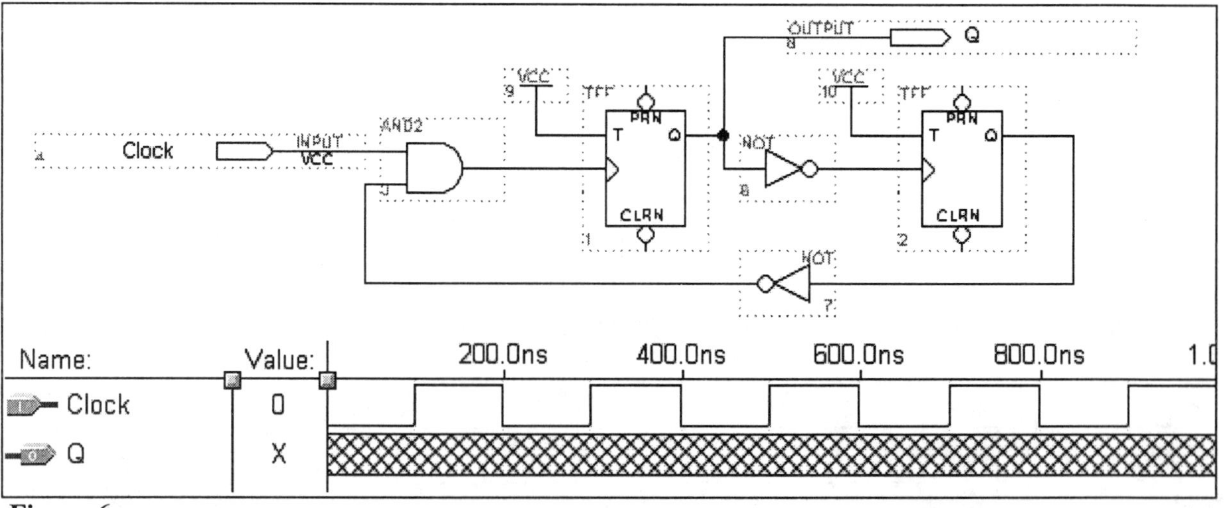

Figure 6

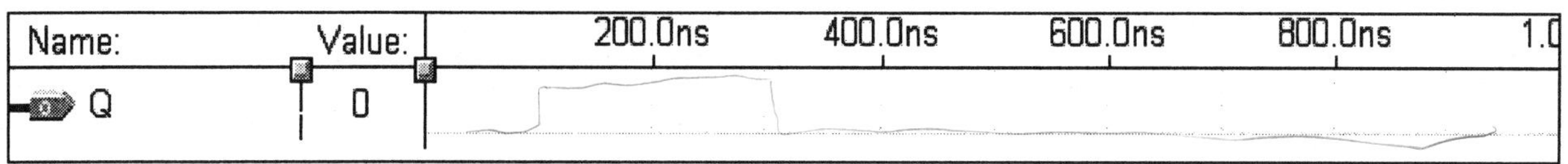

Figure 7

18. Assume both Q outputs are a logic-LOW. Explain how the circuit responds to the first two clock pulses.

____

____

____

19. Save all files to Drive A as **t-fflop**, then exit the Graphic and Waveform Editors.

Part 3 Procedure

The J-K Flip-Flop

Discussion:

The J-K flip-flop turned the "illegal" state of the S-R flip-flop into the productive condition, or state, called the toggle state. This fact makes the J-K a useful flip-flop for making counter circuits. Two J-K flip-flops discussed in this lab are the 7476 and 7476a (the 7476a is Altera's label for the 74LS76a).

1. Open the Max+plus II software. Assign the project name **jk-fflop**.

2. Open a new Graphic Editor, then import the **jkff** symbol from the **Prim** directory and connect input/output terminals with the labels shown in Figure 8.

3. Open a new Waveform Editor and create the waveforms shown in Figure 5.

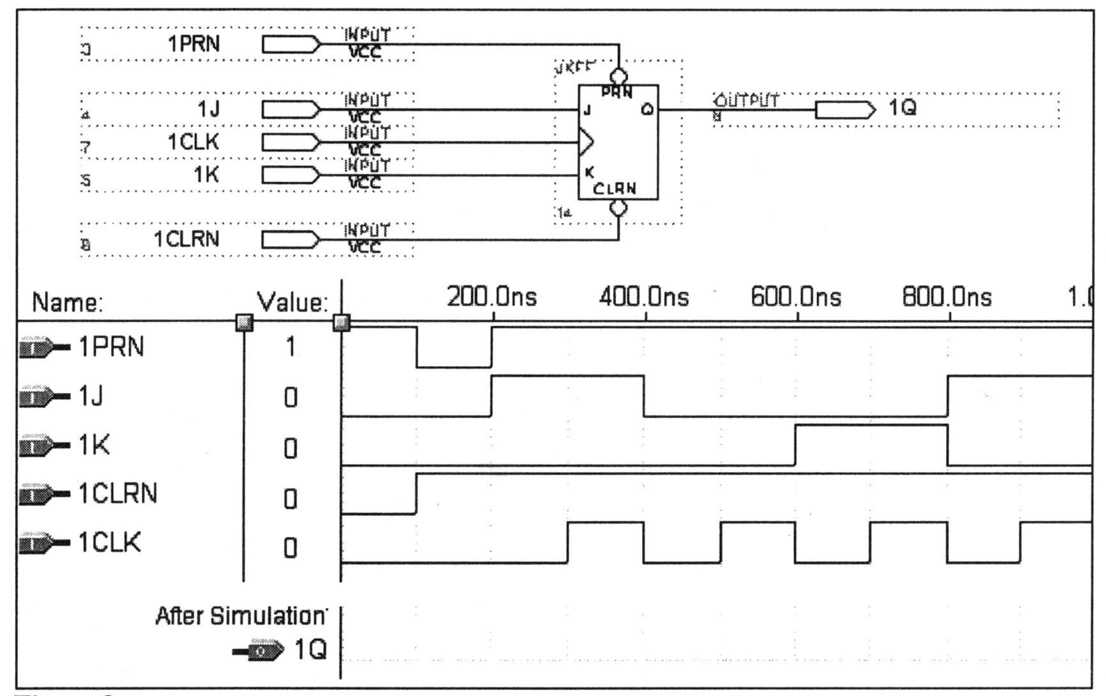

Figure 8

4. After simulating the circuit, draw the Q output waveform for the circuit constructed in the space provided in Figure 8.

5. Evaluate the timing diagram you have just completed. According to the timing diagram, the **jkff** symbol is (leading/trailing) edge triggered.

6. Carefully examine the output waveform of Figure 8 with respect to the input waveforms and complete Table 1. For each time segment, identify the logic levels of the waveforms, and then identify the flip-flop mode of operation.

Time segment	1PRN	1J	1K	1CLRN	1CLK	1Q	MODE
0 to 100 ns	1	0	0	0	0	0	
100 to 200 ns	0	0	0	1	0	1	
200 to 300 ns	1	1	0	1	0	1	
300 to 400 ns	1	1	0	1	1	1	
400 to 500 ns	1	0	0	1	0	1	
500 to 600 ns	1	0	0	1	1	1	
600 to 700 ns	1	0	1	1	0	1	
700 to 800 ns	1	0	1	1	1	0	
800 to 900 ns	1	1	0	1	0	0	
900 ns to 1 μs	1	1	0	1	1	1	

Table 1

7. Revise the timing diagram as shown in Figure 9. Sketch the new output after simulation in Figure 9.

8. At what time did the Q output waveform change from a logic-LOW to a logic-HIGH? _______ ns

9. What condition caused the Q output to change from a logic-LOW to a logic-HIGH? __________

10. Explain why the J-K flip-flop Q output in Figure 5 did not change on either the leading or trailing edge of the clock.

___

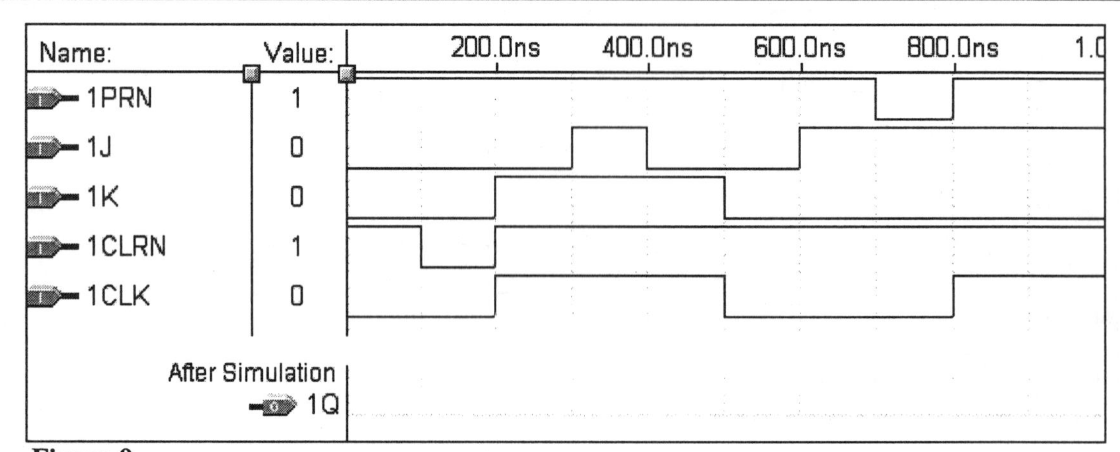

Figure 9

11. According to the results displayed by the waveforms in Figure 14 and Figure 15, preset and clear inputs to the J-K flip-flop are (synchronous/asynchronous).

12. According to the results displayed by the waveforms in Figure 14 and Figure 15, the J, K, and clock inputs to the J-K flip-flop are (synchronous/asynchronous).

13. Which set of inputs to the flip-flop is dominant? (synchronous/asynchronous)

14. Save all files to Drive A as **JK-fflop**, then exit the Graphics and Waveform Editors.

Part 4 Procedure

1. Open the Graphics and Waveform Editors. Assign the project name **7476jk**.

2. Open a new Graphic Editor and import the 7476 and 7476a symbols from the **mf** directory. Connect input/output terminals to both ICs with the labels shown in Figure 10.

3. Open a new Waveform Editor, set Grid Size to 50 ns, and then create the waveforms shown in Figure 10.

4. After compiling and simulating the circuit, draw the Q output waveform for the circuit constructed in the space provided in Figure 10.

5. Evaluate the timing diagram you have just completed. According to the timing diagram, the **7476** is (leading/trailing)-edge triggered and the **7476a** is (leading/trailing)-edge triggered.

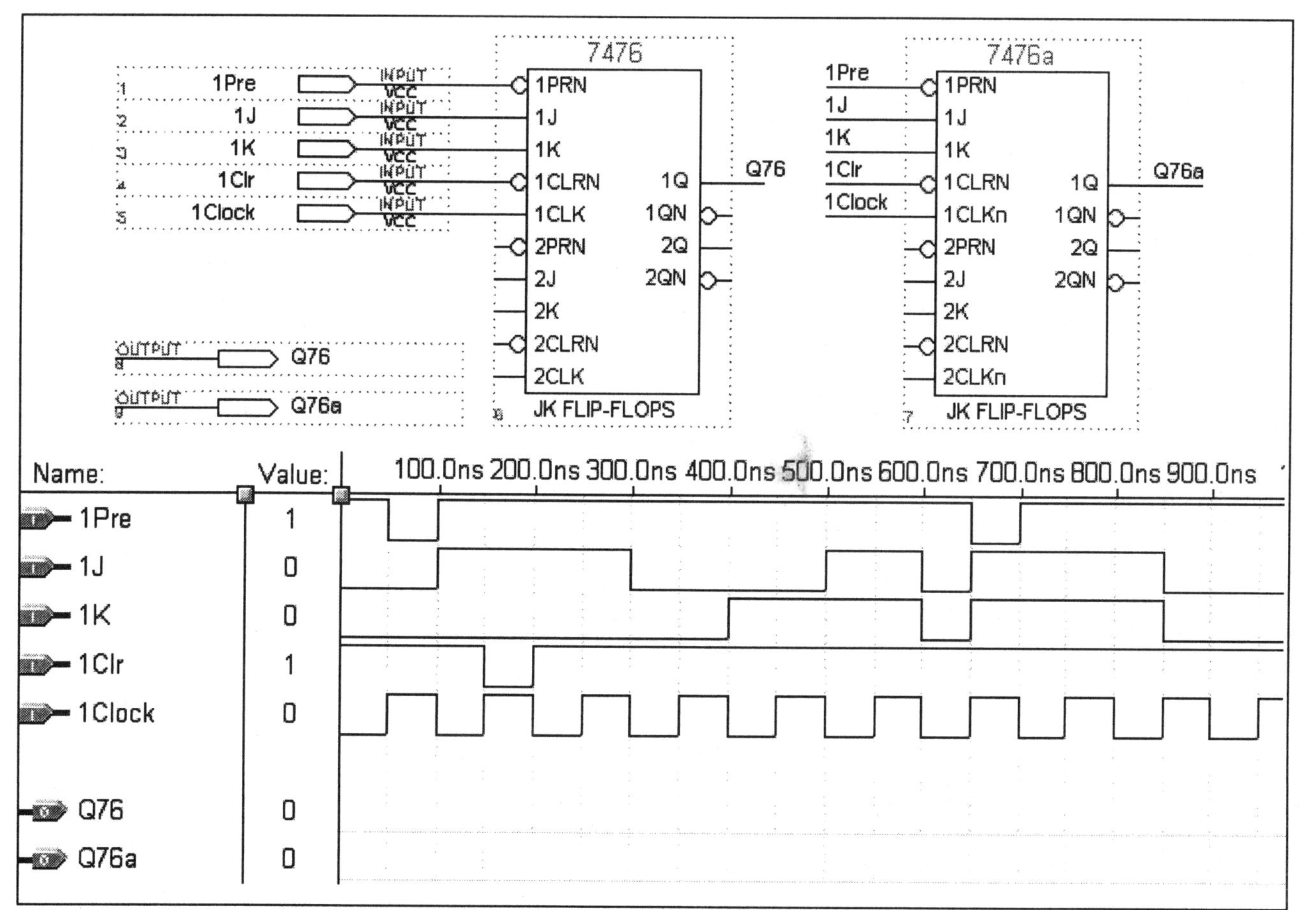

Figure 10

6. Carefully examine the output waveforms in Figure 10 with respect to the input waveforms and complete Table 2. For each time segment, identify the flip-flop mode of operation (Set, Reset, No Change, or Toggle) and the value the Q output will become for each flip-flop.

7. The PRE and CLR inputs to 7476 are (synchronous/asynchronous).

8. The PRE and CLR inputs to 74LS76, the 7476a used in Figure 10, are (synchronous/asynchronous).

9. Write a statement comparing the 7476 and 74LS76 (7476a) based on the results obtained in Part 3 of this lab.

____

____

10. Obtain a hard copy of the Graphic and Waveform Editors used to demonstrate Figure 10 in Part 4 of this lab. Label these hard copies **Part 4, Step 10A** and **Part 4, Step 10B**, respectively.

11. Demonstrate the circuit and waveforms for Part 4 of this lab to the instructor. Obtain the signature of approval directly on the answer page for this lab.

12. Save all files to Drive A as **7476jk**, then exit the Graphic and Waveform Editors.

13. Write a 1- to 2-page summary containing and making reference to at least one embedded figure from this lab.

14. Place all papers for this lab in the following sequence, then submit the lab to your instructor for grading.

    - Cover page
    - Typed summary
    - The completed answer pages for this lab
    - Hard copy of the Graphic Editor, **Part 4, Step 10A**
    - Hard copy of the Waveform Editor, **Part 4, Step 10B**

Lab 16: Asynchronous Counters

Objectives:

1. Create a 3-bit (Modulus 8) asynchronous binary counter using J-K flip-flops
2. Create a 3-bit (Modulus 5) asynchronous binary counter using J-K flip-flops
3. Use the Timing Analyzer to determine propagation delays
4. Construct a 4-bit binary counter using the 7493 integrated circuit
5. Construct a 4-bit decade counter using the 7490 integrated circuit
6. Study the effects of the preset and clear inputs with respect to the clock
7. Change the counter modulus using full or partial decoding

Materials List:

- Max+plus II software by Altera Corporation
- University Board by Altera Corporation (optional)
- Computer requirements:
  - Minimum 486/66 with 8 MB RAM
- Floppy disk

Discussion:

J-K flip-flops with preset and clear inputs can be cascaded to create an *n*-bit binary or BCD counter.

The method of clocking each stage within the counter determines if the counter is asynchronously or synchronously clocked. Asynchronous clocking implies that each stage does *not* share the same clock, nor toggle at the same time, as the other flip-flops in the counter. Each flip-flop in an asynchronous counter is clocked by the Q output of the previous flip-flop. If the first stage toggles, it may cause the second stage to toggle, which in turn may cause the third stage to toggle, and so forth. Asynchronous counters, often called ripple counters, are the focus of this lab.

Synchronous counters are designed with all flip-flops responding to the same clock pulse, simultaneously. Synchronous counters are the focus of Lab 16.

The time it takes the Q output of the flip-flop to change with respect to a trigger event on the clock input is called the propagation delay time, $\mathbf{t_p}$. For asynchronous counters, all propagation delay times for each stage are additive. For instance, if a binary counter has the count of 1111_2, it will change to 0000_2 on the next clock, and the delay for the output to change from a logic-HIGH to a logic-LOW will be 25 ns. Then the delay time for the counter to go from the Fh count to 0h count will be 100 ns.

The inherent characteristics of a J-K flip-flop wired in the toggle mode produces a square ware output frequency one-half the clock frequency. Several toggle flip-flops cascaded in asynchronous mode will create a 2^N frequency divider, where N represents the number of flip-flops, as illustrated in Figure 1. In the example of Figure 1, the total frequency division is $2^N = 2^3 = \div 8$.

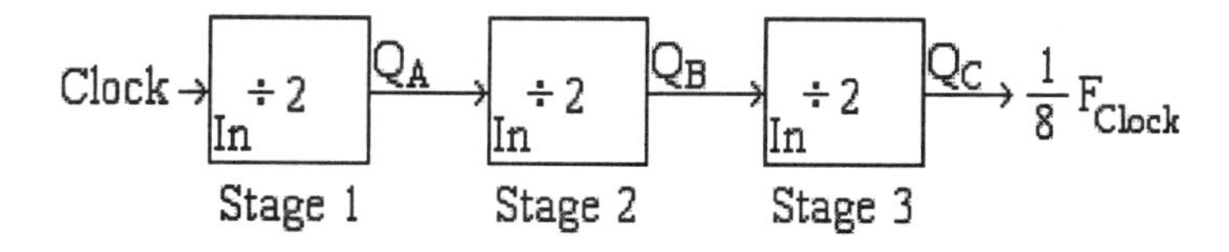

Figure 1

The modulus may be altered by using a decoder gate and wiring its output back to the clear inputs of each flip-flop. Since the clear input is asynchronous, or clock independent, the number the decoder is wired to decode will not be seen as part of the counter count sequence.

Part 1 Procedure

1. Open the Max+plus II software. Assign the project name **counter1** and MAX7000S as the device family.

2. Open a new Graphic Editor and create the 3-bit counter shown in Figure 2 using JKFF symbols.

3. Open a new Waveform Editor, set Grid Size to 50 ns, and create the waveforms shown in Figure 3.

4. Compile and simulate the circuit. Assuming zero errors, draw the resulting output waveforms in the area provided in Figure 3.

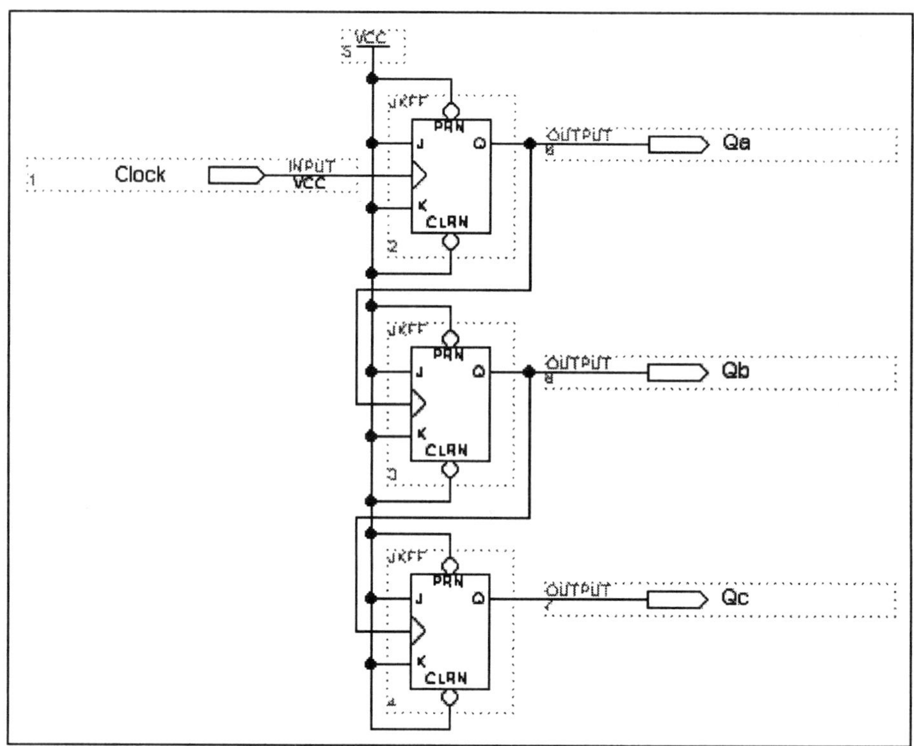

Figure 2

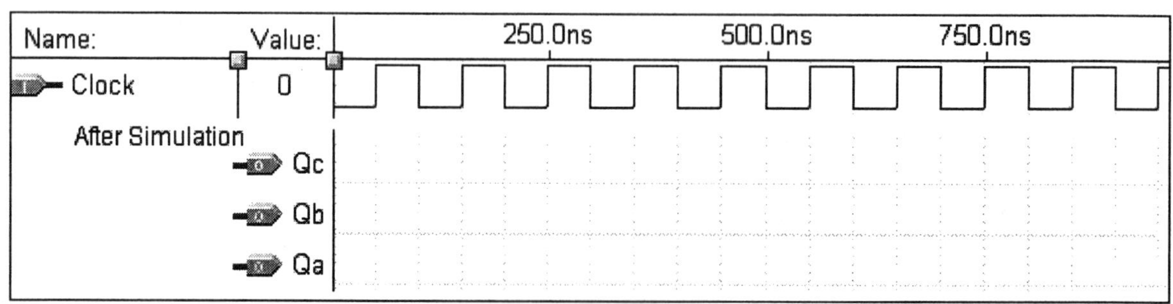

Figure 3

Time Segment	Qc Qb Qa (Bits 0 or 1)	Octal Value
100 to 150 ns	_ _ _	____
200 to 250 ns	_ _ _	____
300 to 350 ns	_ _ _	____
400 to 450 ns	_ _ _	____
500 to 550 ns	_ _ _	____
600 to 650 ns	_ _ _	____
700 to 750 ns	_ _ _	____
800 to 850 ns	_ _ _	____

Table 1

5. Copy the binary bit pattern for each time segment in Figure 3 into the respective cell block in Table 1, then convert the binary bit pattern to its octal equivalent.

6. What direction does the circuit in Figure 2 count? (Up/Down)

7. Based on your observation, the Q outputs of the circuit of Figure 2 change with respect to the (leading/trailing)-edge of the clock applied to the circuit.

8. Turn off the Snap to Grid feature in the Options menu.

9. Click the Zoom-in button (the magnifying glass with a plus sign icon in the Draw tool bar) several times to magnify the waveforms to view only the time segment from 50 ns to 100 ns. Reposition the horizontal scroll bar as necessary to view this segment of the waveform.

10. Position the cursor on the 50 ns mark in the Waveform Editor. When you click on the 50 ns time, a vertical blue line will appear, and the Ref: box in the Waveform Editor (see Figure 4) will show 50 ns. The Time: box in the Waveform Editor shows the position of the mouse as the mouse is moved around the Waveform Editor.

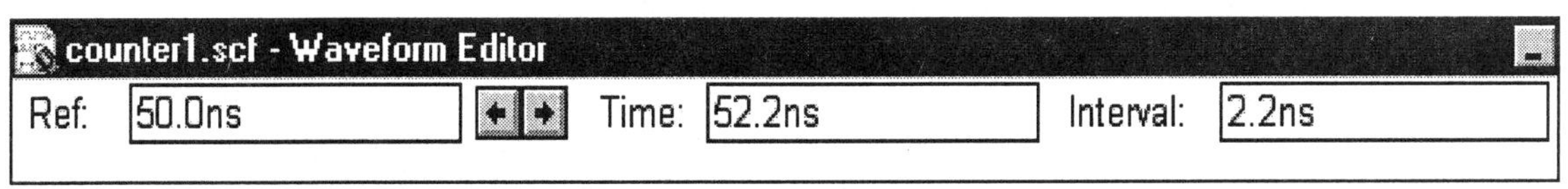

Figure 4

11. Position the mouse pointer right on the blue vertical line on top of the Qa waveform, then click and hold the left mouse button down as you drag the mouse to the right. The section of the waveform you are dragging the mouse across will be highlighted (white waveform on black background) and the Interval: box will show the horizontal time of the highlighted area. Release the mouse button when the mouse pointer just touches the rising edge of the Qa waveform.

12. Record the propagation delay time it takes Qa to go from a logic-LOW to a logic-HIGH, T_{PLH}, with respect to the positive edge of the clock transition.

Qa delay at 50 ns: T_{PLH} = ____________________

13. Repeat Step 11 to determine the delay times for Qb and Qc with respect to the clock. Record the results below.

Qb delay at 50 ns: T_{PLH} = ____________________
Qc delay at 50 ns: T_{PLH} = ____________________

14. Position the cursor at 450 ns and determine the delay times for each waveform with respect to the clock transition.

Qa delay at 450 ns: T_{PLH} = ____________________
Qb delay at 450 ns: T_{PLH} = ____________________
Qc delay at 450 ns: T_{PLH} = ____________________

15. As you may have suspected, there is an easier method to determine delay times. Select the Max+plus II option in the main menu, then select the Timing Analyzer option (Figure 5).

16. Click the Start button in the Timing Analyzer.

17. Click the OK button when the timing analysis is completed.

18. Maximize the Delay Matrix and record the delay times in Table 2.

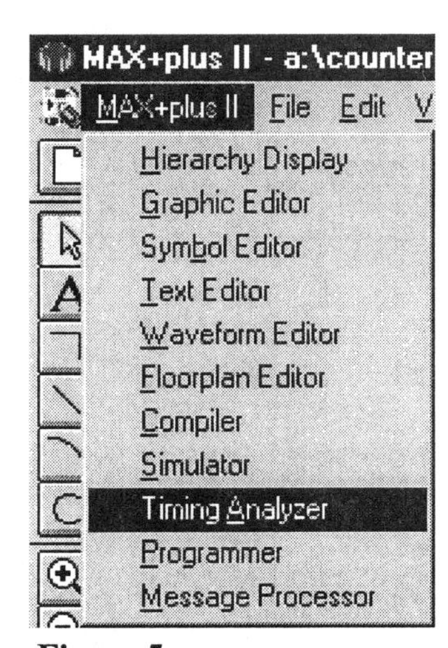

Figure 5

	Qa	Qb	Qc
Clock			

Table 2

19. During the compilation, a report is generated and placed on your diskette. The report may be accessed using a word processor; however, at this time, you will be using the default text editor to view this report.

20. Recompile your **counter1** file. When finished, an icon labeled **rpt** appears directly below the Fitter box in the Compiler window. Double click on this **rpt** icon (see Figure 6). You must have the Compiler window maximized to see the details within the compiler.

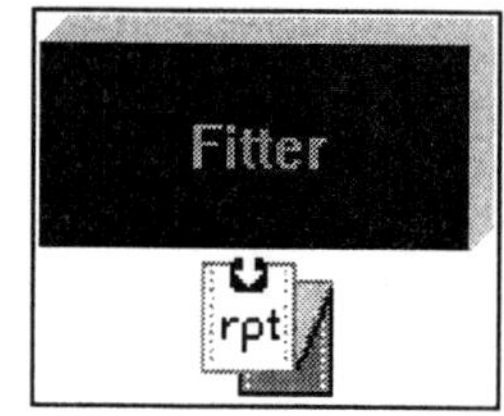

Figure 6

21. Scroll through the report file and answer the following questions.

    A. What pin was assigned to the clock input? __________
    B. How many input pins were used for your circuit? ________
    C. How many output pins were used for your circuit? ________
    D. How much was the IC utilized for the circuit of Figure 2? ________
    E. How many VCCIO pins were assigned to the chip? _______
    F. What is the acceptable voltage range for the VCCIO pins? ______________
    G. What voltage is to be applied to the VCCINT pins? ______________
    H. What name was assigned to the unused pins of the IC? _______________________
    I. What pins were labeled GND? ____
    J. What pin was Qa assigned? ____________
    K. What pin was Qb assigned? ____________
    L. What pin was Qc assigned? ____________
    M. What was the total compilation time for this circuit? ____________

22. Based on the waveforms drawn for Figure 3, the frequency of the clock was ____________.

23. Based on the waveforms drawn for Figure 3, the frequency of Qa was ____________.

24. Based on the waveforms drawn for Figure 3, the frequency of Qb was ____________.

25. Based on the waveforms drawn for Figure 3, the frequency of Qc was ____________.

26. Write a statement commenting on the frequency relationships of the outputs Qa, Qb, and Qc with respect to the clock frequency.

    ____

    ____

27. Save all files to Drive A as **counter1**, then exit the Graphic and Waveform Editors.

Part 2 Procedure

1. Open the Max+plus II software. Assign the project name **counter2**.

2. Open a new Graphic Editor and create the circuit shown in Figure 7.

3. Open a new Waveform Editor, set Grid Size to 75 ns, then construct the waveforms shown in Figure 8.

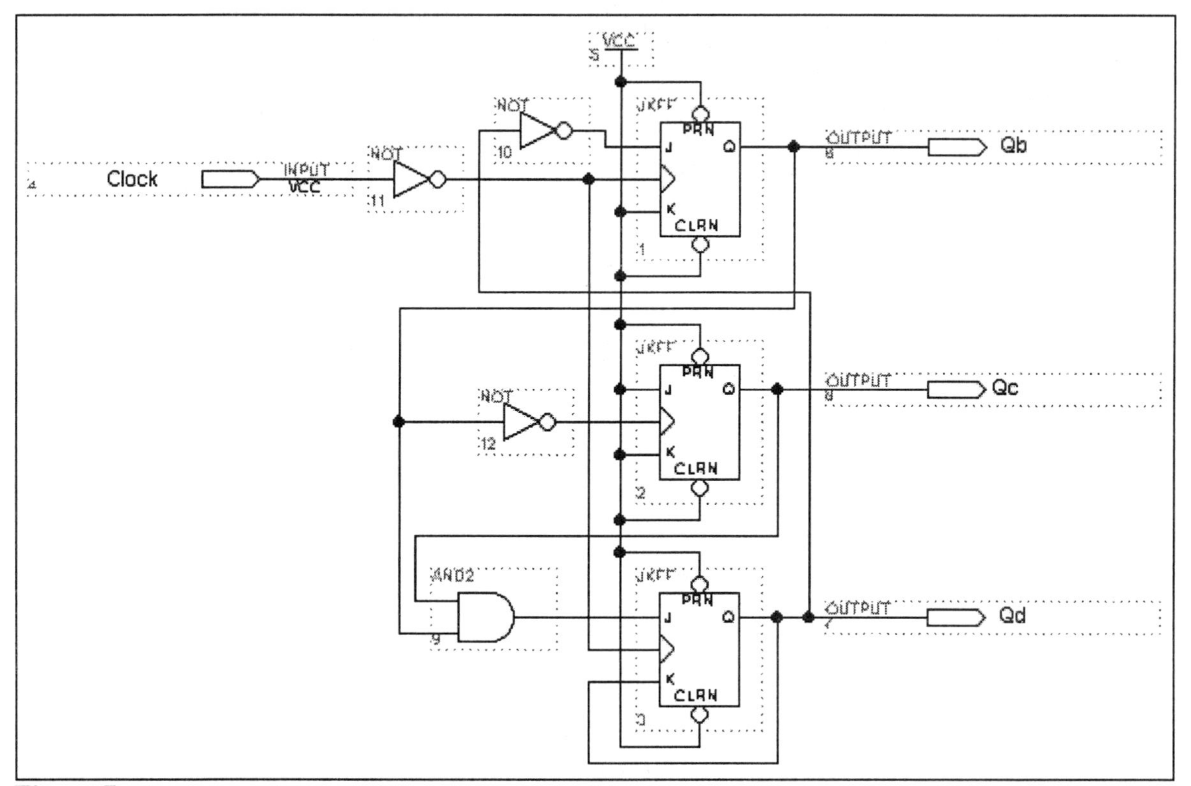

Figure 7

4. Compile and simulate the circuit. Assuming zero errors, sketch the output waveforms in the space provided at the bottom of Figure 8.

5. Based on the resulting waveforms, the counter outputs are changing with respect to the clock's (rising/falling) edge.

6. Complete Table 3, identifying the octal value of the Q outputs, DCB, for the time segments listed.

7. Based on the waveforms and the DCB count sequence list in Table 3, the counter shown in Figure 7 counts (up/down).

8. What is the highest DCB count shown in Table 3?

9. The counter modulus is defined as the total number of clocks it takes for the counter to count a complete count cycle. What is the modulus of the counter shown in Figure 2? ____________

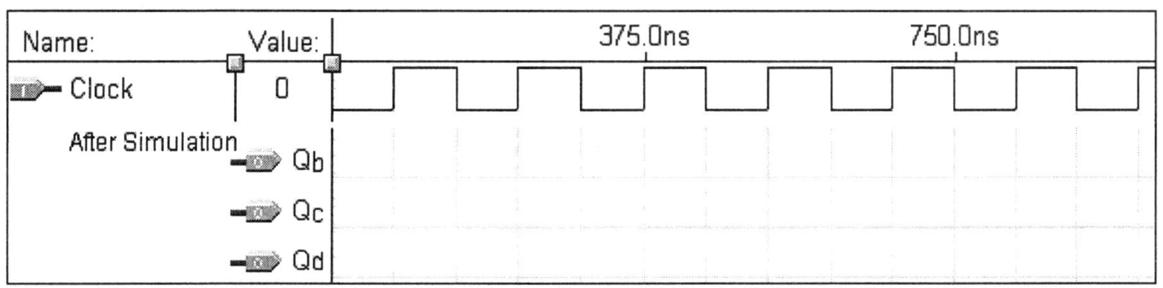

Figure 8

Time (ns)	40	110	180	260	340	400	480	560	640	710	790	860	940
DCB Value													

Table 3

10. What is the modulus of the counter shown in Figure 7? ____________

11. Modify your Graphic Editor schematic to what is shown in Figure 9.

12. Set Grid Size to 25 ns and make the necessary changes to the Waveform Editor file as shown in Figure 10. Change the project name to **mod10** and save both Graphic and Waveform Editor files as **mod10**.

13. Compile and simulate the circuit constructed. Correct all errors before continuing.

14. Neatly sketch the output waveforms in the area provided in Figure 10.

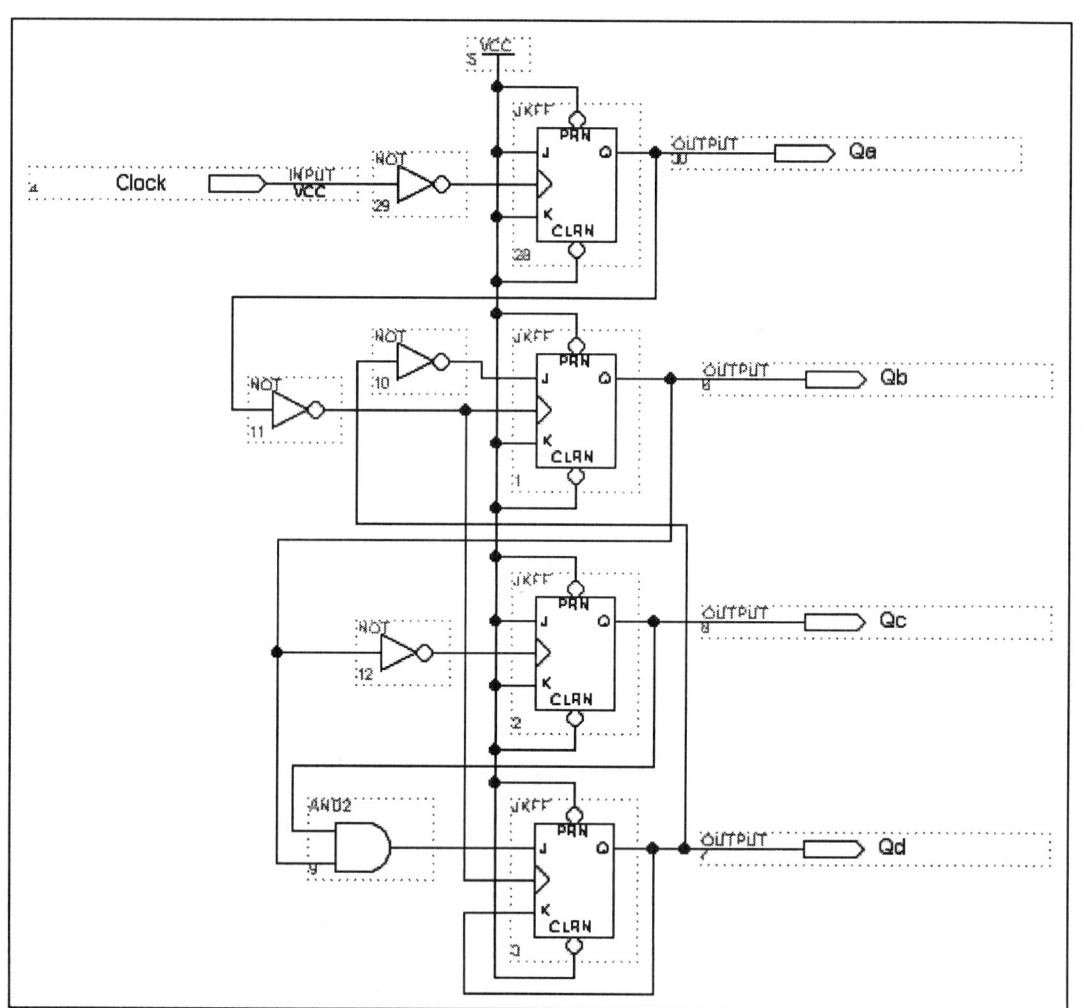

Figure 9

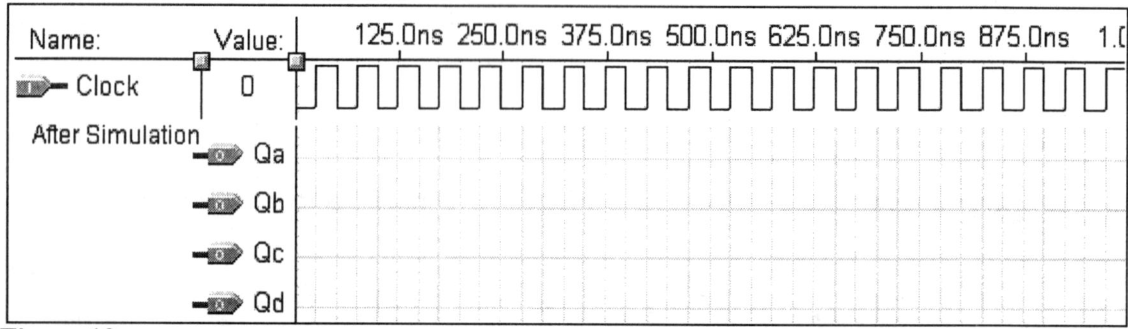

Figure 10

15. Based on the results obtained for Figure 10, the counter outputs change on the (rising/falling) edge of the applied clock.

16. Carefully examine the counter waveforms. What is the count sequence?

____

17. The modulus of an up counter may be determined by subtracting the low count from the high count then adding 1 to the difference. What is the modulus of the counter in Figure 9? _______

18. Save all files to Drive A as **mod10**, then exit both Graphic and Waveform Editors.

Part 3 Procedure

1. Open the Max+plus II software. Assign the project name **7490ctr**.

2. Open a new Graphic Editor and construct the circuit shown in Figure 11.

3. Open a new Waveform Editor, set Grid Size to 40 ns, then create the waveforms shown in Figure 12.

4. Compile, then run the Simulator. Correct all errors before continuing.

5. Draw the resulting output waveforms in the area provided in Figure 12.

6. Based on the waveforms for the 7490, the counter outputs change on the (rising/falling) edge of the applied clock.

7. List the count sequence for the 7490. _______________________

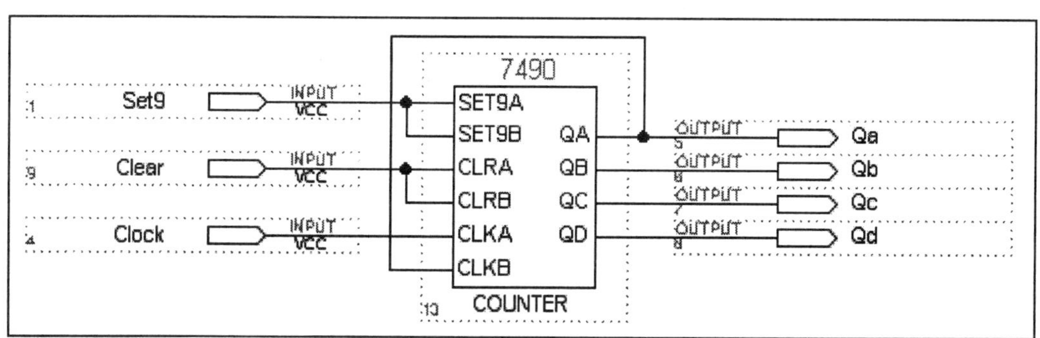

Figure 11

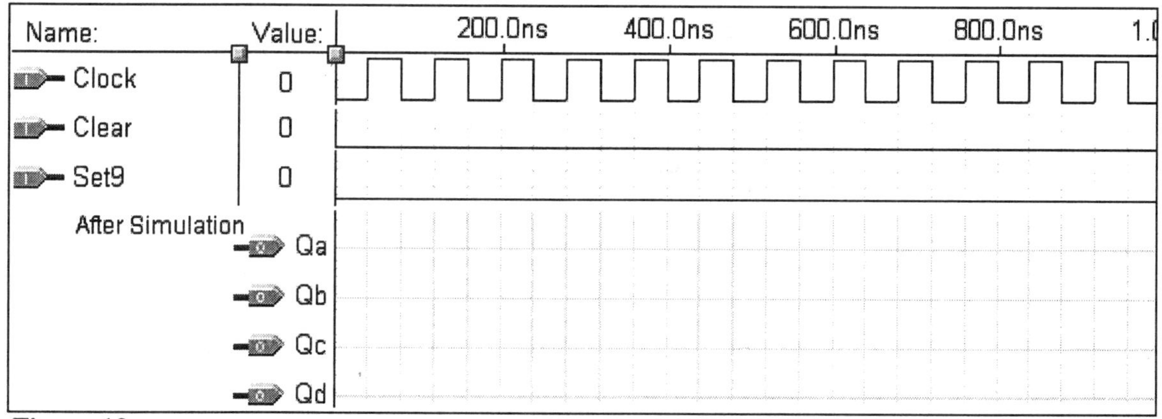

Figure 12

8. Based on the waveforms, what is the modulus of the 7490? _____________ (Note this is called the natural modulus.)

9. Highlight and change the logic level of the clear waveform from 400 ns to 600 ns to a HIGH state.

10. Re-simulate the waveforms.

11. Describe, in appropriate detail, the output waveforms for this counter-based on the logic levels of the clear input.

____

____

____

12. Based on the results of Step 10, the clear inputs are (asynchronous/synchronous) and are active-(LOW/HIGH).

13. Change the SET9 waveform to a constant logic-HIGH.

14. Highlight and set the clear input waveform to a constant logic-LOW level.

15. Click the Simulator button.

16. Based on the results of Step 15, the SET9 inputs are (asynchronous/synchronous) and are active-(LOW/HIGH).

17. Describe, in appropriate detail, the effect the SET9 inputs have on this counter.

____

____

18. Save all files to Drive A as **7490ctr**, then exit both Graphic and Waveform Editors.

Part 4 Procedure

1. Open the Max+plus II software. Assign the project name **7493ctra**.

2. Open a new Graphic Editor and construct the circuit shown in Figure 13.

3. Open a new Waveform Editor, set Grid Size to 30 ns, and then create the waveforms shown in Figure 14.

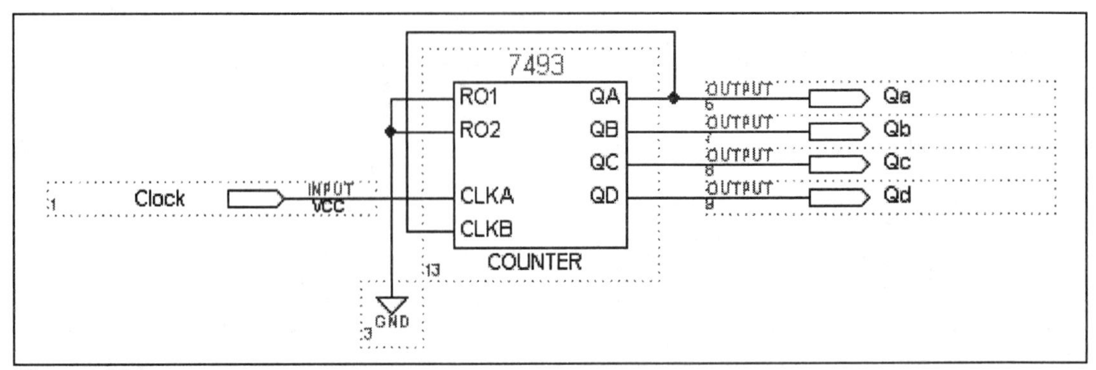

Figure 13

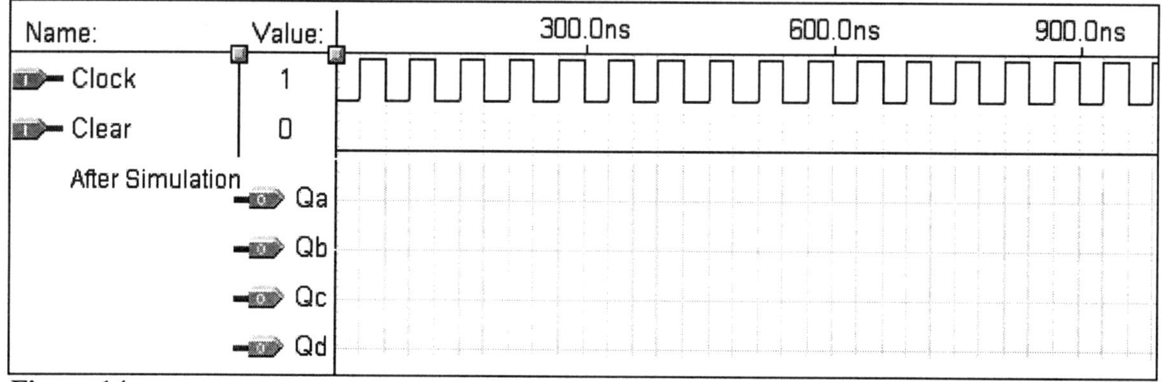

Figure 14

4. Compile and simulate the circuit, then neatly draw the output waveforms in the space provided in Figure 14.

5. The 7493 counter counts (up/down) from ______ to _____.

6. The 7493 counter advances count on the (leading/trailing) edge of the clock signal.

7. Set the clear input to a logic-HIGH from 600 ns to 1 μs.

8. Re-simulate the circuit and describe the effects of the clear input on the Q outputs.

____

9. Based on your observation of the effects of the clear input, the Clear input on the 7493 is (asynchronous/synchronous).

10. Saving all files to Drive A as **7493ctra**, then exit both Graphic and Waveform Editors.

Part 5 Procedure

1. Open the Max+plus II software. Assign the project name **7493ctrb**.

2. Open a new Graphic Editor and construct the circuit shown in Figure 15.

3. Open a new Waveform Editor, set Grid Size to 30 ns, and create the waveforms shown in Figure 16.

4. Draw the output waveform generated by the software after compilation and simulation.

5. The 7493 counter as wired in Figure 15 counts (up/down) from ______ to _____.

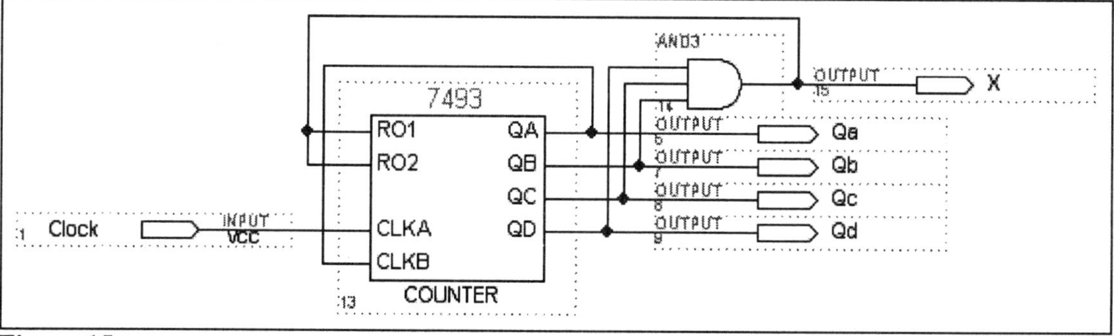

Figure 15

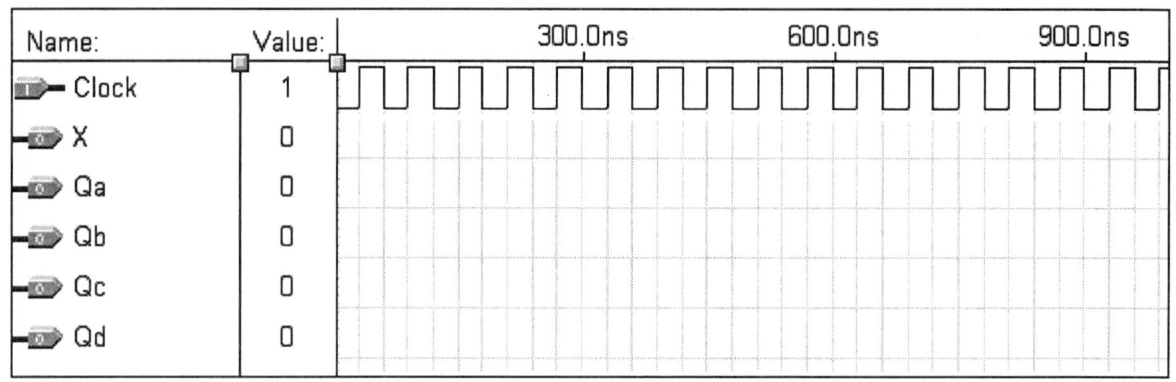

Figure 16

6. What count was detected by the decoder gate? _________

7. Describe what action took place when the decoder output was "active."

   ____

8. What is the natural modulus of the 7493? ___________

9. What is the modulus of the circuit shown in Figure 15? ____________

10. Save all files to Drive A as **7493ctrb**, then exit both Graphic and Waveform Editors.

Part 6 Procedure

1. Open the Max+plus II software. Assign the project name **7493ctrc**.

2. Open a new Graphic Editor. Use the 7493 IC and decoder gate to create a Mod 10 counter that gives a BCD count sequence.

3. Open a new Waveform Editor and create a set of waveforms similar to Figure 16.

4. Compile and simulate the circuit.

5. Assuming zero errors, demonstrate the operational Mod 10 counter to the instructor. Obtain the signature of approval directly on the answer page for this lab.

6. Obtain a hard copy of the Graphic and Waveform Editors. Label these hard copies **Part 6, Step 6a** and **Part 6, Step 6b**, respectively.

7. Save all files to Drive A as **7493ctrc**, then exit both Graphic and Waveform Editors.

8. Write a 1- to 2-page summary pertaining to the results obtained from this lab. Include at least one embedded graphic with corresponding waveforms after compilation and simulation.

9. Staple all papers for this lab in the following sequence, then submit the lab to your instructor for grading.
   - Cover page
   - Typed summary
   - The completed answer page for this lab
   - Hard copy of the Graphic Editor, **Part 6, Step 6a**
   - Hard copy of the Waveform Editor, **Part 6, Step 6b**

Lab 16: Asynchronous Counters Answer Pages

Name:______________________

Part 1

Name: 250.0ns 500.0ns 750.0ns

Qc

Qb

Qa

Figure 3

Time Segment	Qc Qb Qa (Bits 0 or 1)	Octal Value
100 to 150 ns	__ __ __	____
200 to 250 ns	__ __ __	____
300 to 350 ns	__ __ __	____
400 to 450 ns	__ __ __	____
500 to 550 ns	__ __ __	____
600 to 650 ns	__ __ __	____
700 to 750 ns	__ __ __	____
800 to 850 ns	__ __ __	____

Table 1

6. (up/down) 7. (leading/trailing)

12. Qa delay at 50 ns: T_{PLH} = ________

13. Qb delay at 50 ns: T_{PLH} = ________
    Qc delay at 50 ns: T_{PLH} = ________

14. Delay at 450 ns Qa: T_{PLH} = ________

    Qb: T_{PLH} = ________

    Qc: T_{PLH} = ________

	Qa	Qb	Qc
Clock			

Table 2

21. A. ________ B. ________ C. ________ D. ________

    E. ________ F. ________ G. ________ H. ________

    I. ____________________ J. ________ K. ________

    L. ________ M. ________

22. ________ 23. ________ 24 ________ 25. ________

26. ____

____

Part 2

5. (rising/falling) 7. (up/down) 9. ________

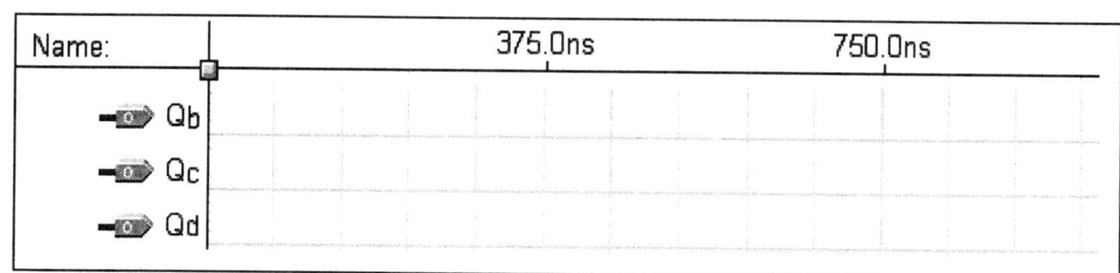

Figure 8

Time (ns)	40	110	180	260	340	400	480	560	640	710	790	860	940
DCB Value													

Table 3

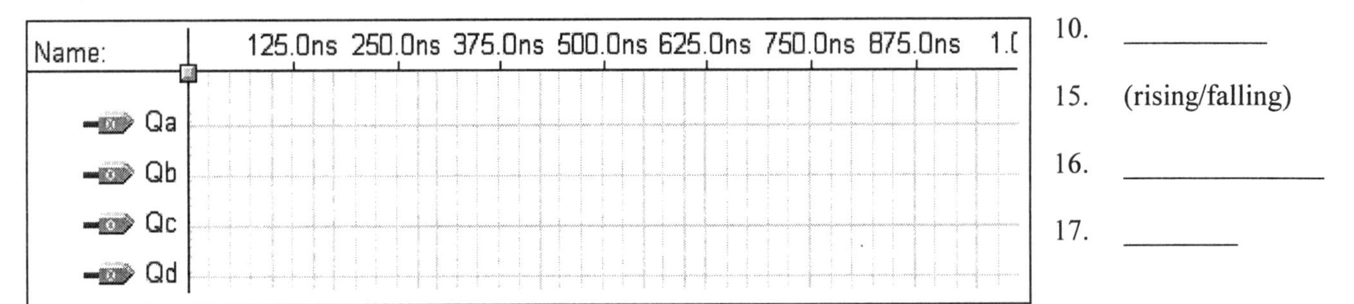

Figure 10

10. __________

15. (rising/falling)

16. ______________

17. ________

Part 3

6. (rising/falling) 7. ___________ 8. ______________

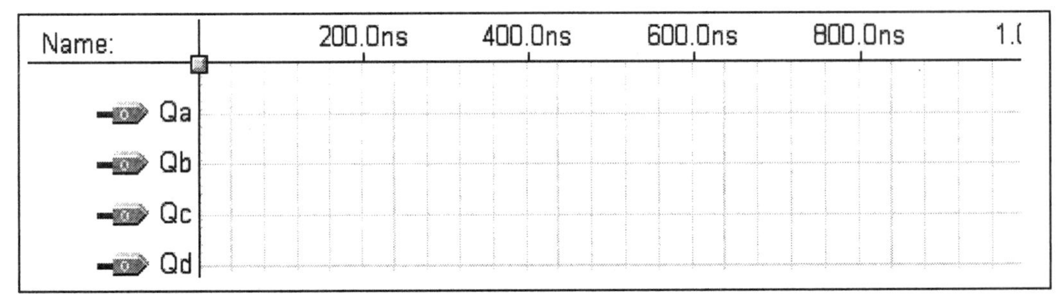

Figure 12

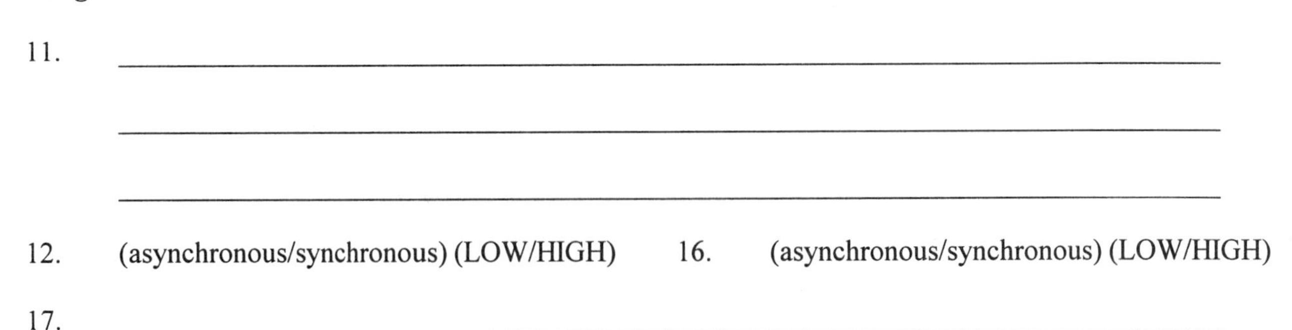

11. ____

____

____

12. (asynchronous/synchronous) (LOW/HIGH) 16. (asynchronous/synchronous) (LOW/HIGH)

17. ____

Part 4

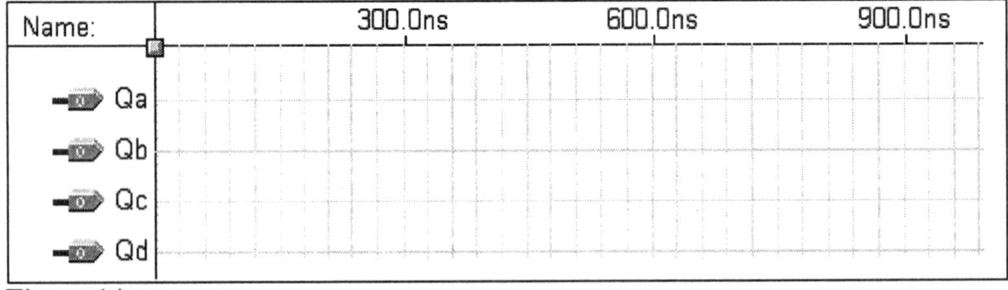

Figure 14

5. (up/down) from ______ to _____ 6. (leading/trailing)

8. ____ 9. (asynchronous/synchronous)

Part 5

5. (up/down) from ______ to _____ 6. ____________

Name:	Value:	300.0ns	600.0ns	900.0ns
X	0			
Qa	0			
Qb	0			
Qc	0			
Qd	0			

Figure 16

7. ____

8. _______________ 9. _______________

Part 6

5. Demonstrated to: ______________________________ Date: _______________

Grade: ___________

Lab 17: Synchronous Counters

Objectives:

1. Create a 4-bit (Modulus 16) synchronous binary counter using J-K flip-flops
2. Use the Timing Analyzer to determine propagation delays
3. Construct a 4-bit binary counter using the 74191 integrated circuit
4. Cascade two counters, making a Mod 256 counter
5. Change the counter modulus using a decoder

Materials List:

- Max+plus II software by Altera Corporation
- University board with CPLD (optional)
- Computer requirements:
- Minimum 486/66 with 8 MB RAM
- Floppy disk

Discussion:

One disadvantage of asynchronous counters is that propagation delays are additive, making asynchronous counters low-frequency devices. Synchronous counters eliminate this problem for all flip-flop clock inputs that are tied to the same source; as a result all flip-flops are triggered at the same time. With asynchronous counters, the clock of each flip-flop was clocked by the previous Q output. With synchronous counters, the Q output of previous stages determines the mode of flip-flop operation as either toggle or no-change when the next clock arrives. For the toggle condition, both J and K inputs are a logic-HIGH. For the no-change condition, J and K inputs are both a logic-LOW.

The preset and clear inputs of the flip-flops may be used to load a number into the counter or to clear the counter to zero. These asynchronous inputs are disabled for Part 1 of this lab by wiring all PRE and CLR inputs to Vcc. The PRE and CLR inputs will be demonstrated once counter ICs are introduced.

The 74190 and 74191 integrated circuits are virtually identical except the 74190 is a 4-bit BCD counter, 0 to 9, and the 74191 is a 4-bit binary counter, 0 to F. Both chips have identical pin assignments and identical control functions. The counter enable, CTEN, input is active-LOW and is considered to be a synchronous input since CTEN affects the synchronous J and K inputs to the first flip-flop in the counter. The pre-settable inputs, D, C, B, and A, are controlled by the asynchronous LOAD input, which is also active-LOW. The counters may count UP or DOWN as determined by the logic state applied to the D/U input. Both CTEN and LOAD are active-LOW. Because of these similarities, only the 74191 will be the focus of this lab.

As with the asynchronous counter, the modulus of the 74191 counter will be changed using a decoder gate. A decoder glitch in the output waveforms will appear any time a decoder is used to change the modules.

Part 1 Procedure

1. Open the Max+plus II software. Assign the project name **sync-ctr**.
2. Open a new Graphic Editor and construct the circuit shown in Figure 1.
3. Open a new Waveform Editor, set Grid Size to 25 ns, and then create the waveforms shown in Figure 2.
4. Run the Compiler and Simulator. Sketch the resulting output waveforms in the area provided in Figure 2.
5. What direction does the circuit in Figure 2 count? (Up/Down)

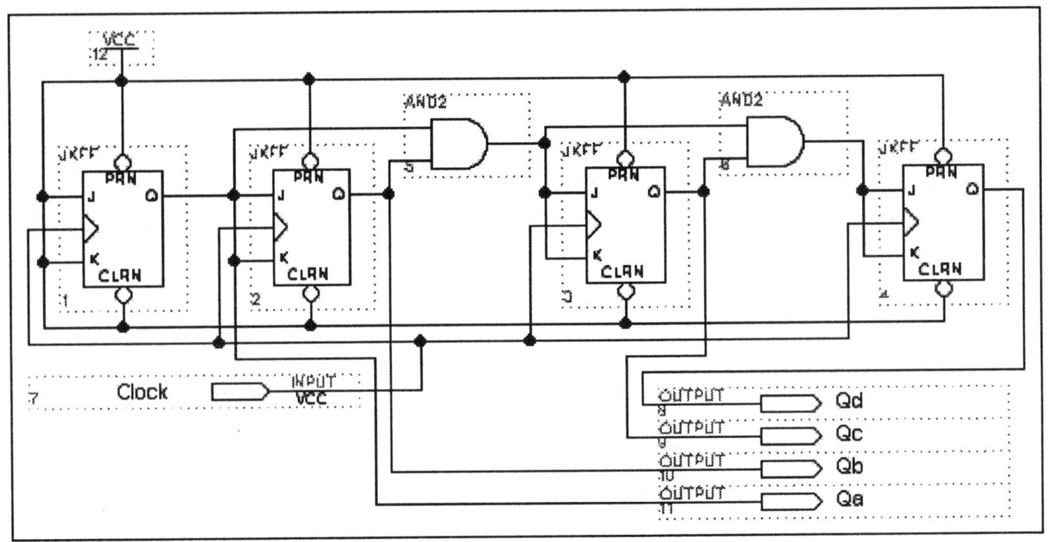

Figure 1

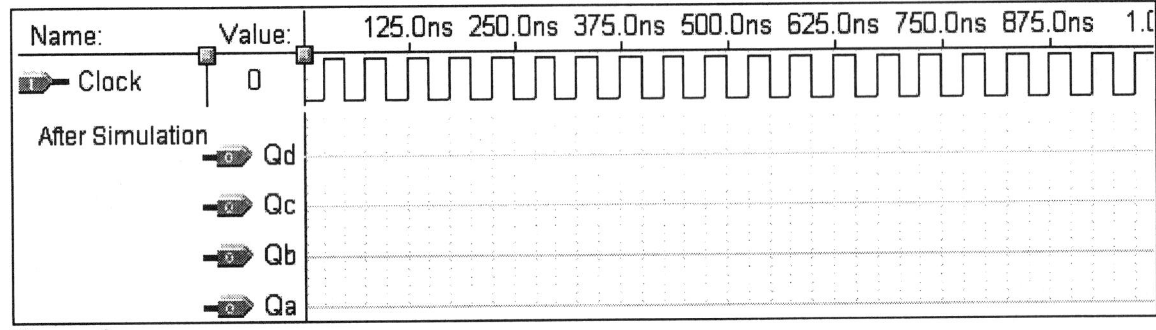

Figure 2

6. Based on your observation, the Q outputs of the circuit of Figure 2 change with respect to the (leading/trailing) edge of the clock applied to the circuit.

7. Copy the binary bit pattern for each time segment in Figure 3 into the respective cell block in Table 1, then convert the binary bit pattern to its hexadecimal equivalent.

Time Segment	Qd Qc Qb Qa Bits (1 or 0)	Hex Value
0 to 25 ns	_ _ _ _	____
25 to 75 ns	_ _ _ _	____
75 to 125 ns	_ _ _ _	____
125 to 175 ns	_ _ _ _	____
175 to 225 ns	_ _ _ _	____
225 to 275 ns	_ _ _ _	____
275 to 325 ns	_ _ _ _	____
325 to 375 ns	_ _ _ _	____

Time Segment	Qd Qc Qb Qa Bits (1 or 0)	Hex Value
375 to 425 ns	_ _ _ _	____
425 to 475 ns	_ _ _ _	____
475 to 425 ns	_ _ _ _	____
525 to 475 ns	_ _ _ _	____
575 to 525 ns	_ _ _ _	____
625 to 675 ns	_ _ _ _	____
675 to 625 ns	_ _ _ _	____
725 to 775 ns	_ _ _ _	____

Table 1

8. Open the Timing Analyzer and click the Start button in the Delay Matrix window.

9. Record the delay times in Table 2 for the Q outputs with respect to the clock signal.

Qa Delay	Qb Delay	Qc Delay	Qd Delay

Table 2

10. Recompile your **sync-ctr** file. When finished, an icon labeled **rpt** appears directly below the Fitter box in the Compiler window. Double click on this **rpt** icon (see Figure 3). You must have the Compiler window maximized to see the details within the compiler.

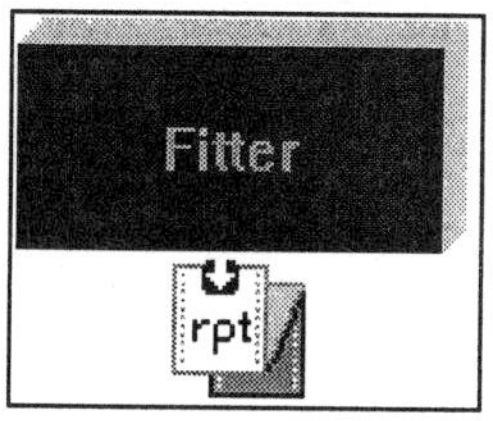

Fitter

rpt

Figure 3

11. Scroll through the report file and answer the following questions.

    A. What pin was assigned to the clock input? __________
    B. How many input pins were used for your circuit? ________
    C. How many output pins were used for your circuit? ________
    D. How much was the IC utilized for the circuit of Figure 2? ________
    E. How many VCCIO pins were assigned to the chip? _______
    F. What is the acceptable voltage range for the VCCIO pins? _____________
    G. What voltage is to be applied to the VCCINT pins? ______________
    H. What name was assigned to the unused pins of the IC? ______________________
    I. What pins were labeled GND? ____
    J. What pin was Qa assigned? ____________
    K. What pin was Qb assigned? ____________
    L. What pin was Qc assigned? ____________
    M. What pin was Qd assigned? ____________
    N. What was the total compilation time for this circuit? ____________

12. Based on the waveforms drawn for Figure 2, the frequency of the clock was ____________.

13. Based on the waveforms drawn for Figure 2, the frequency of Qa was ____________.

14. Based on the waveforms drawn for Figure 2, the frequency of Qb was ____________.

15. Based on the waveforms drawn for Figure 2, the frequency of Qc was ____________.

16. Based on the waveforms drawn for Figure 2, the frequency of Qd was ____________.

17. Write a statement commenting on the frequency relationships of the outputs Qa, Qb, Qc, and Qd with respect to the clock frequency.

    ____

    ____

18. Save all files to Drive A as **sync-ctr**, then exit the Graphic and Waveform Editors.

Part 2 Procedure

1. Open the Max+plus II software. Assign the project name **mod16ic**.

2. Open the Graphic Editor and construct the circuit shown in Figure 5 using the 74191 symbol.

3. Open the Waveform Editor and create the waveforms as shown in Figure 6.

4. Run the Compiler and Simulator. Correct all errors before continuing.

5. What count appears on the Q[3..0] output? ________

6. Highlight the D[3..0] waveform and click the Group Waveform Overwrite icon (see Figure 4) in the Draw tool bar.

XG

Figure 4

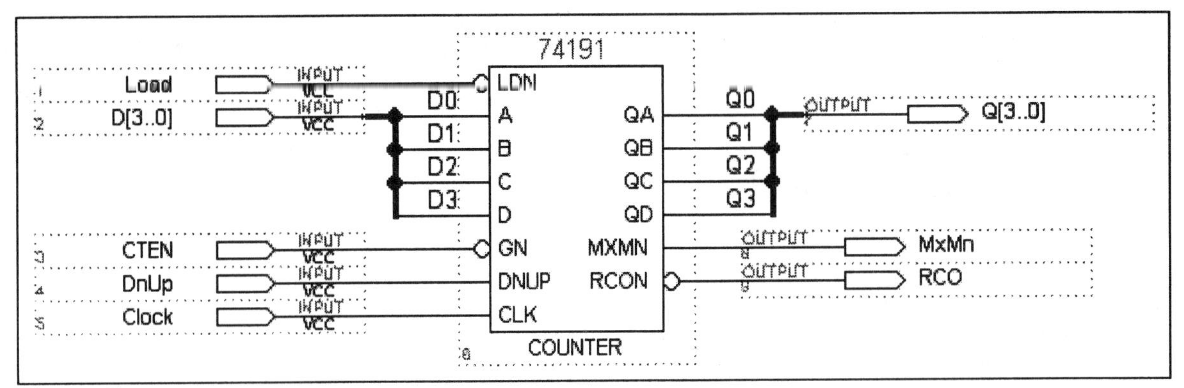

Figure 5

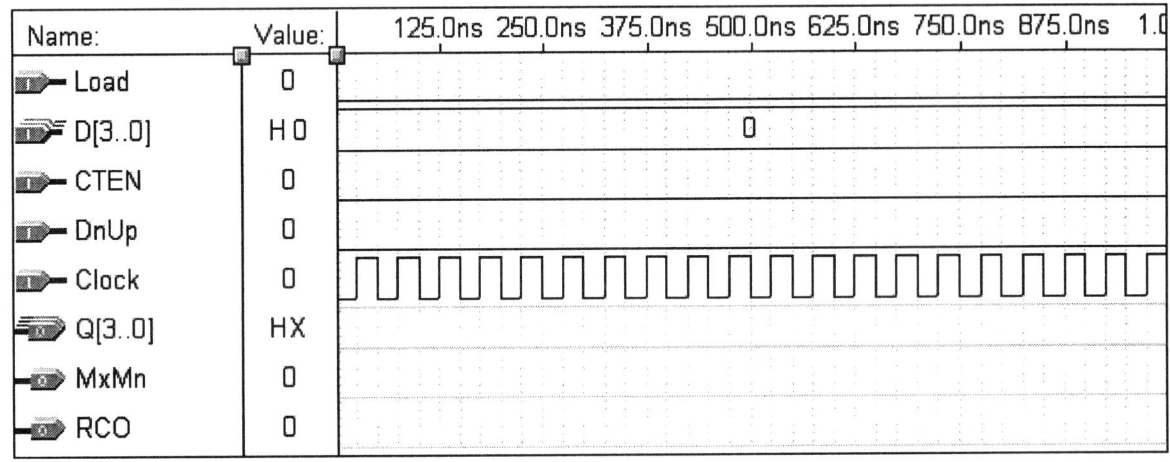

Figure 6

7. Assign the Group Value: 3.

8. Click the Simulate button.

9. What count appears on the Q[3..0] output? _________

10. Change the section of the load waveform from 500 ns to 1 μs to a logic-HIGH level.

11. Click the Simulate button.

12. What count appears on the Q[3..0] output from 0 to 500 ns? _________

13. What count appears on the Q[3..0] output from 500 ns to 1 μs? _________________________

14. Based on the results obtained from Steps 4 through 13, the load input to the 74191 is (synchronous/asynchronous).

15. Based on the results obtained from Steps 4 through 13, the load input to the 74191 is active-(LOW/HIGH).

16. Change the load waveform to a constant logic-HIGH level from 0 ns to 1 μs.

17. Click the Simulate button.

18. Sketch the resulting output waveforms in the area provided in Figure 6.

19. Save all files to Drive A as **mod16ic**, then exit the Graphic and Waveform Editors.

Part 3 Procedure

1. Open the Max+plus II software. Assign the project name **mod-x**.
2. Open the Graphic Editor and construct the circuit shown in Figure 7 using the 74191 symbol.
3. Open the Waveform Editor and create the waveforms as shown in Figure 8.
4. Run the Compiler and Simulator. Correct all errors before continuing.
5. What count appears on the Q[3..0] output? _________
6. What number (in hex) is the decoder wired for? This is the number that makes the load active: _________
7. What is the high count of Figure 7 when D[3..0] is 0_H? ______
8. What is the low count of Figure 7 when D[3..0] is 0_H? ______
9. What is the modulus of the counter as wired in Figure 7? ______
10. Change the data bus to 5_H and re-simulate the circuit. How does this change affect the high count?

    ____

11. How does this change affect the high count? ____________________________

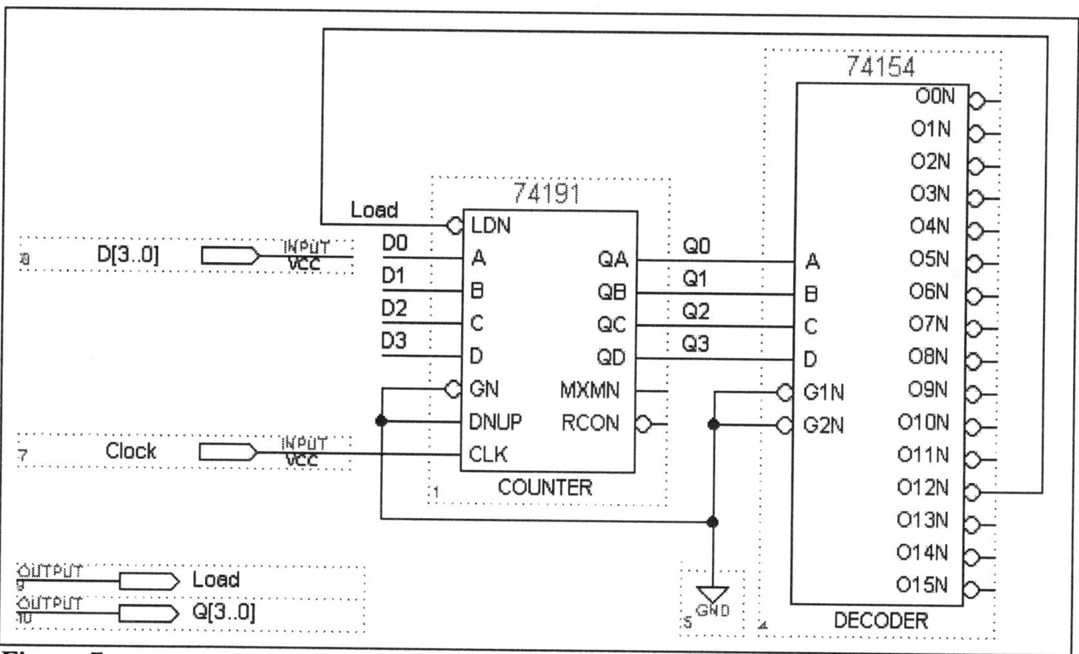

Figure 7

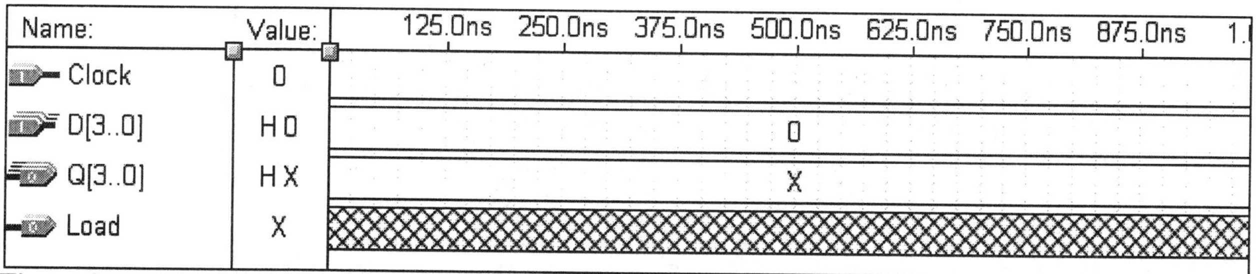

Name:	Value:	125.0ns 250.0ns 375.0ns 500.0ns 625.0ns 750.0ns 875.0ns 1.
Clock	0	
D[3..0]	H 0	0
Q[3..0]	H X	X
Load	X	

Figure 8

12. How does this change affect the low count? ______________________

13. How does this change affect the modulus of the counter? ______________________

14. Delete the horizontal wire at the O12N output of the decoder, then wire output O10N of the decoder to the right-most vertical wire going to LDN of the counter. Re-compile and re-simulate the circuit.

15. Explain how using O10N of the decoder affects the high count, low count, and modulus of the counter.

____

____

16. Save all files to Drive A as **mod-x**, then exit the Graphic and Waveform Editors.

Part 4 Procedure

1. Open the Max+plus II software. Assign the project name **mod-256**.

2. Open a new Graphic Editor and construct the circuit shown in Figure 9.

3. Open a new Waveform Editor, set Grid Size to 25 ns, and then create the waveforms shown in Figure 10.

4. Since the modulus of this circuit is 256, it will take 256 clock pulses to see the entire count sequence. Multiplying the grid size by 2 gives us the period for one cycle of the clock. Multiplying the period by 256 gives us the End Time required to see the entire count sequence. The result is 12800 ns, or 12.8 μs.

5. Set End Time to 12.8 μs (File – End Time), and then create the clock signal using Overwrite Count Value.

6. Run the Compiler and Simulator. You will encounter several warnings regarding D[7..0]. Since the load input is wired to Vcc, D[7..0] is currently not used. Do not worry about the warnings, but do correct all errors before continuing.

7. Zoom out to view the entire 12.8 μs of time. The frequency relationship of several high order Q outputs can be seen; however, the detail of the clock and low-order Q outputs cannot be determined without zooming in.

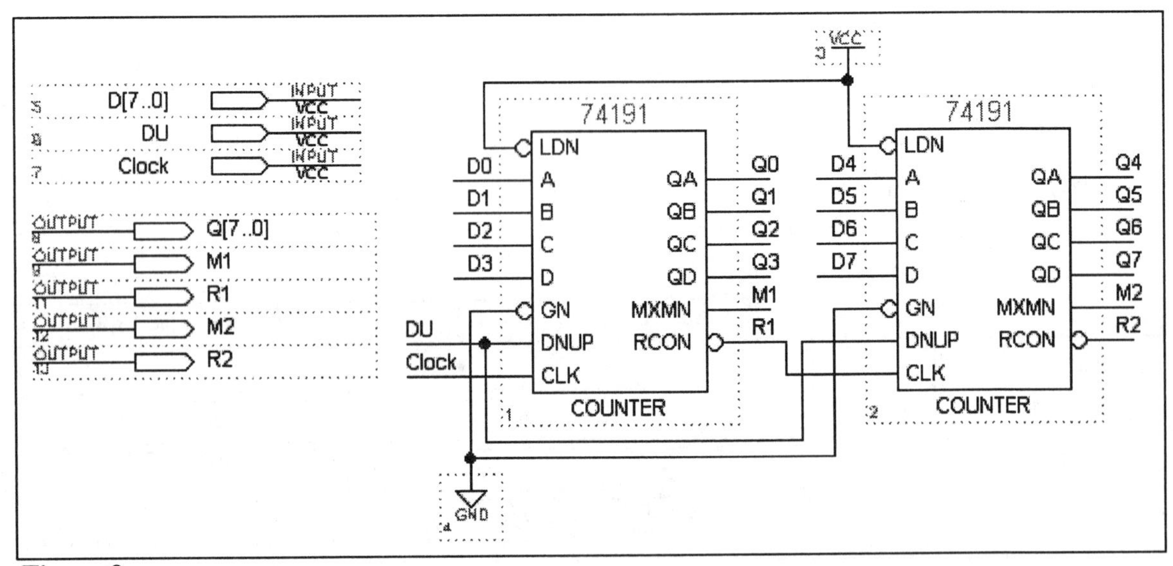

Figure 9

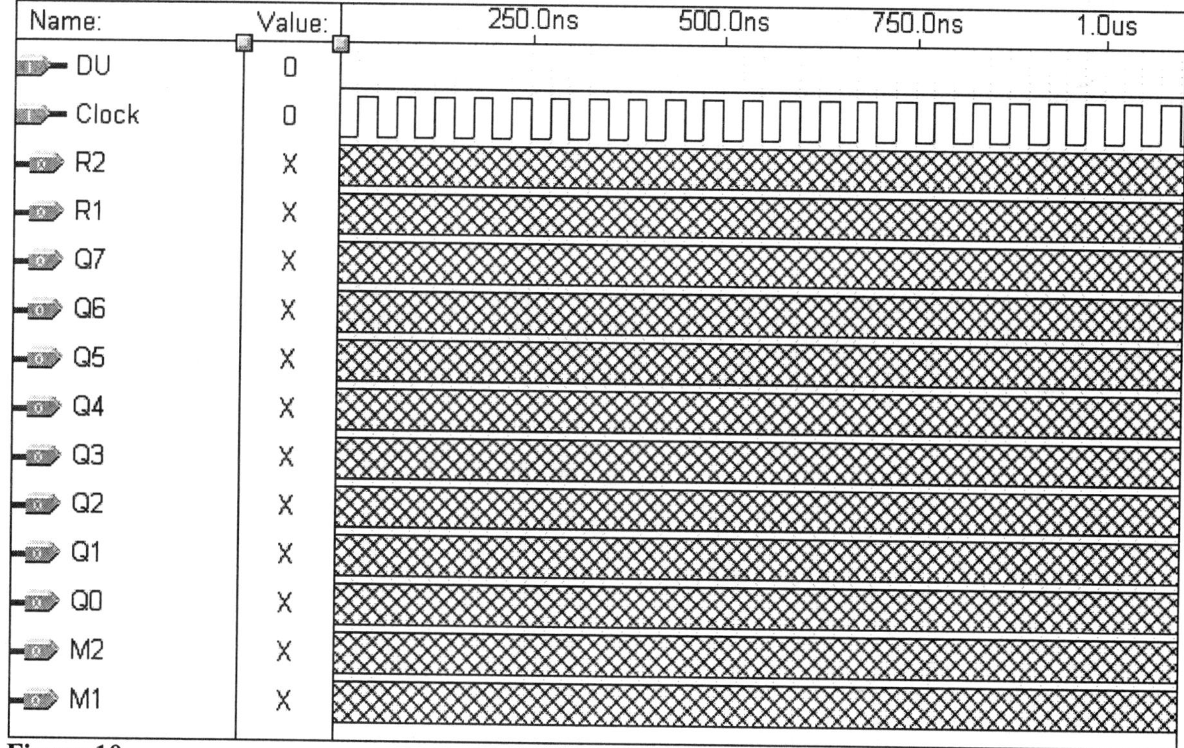

Figure 10

8. With the 12.8 μs of time displayed, how many negative spikes occur on R2? _______; on R1? _______

9. M2 is the output of the most significant of the two 74191 ICs. What count is on Q[7..4] when M2 is a logic-HIGH? _____

10. M1 is the output of the least significant of the two 74191 ICs. What count is on Q[3..0] when M1 is a logic-HIGH? _____ (Remember to zoom in to see the details.)

11. Zoom in until you see only 2 pulses on R1. How many clock pulses does it take to make pulses R1 low? _____

12. Since the clock input of the most significant IC is dependent on the RCON output of the least significant IC, inter-stage clocking for this counter is (synchronous/asynchronous).

13. Save all files to Drive A as **mod-256**, then exit the Graphic and Waveform Editors.

Part 5 Procedure

1. Open the Max+plus II software. Assign the project name **mod-xxx**.

2. Modify the circuit shown in Figure 11.

3. Open a new Waveform Editor, set Grid Size to 25 ns, and then create the waveforms shown in Figure 12.

4. Highlight the D[7..0] bus and assign a group value of $B0_H$.

5. Change End Time to 12800 ns. Highlight and assign the clock wave a count sequence using Overwrite Count Value.

6. Click the Compile and Simulate buttons. Correct all errors before continuing.

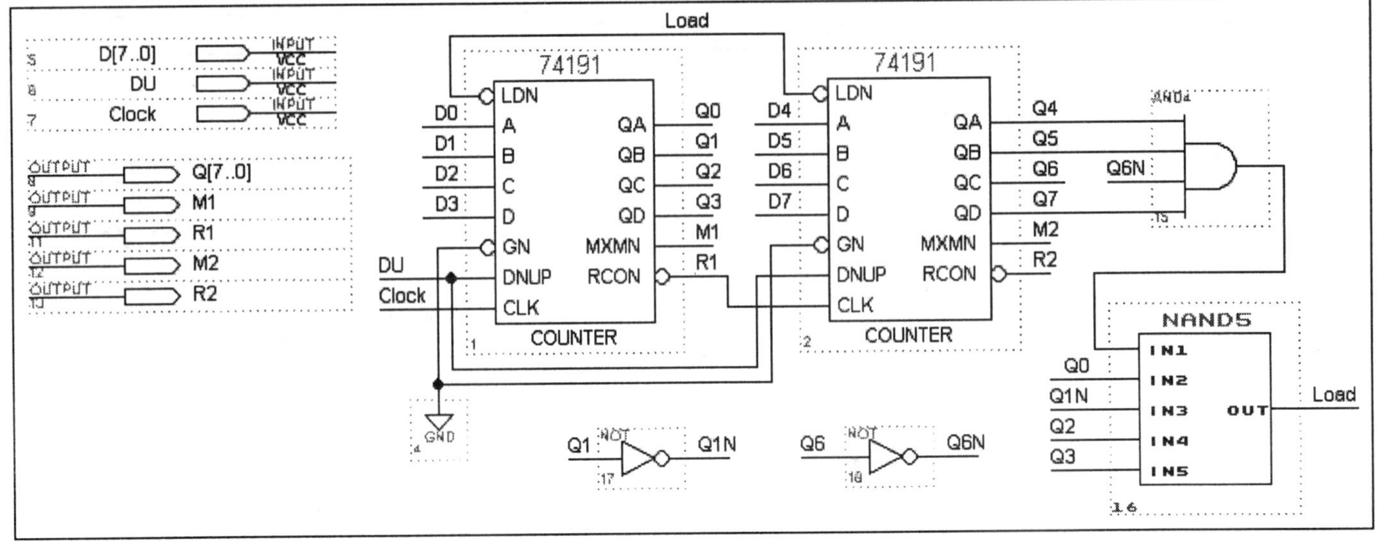

Figure 11

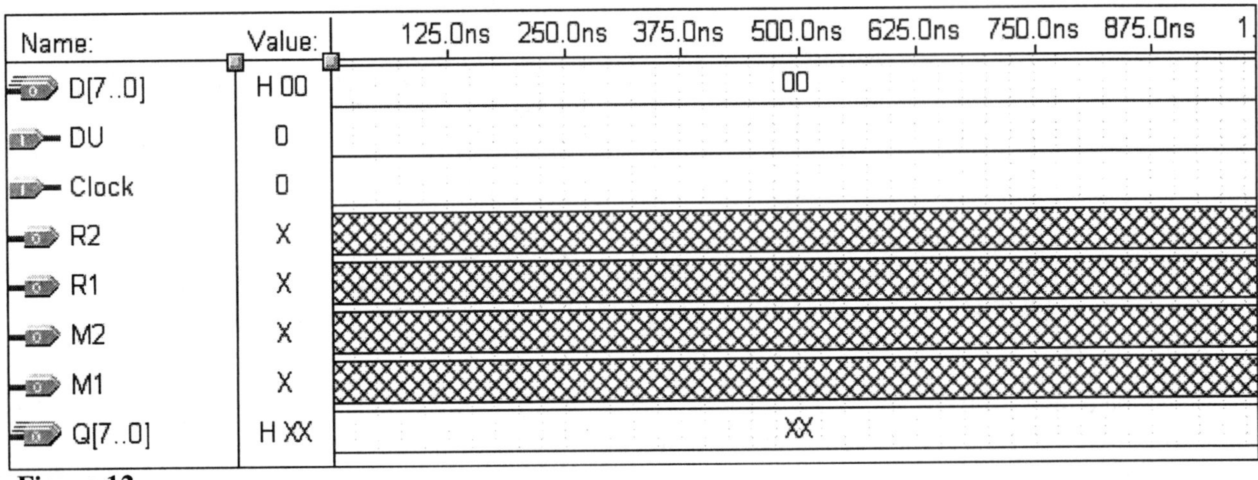

Figure 12

7. Zoom out to see the entire 12.8 μs of time.

8. Place the mouse pointer on the last M1 pulse, then press the left mouse button.

9. A blue vertical line should appear at about 8.7324 μs. Zoom in until the details of the count sequence on Q[7..0] are visible.

10. The cursor should be very close to the AF to B0 count. Starting at B0, scroll to the right, recording the count sequence until the pattern starts to repeat.

    B0, ____

11. What is the highest count? ______

12. Once the counter develops the count sequence, the counter modulus is _________.

13. Based on the schematic, what number on the output of the most significant counter IC will make the output of the AND4 gate “active”? _________

14. Based on the schematic, what number on the output of the least significant counter IC will make the output of the NAND5 gate “active”? _________

15. The AND4 and NAND5 gates make up a full decoder network and are wired to decode the number ________$_H$. This number is not part of the count sequence but asynchronously loads ______ into the counter.

16. Demonstrate the circuit you created to your instructor. Obtain the signature of approval directly on the answer page for this lab.

17. Obtain a hard copy of the Graphic and Waveform files. Label these hard copies **Part 3, Step 21A** and **Part 3, Step 21B**.

18. Save all files to Drive A as **mod-xxx**, then exit the Graphic and Waveform Editors.

19. Write a 1- to 2-page summary pertaining to the results obtained from this lab. Include embedded graphics as part of your summary.

20. Place all papers for this lab in the following sequence, then submit the lab to your instructor for grading.

    - Cover page
    - Typed summary
    - The completed answer page for this lab
    - Hard copy of the Graphic Editor, **Part 5, Step 17A**
    - Hard copy of the Waveform Editor, **Part 5, Step 17B**

Lab 17: Synchronous Counters Answer Pages

Name: ______________________________

Part 1

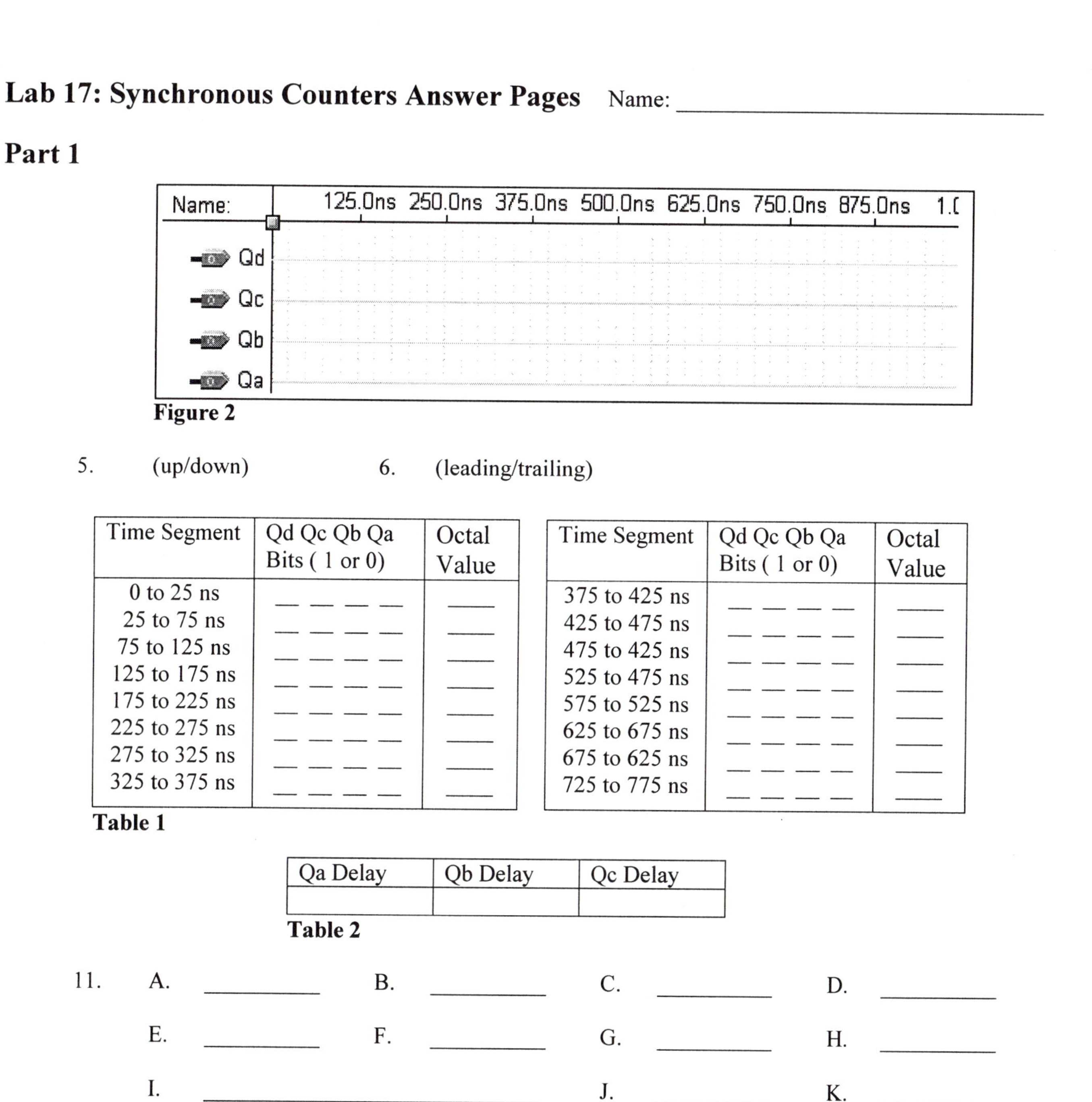

Figure 2

5. (up/down) 6. (leading/trailing)

Time Segment	Qd Qc Qb Qa Bits (1 or 0)	Octal Value
0 to 25 ns	__ __ __ __	____
25 to 75 ns	__ __ __ __	____
75 to 125 ns	__ __ __ __	____
125 to 175 ns	__ __ __ __	____
175 to 225 ns	__ __ __ __	____
225 to 275 ns	__ __ __ __	____
275 to 325 ns	__ __ __ __	____
325 to 375 ns	__ __ __ __	____

Time Segment	Qd Qc Qb Qa Bits (1 or 0)	Octal Value
375 to 425 ns	__ __ __ __	____
425 to 475 ns	__ __ __ __	____
475 to 425 ns	__ __ __ __	____
525 to 475 ns	__ __ __ __	____
575 to 525 ns	__ __ __ __	____
625 to 675 ns	__ __ __ __	____
675 to 625 ns	__ __ __ __	____
725 to 775 ns	__ __ __ __	____

Table 1

Qa Delay	Qb Delay	Qc Delay

Table 2

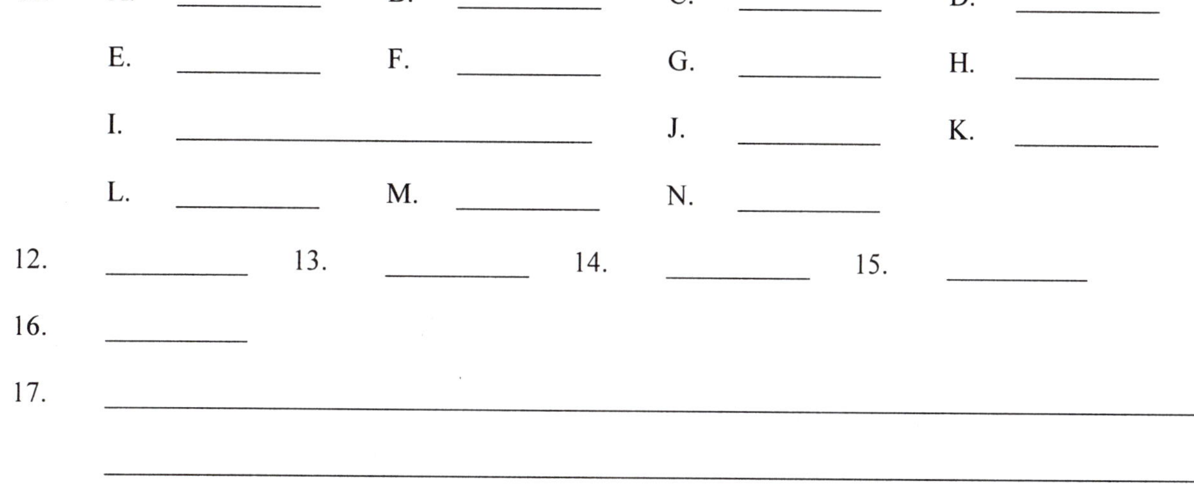

11. A. ________ B. ________ C. ________ D. ________

E. ________ F. ________ G. ________ H. ________

I. ______________________ J. ________ K. ________

L. ________ M. ________ N. ________

12. ________ 13. ________ 14. ________ 15. ________

16. ________

17. ____

____

Part 2

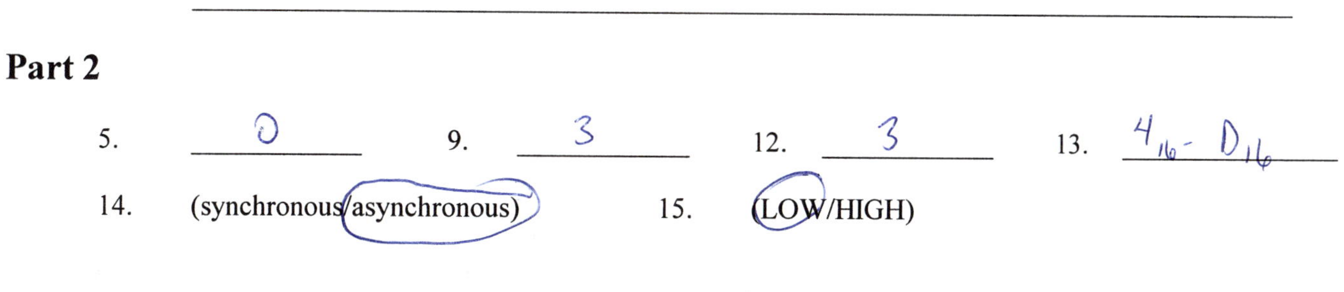

5. 0 9. 3 12. 3 13. 4_{16} - D_{16}

14. (synchronous/asynchronous) [asynchronous circled] 15. (LOW/HIGH) [LOW circled]

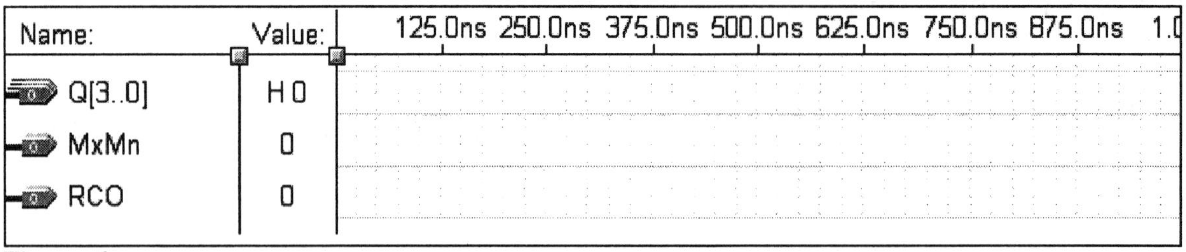

Name:	Value:	125.0ns	250.0ns	375.0ns	500.0ns	625.0ns	750.0ns	875.0ns	1.0
Q[3..0]	H 0								
MxMn	0								
RCO	0								

Figure 6

Part 3

5. _______ 6. _______ 7. _______ 8. _______ 9. _______

10. ________________________________ 11. ________________________________

12. ________________________________ 13. ________________________________

15. ____

____

Part 4

8. R2: ________ R1: ________ 9. ___________

10. ___________ 11. ___________ 12. (synchronous/asynchronous)

Part 5

10. ____

11. __________ 12. __________ 13. __________ 14. __________

15. __________

16. Demonstrated to: ____________________________________ Date: ________________

Grade: __________

Lab 18: Shift Registers

Objectives:

1. Construct and demonstrate the serial shifting of data through D-type flip-flops
2. Analyze the 74194 constructed as a serial right or serial left shift register
3. Construct and demonstrate parallel shifting of data through several 74194s
4. Construct and analyze a parallel in, serial out shift register
5. Construct and analyze a serial in, parallel out shift register

Materials List:

- Max+plus II software by Altera Corporation
- University Board with CPLD (optional)
- Computer requirements:
  Minimum 486/66 with 8 MB RAM
- Floppy disk

Discussion:

Figure 1 shows a basic 4-bit shift register using four transparent D-type flip-flops with synchronous clocking. Data applied to the D input of flip-flop A will appear on the Q output of flip-flop A when a rising edge of the clock appears. Since Qa is the data input to flip-flop B, the content that was in flip-flop A will be transferred to flip-flop B at the same time data was entered into flip-flop A. This data shifting is illustrated in Figure 2.

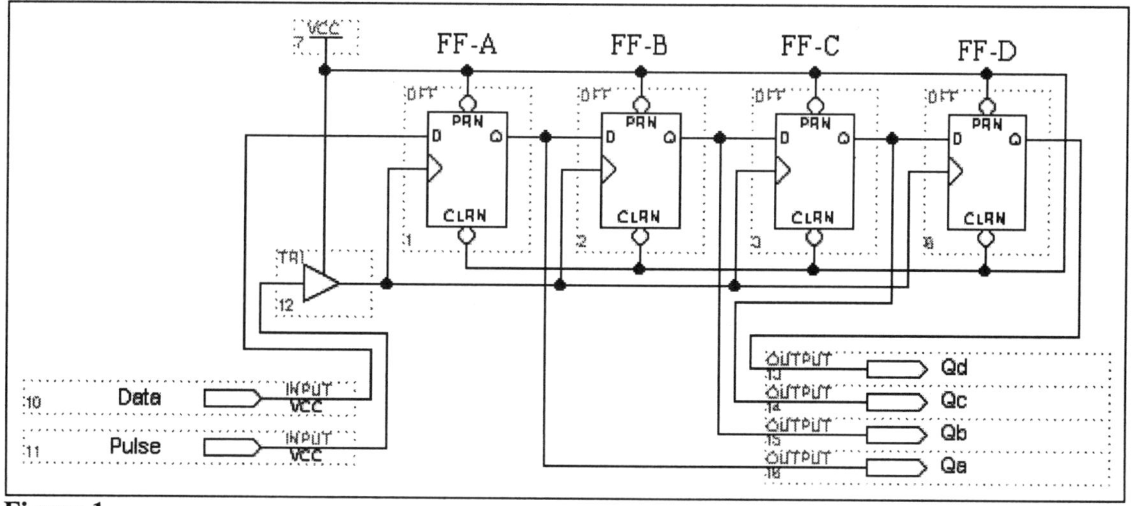

Figure 1

Initially, at time t_0, all flip-flops contain logic-LOWs and the data is a logic-HIGH. When the rising edge of the clock occurs at time t_1, the high state of the data is moved into flip-flop A. The content that was in flip-flop A was transferred to flip-flop B, the content that was in flip-flop B was transferred to flip-flop C, the content that was in flip-flop C was transferred to flip-flop D, and the content that was in flip-flop D was shifted out to the next stage, if one exists. Since all flip-flops initially contained zeros, the shifting may not be obvious, but now the flip-flops contain 1_H, 0001_2.

Notice the data shifting that takes place at the second rising edge of the clock at time t_2. Since the data input is a logic-HIGH, that data is shifted into flip-flop A. The logic-HIGH that was in flip-flop A shifted to flip-flop B. Of course, the contents of flip-flop C moved to flip-flop D and the contents in flip-flop D moved on to the next stage. The flip-flops now contain 3_H, 0011_2.

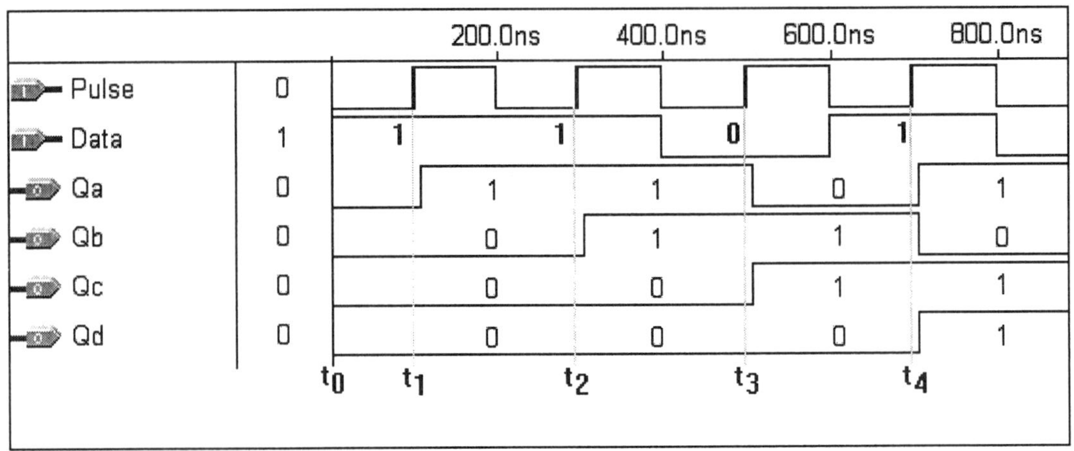

Figure 2

On the third clock pulse at time t_3, the logic-LOW at the data input is transferred to flip-flop A. As the original contents of each flip-flop are transferred to the next, the flip-flops now contain 6_H, 0110_2. At t_4, flip-flop A receives a logic-HIGH and the data in each flip-flop is shifted to the next flip-flop, resulting in D_H, 1101_2, now stored in "memory." It takes one clock pulse to shift one data bit one flip-flop position. This equates to four clock pulses to shift one nibble of data through four flip-flops.

Note the final contents of the four flip-flops contain 1101_2, the same sequence of the bits applied on the data input. Data shifting from the least-significant flip-flop (A) to the most-significant flip-flop (D in this case) is called right shifting.

The 74194 is a 4-bit shift register that may be wired to serial shift data either right or left, or parallel shift four bits, simultaneously. Various shifting techniques will be studied using the 74194.

Part 1 Procedure

1. Open the Max+plus II software. Assign the project name **reg-1**.

2. Open a new Graphic Editor and construct the circuit shown in Figure 1.

3. Open a new Waveform Editor, set Grid Size to 50 ns, and create the waveforms shown in Figure 3.

4. Run the Compiler and Simulator. Correct all errors before continuing.

5. Neatly and accurately sketch the output waveforms in the area provided in Figure 3.

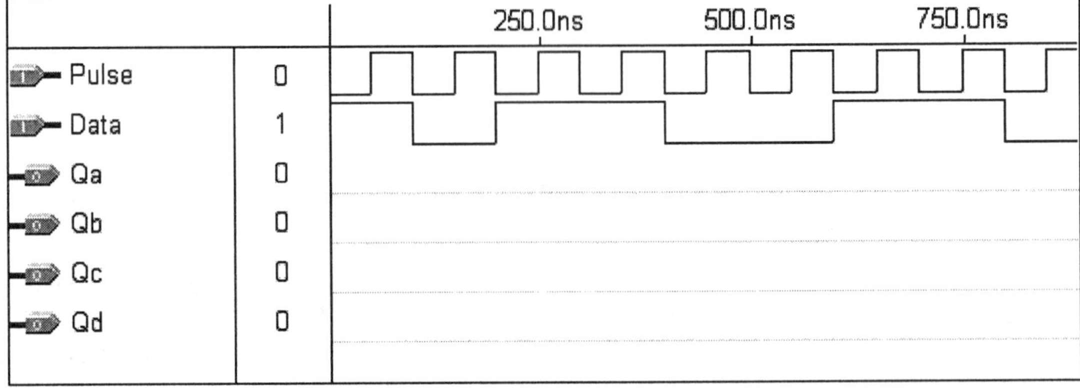

Figure 3

6. Record the bit pattern of the Data waveform, copying the logic level of the data that is present just before the rising edge of each clock pulse (eight bits required).

   Data sequence applied: ____________________$_2$

7. Complete Table 1, showing the data shifting through the flip-flops after each rising edge of the clock pulse. Read "Bit D" as the most-significant bit when converting the bit pattern to hexadecimal.

Data Applied	Flip-flop A B C D	Clock Pulse	Contents (in HEX)
____________$_2$	0 0 0 0	Initial	0_H
(from Step 6)	_ _ _ _	1st	____
	_ _ _ _	2nd	____
	_ _ _ _	3rd	____
	_ _ _ _	4th	____
	_ _ _ _	5th	____
	_ _ _ _	6th	____
	_ _ _ _	7th	____
	_ _ _ _	8th	____

Table 1

8. Save all files to Drive A as **reg-1**, then exit the Graphic and Waveform Editors.

Part 2 Procedure

1. Open the Max+plus II software. Assign the project name **reg-2**.

2. Open a new Graphic Editor and construct the circuit shown in Figure 4.

3. Open the Waveform Editor, set Grid Size to 50 ns, and create the waveforms shown in Figure 5.

4. Run the Compiler and Simulator. Correct all errors before continuing.

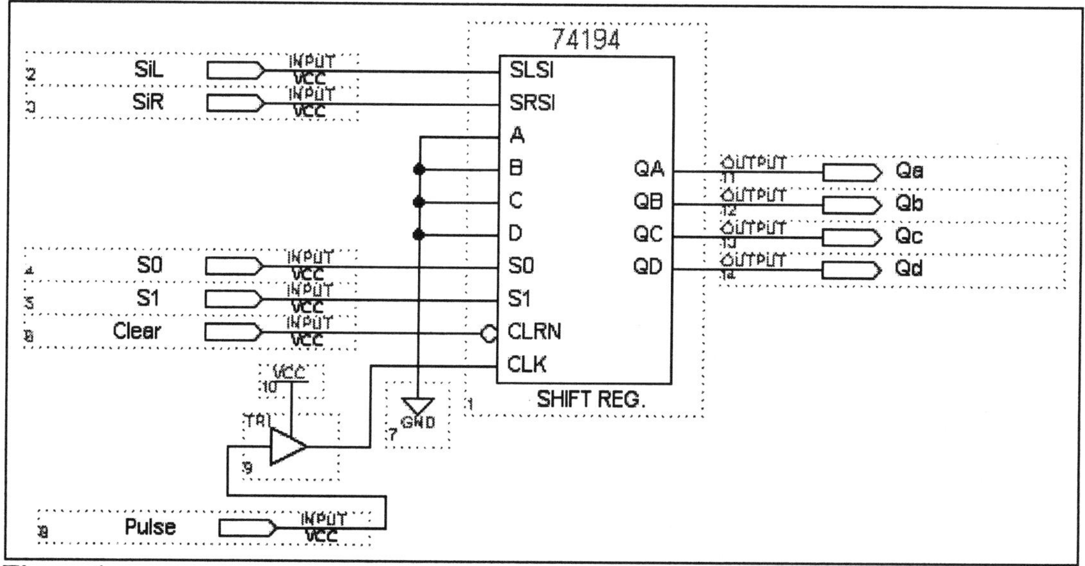

Figure 4

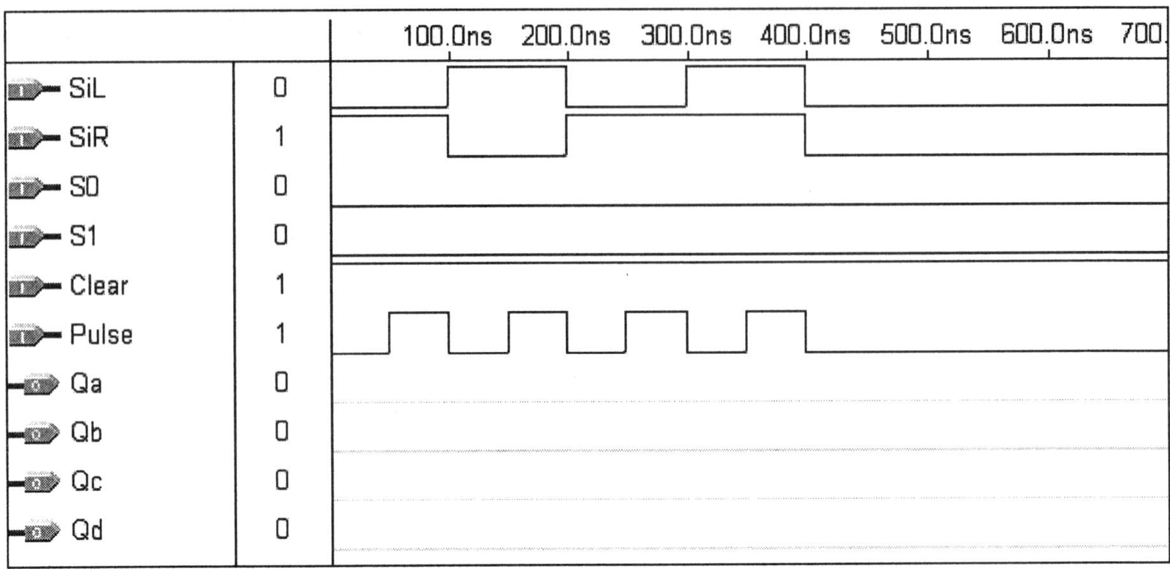

Figure 5

5. Refer to the function table for the 74194, Table 2, for the shift register and explain why the outputs are a constant logic-LOW. ____

____

6. Highlight and set the S0 waveform to a logic-HIGH.

7. Click the Simulate button.

8. Refer to the function table for the 74194 and explain how the inputs are affecting the outputs for the 74194 register. ____

____

____

9. Neatly and accurately sketch the resulting output waveforms in the area provided in Figure 5.

10. Complete Table 3 showing the shifting pattern for the register when S1 = 0 and S0 = 1.

11. Highlight and set the S0 waveform to a logic-LOW.

12. Highlight and set the S1 waveform to a logic-HIGH.

13. Click the Simulate button.

Inputs										Outputs			
Clear	Mode		Clock	Serial		Parallel				Qa	Qb	Qc	Qd
	S1	S0		Left	Right	A	B	C	D				
L	X	X	X	X	X	X	X	X	X	L	L	L	L
H	X	X	L	X	X	X	X	X	X	Qao	Qbo	Qco	Qdo
H	X	X	↑	X	X	a	b	c	d	a	b	c	d
H	L	H	↑	X	H	X	X	X	X	H	Qan	Qbn	Qcn
H	L	H	↑	X	L	X	X	X	X	L	Qan	Qbn	Qcn
H	H	L	↑	H	X	X	X	X	X	Qbn	Qcn	Qdn	H
H	H	L	↑	L	X	X	X	X	X	Qbn	Qcn	Qdn	L
H	L	L	X	X	X	X	X	X	X	Qao	Qbo	Qco	Qdo

Table 2 Function table for the 74194

Clock	SiR Data	A B C D	Hex value
Initial		0 0 0 0	0_H
1st	___	_ _ _ _	___
2nd	___	_ _ _ _	___
3rd	___	_ _ _ _	___
4th	___	_ _ _ _	___

Table 3

		100.0ns 200.0ns 300.0ns 400.0ns 500.0ns 600.0ns 700
Pulse	1	
Qa	0	
Qb	0	
Qc	0	
Qd	0	

Figure 6

14. Refer to the function table for the 74194 and explain how the inputs are affecting the outputs for the 74194 register. ____

____

15. Neatly and accurately sketch the resulting output waveforms in the area provided in Figure 6.

16. Complete Table 4 showing the shifting pattern for the register when S1 = 1 and S0 = 0.

Clock	SiL Data	D C B A	Hex value
Initial		0 0 0 0	0_H
1st	___	_ _ _ _	___
2nd	___	_ _ _ _	___
3rd	___	_ _ _ _	___
4th	___	_ _ _ _	___

Table 4

17. Based on the results shown in Figure 6, the shift register is wired for (Right/Left) shift.

18. Highlight and set the clear waveform to a logic-LOW.

19. Click the Simulate button.

20. Based on the results obtained and the function table, which input is dominant? (SiR , SiL, Clear)

21. Save all files to Drive A as **reg-2**, then exit the Graphic and Waveform Editors.

Part 3 Procedure

1. Open the Max+plus II software. Assign the project name **reg-3**.

2. Open a new Graphic Editor and construct the circuit shown in Figure 7.

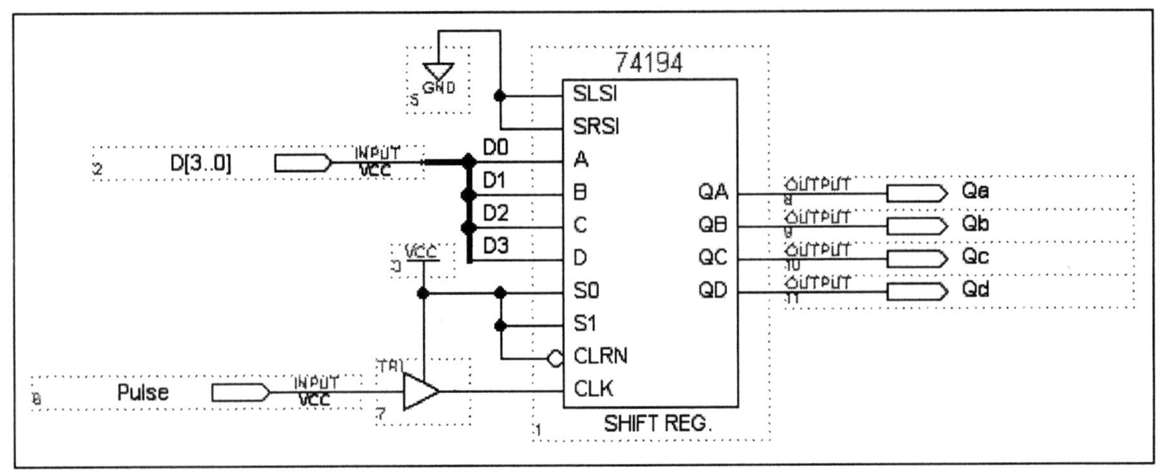

Figure 7

Name:	Value:	200.0ns	400.0ns	600.0ns	800.0ns
Pulse	0				
D[3..0]	H 7	7			B
Qa	0				
Qb	0				
Qc	0				
Qd	0				

Figure 8

3. Open the Waveform Editor, set Grid Size to 100 ns, and create the waveforms shown in Figure 8.

4. Run the Compiler and Simulator. Correct all errors before continuing.

5. Neatly and accurately sketch the resulting output waveforms in the area at the bottom of Figure 8.

6. At what time(s) did the data on the DCBA inputs get loaded into the 74194? _____________ ns

7. What mode of operation is the 74194 shown in Figure 7 wired for? ______________________

8. How many clock pulses did it take to load 7_H into the four flip-flops contained within the 74194? _________

9. Save all files to Drive A as **reg-3**, then exit the Graphic and Waveform Editors.

Part 4 Procedure

1. Open the Max+plus II software. Assign the project name **reg-4**.

2. Open a new Graphic Editor and construct the circuit shown in Figure 9.

3. Open a new Waveform Editor, set Grid Size to 50 ns, and create the waveforms shown in Figure 10.

4. Run the Compiler and Simulator. Correct all errors before continuing.

5. Neatly and accurately sketch the resulting waveforms in the area provided at the bottom of Figure 10.

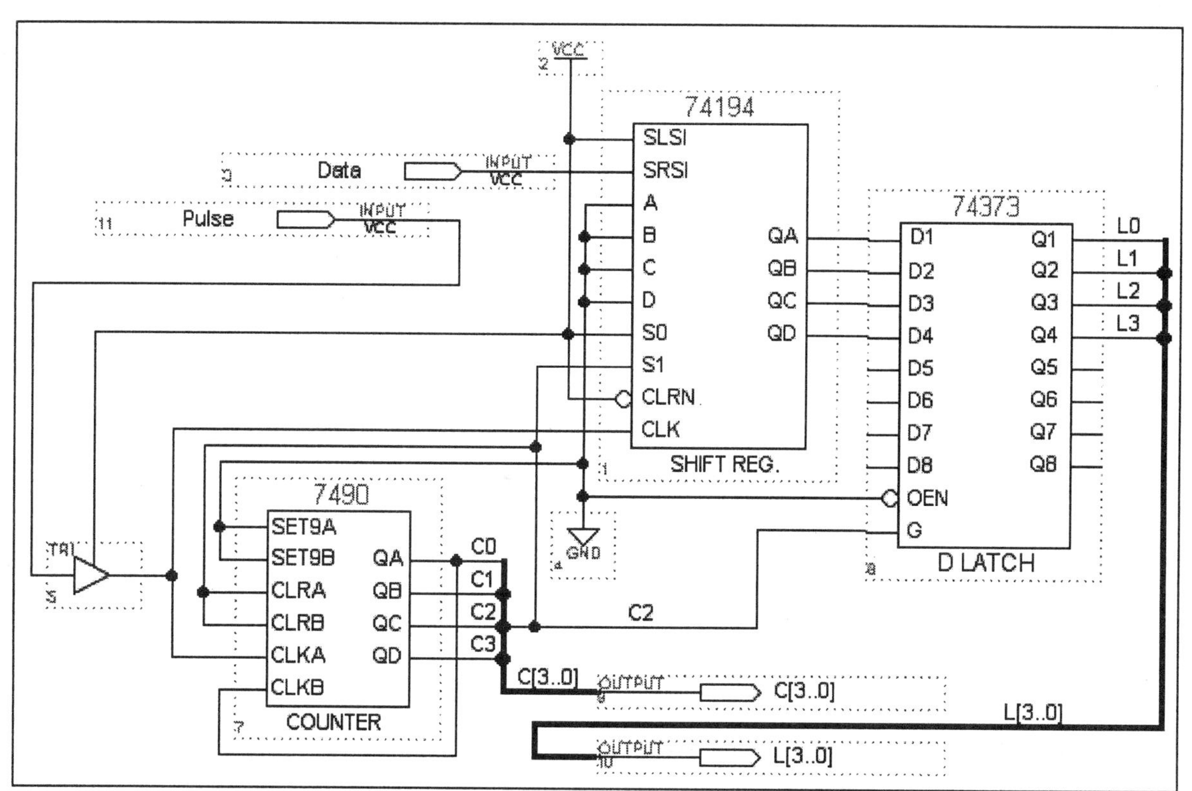

Figure 9

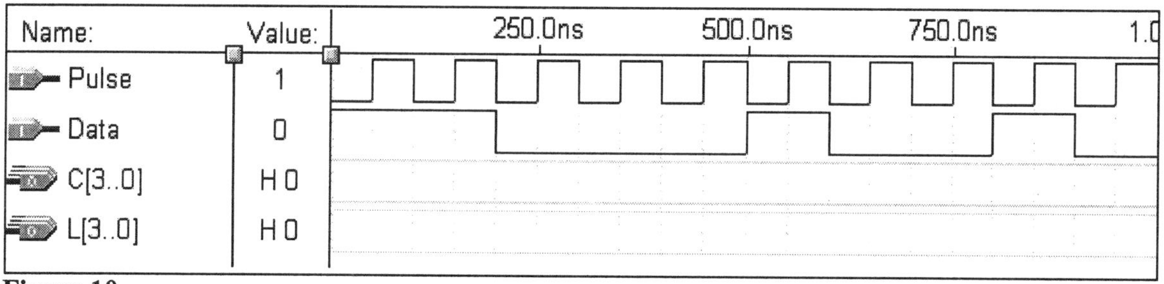

Figure 10

6. Identify the count sequence of the counter. ____________________________________

7. What number is decoded by the counter? __________

8. When C2 of the counter is a logic-LOW, the 74194 is wired as a ___ register.
   A. serial shift right
   B. serial shift left
   C. parallel in

9. When C2 of the counter goes to a logic-HIGH, the 74373
   A. shuts off.
   B. latches onto the register data.
   C. passes the data from the register.

10. When C2 of the counter goes to a logic-LOW due to counter reset, the 74373
    A. shuts off.
    B. latches onto the register data.
    C. passes the data from the register.

11. The circuit shown in Figure 9 is a
    A. serial in/serial out right shift.
    B. serial in/serial out left shift.
    C. serial in right/parallel out.
    D. serial in left/parallel out.
    E. parallel in/serial out right shift.
    F. parallel in/serial out left shift.

12. Save all files to Drive A as **reg-4**, then exit the Graphic and Waveform Editors.

Part 5 Procedure

1. Open the Max+plus II software. Assign the project name **reg-5**.

2. Open a new Graphic Editor and construct the circuit shown in Figure 11.

3. Open the Waveform Editor, set Grid Size to 50 ns, and create the waveforms shown in Figure 12.

4. Run the Compiler and Simulator. Correct all errors before continuing.

5. Neatly and accurately sketch the output waveforms in the area provided at the bottom of Figure 12.

6. Based on the waveforms, the counter advances on the (positive/negative) edge of the pulse.

7. Based on the waveforms, the register shifts data on the (positive/negative) edge of the pulse.

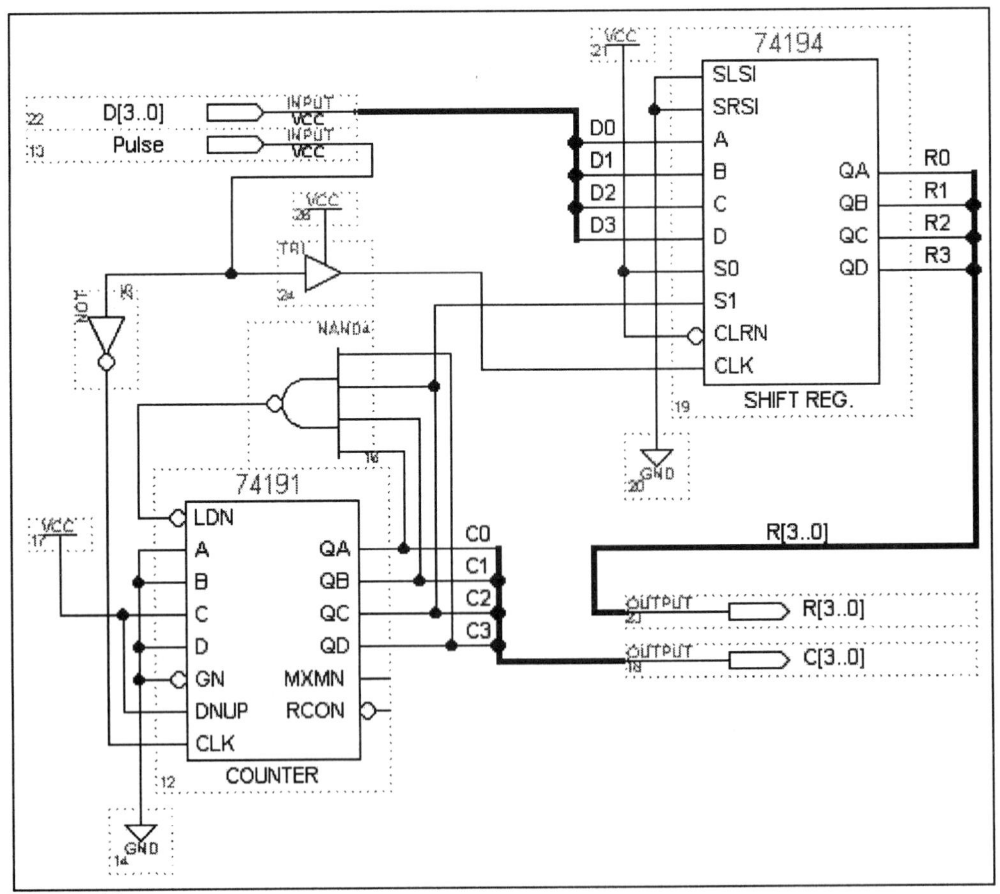

Figure 11

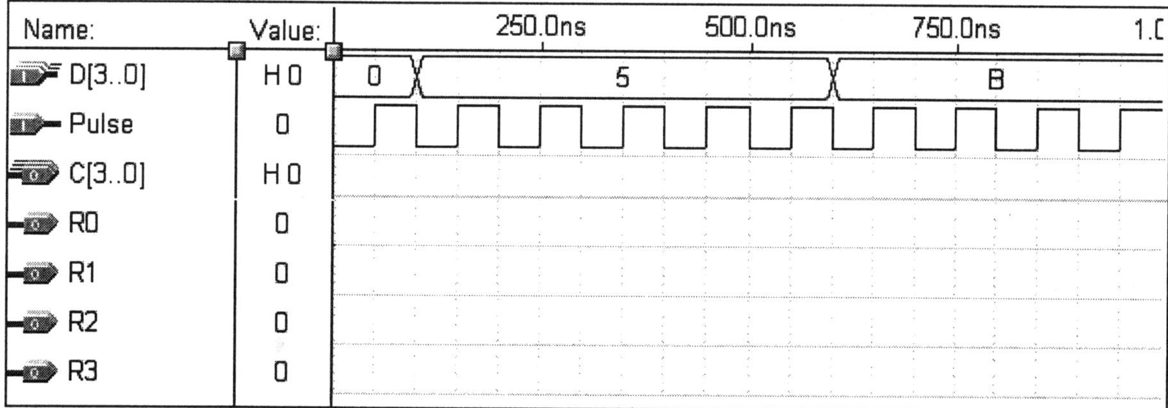

Figure 12

8. What number is decoded in the counter circuit? ____________

9. What number is loaded in the counter circuit? ____________

10. What is the modulus of the counter? ______________

11. What is the mode of operation for the register when the counter is on the 4 count?
    A. Si/So right shift
    B. Si/So left shift
    C. Parallel shift

12. What mode of operation is the register when the counter is not on the 4 count?
    A. Si/So right shift
    B. Si/So left shift
    C. Parallel shift

13. What is the data inputted to the register when the counter is counting from 3 to 0? __________

14. The majority of the time, the register is serially shifting data. Which register output would be used during serial shift mode?
    A. Qa B. Qb C. Qc D. Qd

15. Demonstrate the circuit you created to your instructor. Obtain the signature of approval directly on the answer page for this lab.

16. Obtain a hard copy of the Graphic and Waveform files. Label these hard copies **Part 5, Step 16A** and **Part 5, Step 16B**.

17. Save all files to Drive A as **reg-5**, then exit the Graphic and Waveform Editors.

18. Write a 1- to 2-page technical summary pertaining to the results obtained from this lab. Explain the operation of Figure 11 and include the circuit and simulated waveforms as embedded graphics in your summary.

19. Place all papers for this lab in the following sequence, then submit the lab to your instructor for grading.

    - Cover page
    - Typed summary
    - The completed answer page for this lab
    - Printout of the Graphic Editor, **Part 5, Step 16A**
    - Printout of the Waveform Editor, **Part 5, Step 16B**

Lab 18: Shift Registers Answer Pages

Name: ______________________________

Part 1

250.0ns 500.0ns 750.0ns

Qa

Qb

Qc

Qd

Figure 3

6. ________________ $_2$

Data Applied	Flip-flop A B C D	Clock Pulse	Contents (in HEX)
________ $_2$ (from Step 6)	0 0 0 0	Initial	0_H
	_ _ _ _	1st	______
	_ _ _ _	2nd	______
	_ _ _ _	3rd	______
	_ _ _ _	4th	______
	_ _ _ _	5th	______
	_ _ _ _	6th	______
	_ _ _ _	7th	______
	_ _ _ _	8th	______

Table 1

Part 2

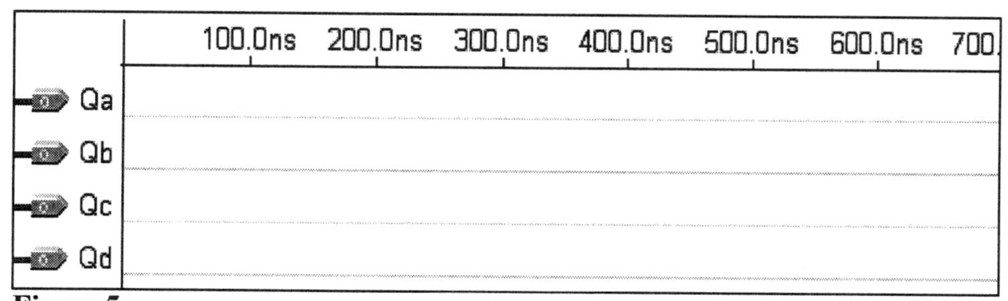

Figure 5

5. ____

____

____

Clock	SiR Data	A B C D	Hex value
Initial		0 0 0 0	0_H
1st	___	_ _ _ _	______
2nd	___	_ _ _ _	______
3rd	___	_ _ _ _	______
4th	___	_ _ _ _	______

Table 3

8. ____

100.0ns 200.0ns 300.0ns 400.0ns 500.0ns 600.0ns 700.

Qa 0

Qb 0

Qc 0

Qd 0

Figure 6

14. ____

Clock	SiL Data	D C B A	Hex value
Initial		0 0 0 0	0_H
1st	___	__ __ __ __	_____
2nd	___	__ __ __ __	_____
3rd	___	__ __ __ __	_____
4th	___	__ __ __ __	_____

Table 4

17. (RIGHT/LEFT)

20. (SiR , SiL, Clear)

Part 3

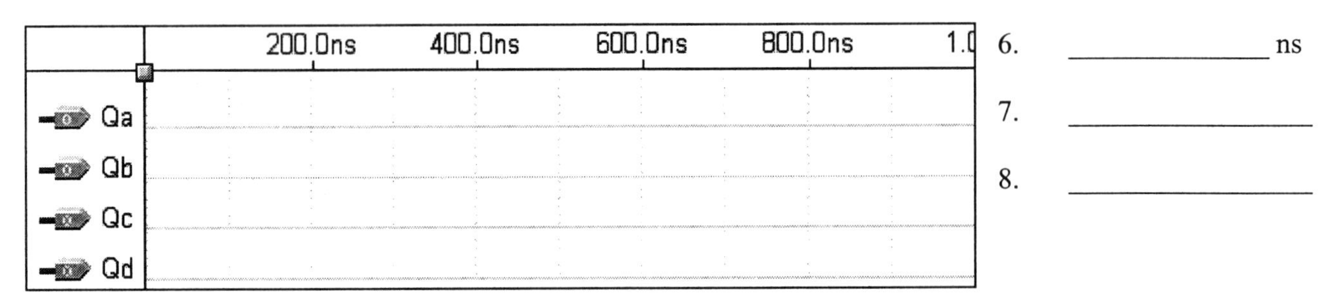

Figure 8

6. ______________ ns

7. ______________

8. ______________

Part 4

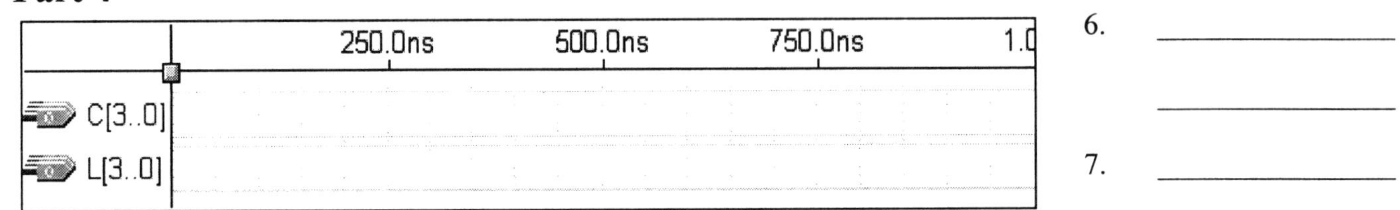

Figure 10

6. ______________

____

7. ______________

8. A B C 9. A B C 10. A B C 11. A B C D E F

Part 5

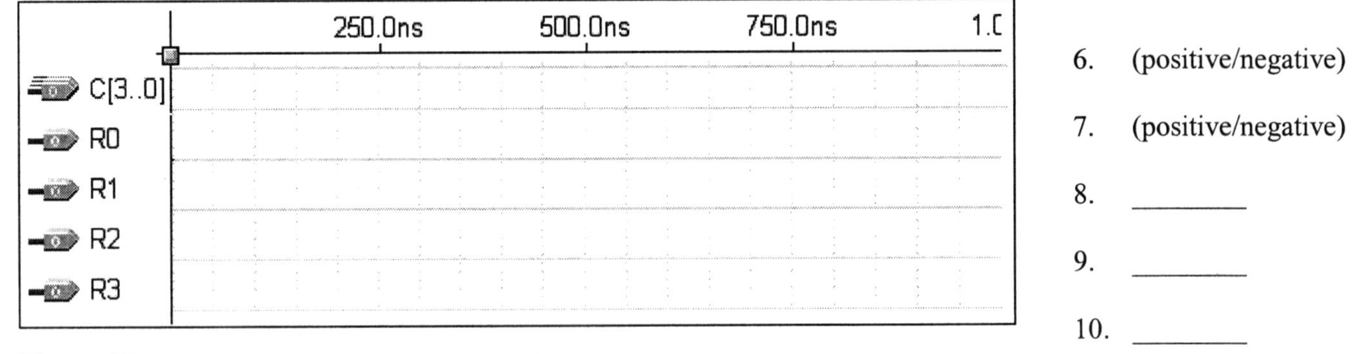

Figure 12

6. (positive/negative)

7. (positive/negative)

8. ________

9. ________

10. ________

11. A B C 12. A B C 13. ________ 14. Qa Qb Qc Qd

15. Demonstrated to: ______________________________ Date: __________

Grade: ________

Lab 19: Johnson and Ring Counters

Objectives:

1. Construct and evaluate the bit shift pattern of a ring counter
2. Use a ring counter to multiply and divide by 2
3. Construct and evaluate the shift pattern of a Johnson counter

Materials list:

- Max+plus II software by Altera Corporation
- University board with a CPLD (optional)
- Computer requirements:
  Minimum 486/66 with 8 MB RAM
- Floppy disk

Discussion:

Johnson and ring counters are circuits made from shift registers. To make a ring counter, wire the serial output back to the serial input. To make a Johnson counter, wire the serial output through an inverter, then the output of the inverter is wired back to the serial input of the register. Of course, if the serial input is used to load new data into the register, then the feedback signal and the serial input signal are to be multiplexed so only one signal source will be used at a time. This lab will use a 2-line to 1-line multiplexer for data selection. If data are first parallel (not serial) loaded into the register, then the multiplexer would not be needed.

Ring counters allow one to read the contents of a shift register without destroying the contents of the register. As the data is serially outputted, it is also fed back to the serial input. Since a register shifts data one flip-flop position on each clock, a 4-bit shift register will take four clock pulses to load (and output) four bits of data. This nondestructive readout of the ring counter may be a useful feature when programming in assembly language.

Right-shift ring counters may also be used to multiply the contents of the register by 2, as long as the most significant bit in the register is a logic-LOW. Left-shift ring counters may divide the contents of the register by 2, as long as the least significant bit is a logic-LOW. The respective bits should be checked before multiplying or dividing, otherwise an overflow may occur. An overflow condition exists if the answer is too large to fit into the register. For instance, the maximum number that an 8-bit register can hold is FF_H. Multiplying $7F_H$ (01111111_2) by 2 (10_H) will give FE_H (11111110_2) as an answer; 80_H times 2 will yield 01_H due to feeding back a logic-HIGH. This will become more apparent while completing the lab.

The bit pattern of Johnson and ring counters will be examined, as well as each counter module.

Part 1 Procedure

1. Open the Max+plus II software. Assign the project name **ring1**.

2. Open the Graphic Editor and construct the circuit shown in Figure 1.

3. Open the Waveform Editor, set Grid Size to 30 ns, and create the waveforms shown in Figure 2.

4. Click the Compiler and Simulator buttons. Correct all errors before continuing.

5. Sketch the output waveforms in the area provided in Figure 2.

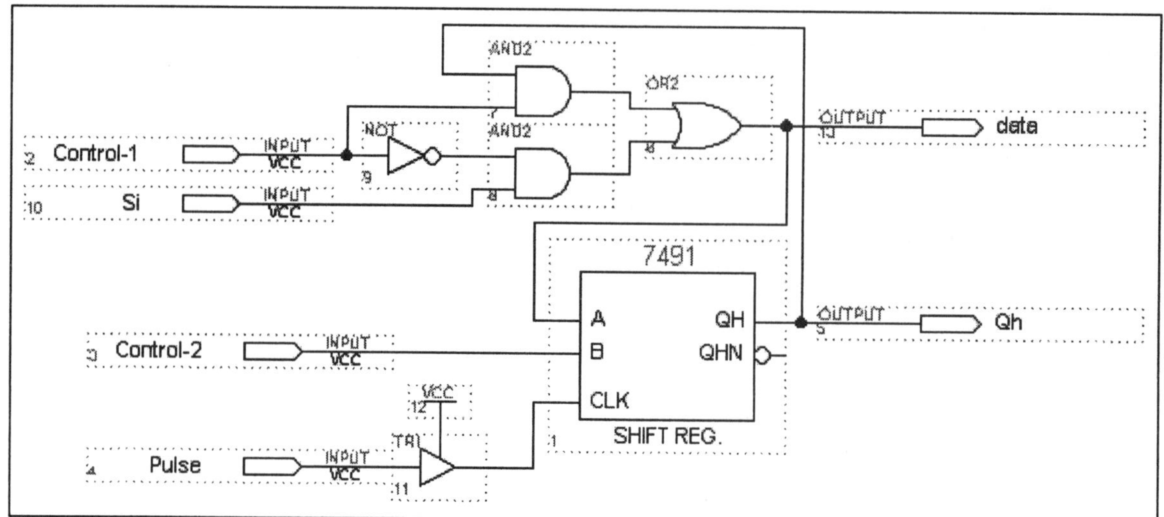

Figure 1

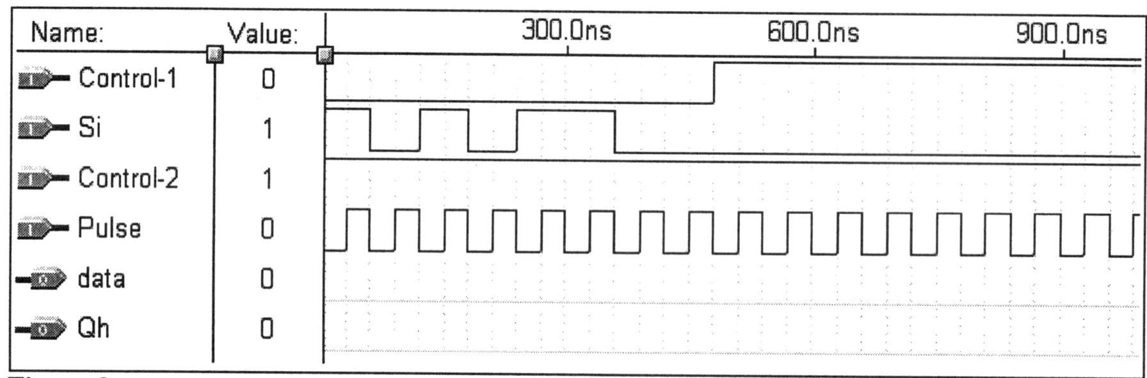

Figure 2

6. The shift register initial contents that can be observed as data are shifted in and out of the register. How many clock pulses did it take to load new data into the register in Figure 1? __________

7. What was the initial contents of the register? ___________

8. What was the data bit sequence loaded into the register? (answer in binary) ________________$_2$

9. When the control-1 input is a logic-LOW, the data entering the "A" input of the register is from

   A. the serial output, Qh. B. the serial input, Si.

10. At what time in Figure 1 does the control-1 input go to a logic-HIGH? __________ns

11. When the control input goes to a logic-HIGH, the data entering the "A" input of the register is from
   A. the serial output, Qh. B. the serial input, Si.

12. Once data has been serial loaded into the register, how many clock pulses does it take before the data reappears in the register as a ring counter? __________

13. Assume the register of Figure 1 contains $b7_H$. Complete Table 1 showing the data shifting through the register wired as a ring counter.

14. Save all files to Drive A as **ring1**, then exit the Graphic and Waveform Editors.

Clock	Binary Pattern H G F E D C B A	Hex Value
Initial contents	_ _ _ _ _ _ _ _	$b7_H$
1st	_ _ _ _ _ _ _ _	______H
2nd	_ _ _ _ _ _ _ _	______H
3rd	_ _ _ _ _ _ _ _	______H
4th	_ _ _ _ _ _ _ _	______H
5th	_ _ _ _ _ _ _ _	______H
6th	_ _ _ _ _ _ _ _	______H
7th	_ _ _ _ _ _ _ _	______H
8th	_ _ _ _ _ _ _ _	______H
9th	_ _ _ _ _ _ _ _	______H

Table 1

Part 2 Procedure

1. Open the Max+plus II software. Assign the project name **ring2**.

2. Open the Graphic Editor and construct the circuit shown in Figure 3.

3. Open the Waveform Editor, set Grid Size to 30 ns, and create the waveforms shown in Figure 4.

4. Click the Compiler and Simulator buttons. Correct all errors before continuing.

5. Sketch the resulting output waveforms in the area provided at the bottom of Figure 4.

6. The 74198 shifts data on the (leading/trailing) edge of the clock pulse.

7. What was the initial contents of the 74198 register before the first clock transition? __________

8. Initially at 0 ns, S0 = ____ and S1 = ____. According to the function table for the 74198, the mode inputs are for (serial shift right/serial shift left/parallel input).

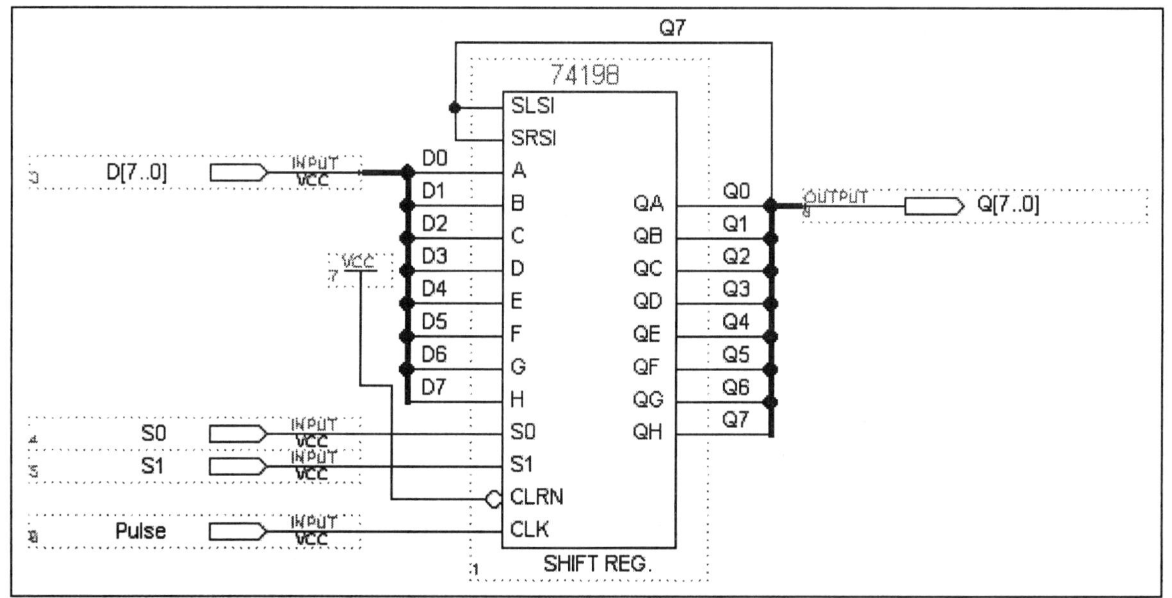

Figure 3

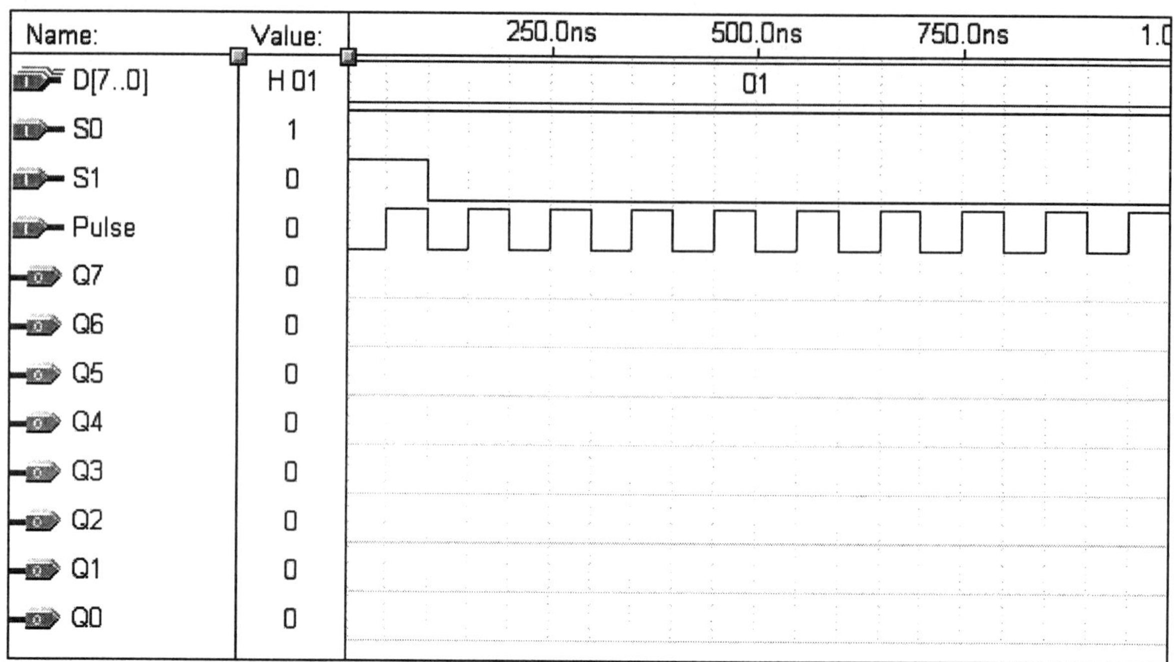

Figure 4

9. On the first clock transition, the data loaded into the register was _________.

10. At what time did S1 go to a logic-LOW? _________

11. After S1 went to a logic-LOW, what was the mode of operation?
(serial shift right/serial shift left/parallel input)

12. Complete Table 2 showing the bit pattern (and hexadecimal equivalents) as data of Figure 4 was shifted through the 74198 register wired as a ring counter (Figure 5).

Clock	Binary Pattern H G F E D C B A	Hex Value
1st	_ _ _ _ _ _ _ _	01_H
2nd	_ _ _ _ _ _ _ _	____H
3rd	_ _ _ _ _ _ _ _	____H
4th	_ _ _ _ _ _ _ _	____H
5th	_ _ _ _ _ _ _ _	____H
6th	_ _ _ _ _ _ _ _	____H
7th	_ _ _ _ _ _ _ _	____H
8th	_ _ _ _ _ _ _ _	____H
9th	_ _ _ _ _ _ _ _	____H
10th	_ _ _ _ _ _ _ _	____H

Table 2

Q_H

$Q_H Q_G Q_F Q_E Q_D Q_C Q_B Q_A$

Figure 5

13. Note the pattern developed by the hexadecimal count. As long as the MSB (most significant bit) was a logic-LOW, this serial shifting pattern demonstrates a ___ function.
A. times 2 B. divide by 2 C. logic

14. Change the data to 24_H , then click the Simulate button.

15. Complete Table 3 showing the binary and hexadecimal pattern based on 24_H.

16. On the first clock pulse, 24_H was loaded into the register. How many clock pulses did it take to swap the two nibbles of date (the 2 and the 4) to get 42_H? _______

Clock	Binary Pattern H G F E D C B A	Hex Value
1st	_ _ _ _ _ _ _ _	24_H
2nd	_ _ _ _ _ _ _ _	______H
3rd	_ _ _ _ _ _ _ _	______H
4th	_ _ _ _ _ _ _ _	______H
5th	_ _ _ _ _ _ _ _	______H
6th	_ _ _ _ _ _ _ _	______H
7th	_ _ _ _ _ _ _ _	______H
8th	_ _ _ _ _ _ _ _	______H
9th	_ _ _ _ _ _ _ _	______H
10th	_ _ _ _ _ _ _ _	______H

Table 3

17. Add Q[7..0] as an output waveform to the Waveform Editor file. Compare the pattern shown in Q[7..0] waveform to your recorded results in Table 3. Are they identical? (Yes/No)

18. As long as the MSB (most significant bit) in Table 3 was a logic-LOW, this serial shifting pattern demonstrated a ___ function.
A. times 2 B. divide by 2 C. logic

19. Save all files to Drive A as **ring2**, then exit the Graphic and Waveform Editors.

Part 3 Procedure

1. Open the Max+plus II software. Assign the project name **johnson1**.

2. Open the **ring1.gdf** file that is located on your disk in Drive A.

3. Modify your Graphic Editor file to match Figure 6.

4. Save the file as **johnson1** to your disk in Drive A.

5. Open the Waveform Editor, set Grid Size to 20 ns, and create the waveforms shown in Figure 7.
   - The control-1 line goes high at 320 ns.
   - Si is a logic-HIGH from 0 to 40 ns, 80 to 120 ns, and 160 to 240 ns.

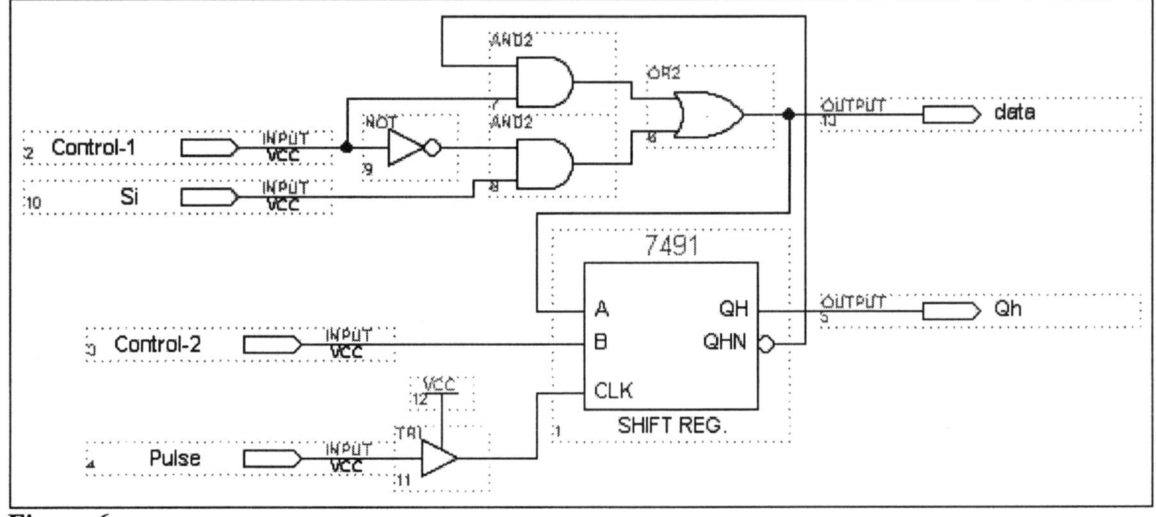

Figure 6

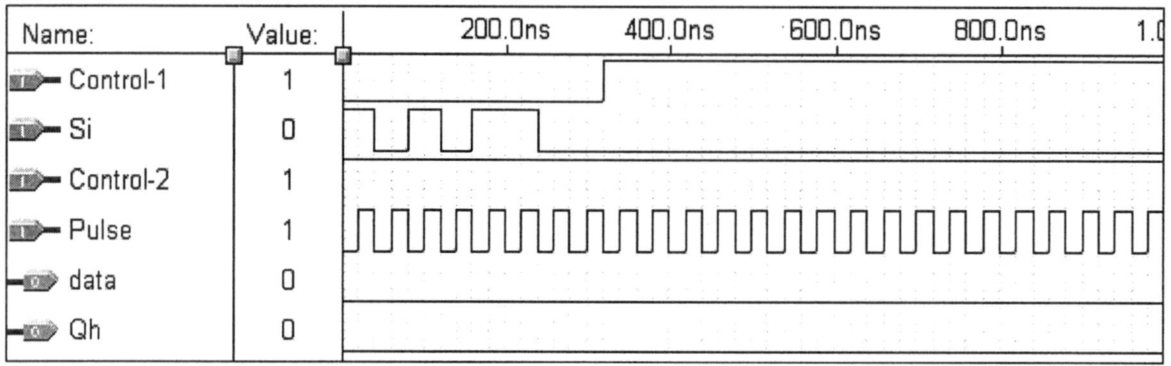

Figure 7

6. Compile and simulate the circuit. Correct all errors before continuing.

7. The waveforms you obtain after simulation should look similar to the waveforms in Figure 8. The data (highlighted) appearing on input A of the register won't start appearing on the output, Q_H, until 300 ns later. At 300 ns when the control-1 input goes to a logic-HIGH, the inverted Q_H output supplies the data and won't be seen on Q_H until 640 ns.

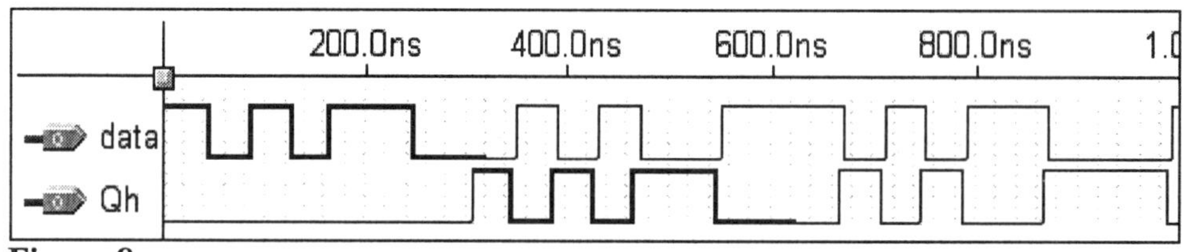

Figure 8

8. Since the 7491 only has one output, the shifting pattern may be difficult to see from the waveforms. Complete Table 4 showing the data being loaded into the 7491 as clock pulses are applied. On the first clock pulse, bit 7 of the data will appear on Q_A, what was on Q_A will be moved to Q_B, and so on. On the second clock pulse, bit 6 will appear on Q_A, moving the remaining bits in the IC one flip-flop position towards Q_H.

Clock	Q_H Q_G Q_F Q_E Q_D Q_C Q_B Q_A	10101100_2
Initial	0 0 0 0 0 0 0 0	
1st	_ _ _ _ _ _ _ _	⇐ bit 7
2nd	_ _ _ _ _ _ _ _	⇐ bit 6
3rd	_ _ _ _ _ _ _ _	⇐ bit 5
4th	_ _ _ _ _ _ _ _	⇐ bit 4
5th	_ _ _ _ _ _ _ _	⇐ bit 3
6th	_ _ _ _ _ _ _ _	⇐ bit 2
7th	_ _ _ _ _ _ _ _	⇐ bit 1
8th	_ _ _ _ _ _ _ _	⇐ bit 0

Table 4

9. Once the IC has been serially loaded, the control-1 input goes to a logic-HIGH, switching the circuit to a Johnson counter. The inverted Q_H output becomes the serial input data to flip-flop A. Complete Table 5 showing the data shifting through the Johnson counter (Figure 9) as clock pulses are applied. Assume the IC contains AC_H at the start of this count sequence.

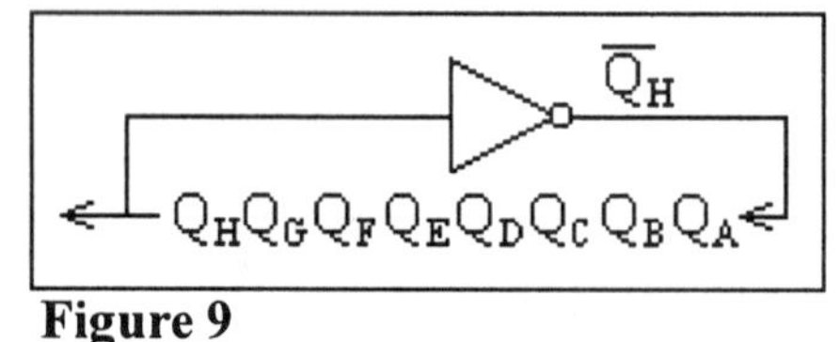

Figure 9

Clock	Q_H Q_G Q_F Q_E Q_D Q_C Q_B Q_A	Hex Equiv.
Initial	_ _ _ _ _ _ _ _	AC_H
1st	_ _ _ _ _ _ _ _	____H
2nd	_ _ _ _ _ _ _ _	____H
3rd	_ _ _ _ _ _ _ _	____H
4th	_ _ _ _ _ _ _ _	____H
5th	_ _ _ _ _ _ _ _	____H
6th	_ _ _ _ _ _ _ _	____H
7th	_ _ _ _ _ _ _ _	____H
8th	_ _ _ _ _ _ _ _	____H
9th	_ _ _ _ _ _ _ _	____H
10th	_ _ _ _ _ _ _ _	____H
11th	_ _ _ _ _ _ _ _	____H
12th	_ _ _ _ _ _ _ _	____H
13th	_ _ _ _ _ _ _ _	____H
14th	_ _ _ _ _ _ _ _	____H
15th	_ _ _ _ _ _ _ _	____H
16th	_ _ _ _ _ _ _ _	____H

Table 5

10. Compare the bit pattern after the 8th clock pulse to the initial contents, AC_H. Compare the bit pattern after the 1st clock pulse to the pattern after the 9th clock pulse. Write a brief statement describing the relationship of the bit pattern the IC will contain after 8 clock pulses, regardless of the initial content of the IC.

____

____

11. Save all files to Drive A as **johnson1**, then exit the Graphic and Waveform Editors.

Part 4 Procedure

1. Open the Max+plus II software. Assign the project name **johnson2**.

2. Open the **ring2.gdf** file located on your diskette in Drive A. Modify the circuit as shown in Figure 10, then save the file as **johnson2.gdf**.

3. Open the **ring2.scf** file located on your disk in Drive A. Modify the waveforms shown in Figure 11, then save the file as **johnson2**.scf.

4. Click the Compiler and Simulator buttons. Correct all errors before continuing.

5. Draw the resulting output waveforms in the area provided at the bottom of Figure 11.

6. What was the number loaded into the resister during the Pi mode? __________

7. Complete Table 6 showing the bit pattern and hex equivalents as clock pulses are applied to the Johnson counter of Figure 10.

8. Modify the waveforms and circuit as described below:
   - Set S0 from 50 ns to 1 μs to a logic-LOW
   - Set S1 to a constant logic-HIGH from 0 ns to 1 μs
   - Change the Q7 label on the wire to the inverter to Q0

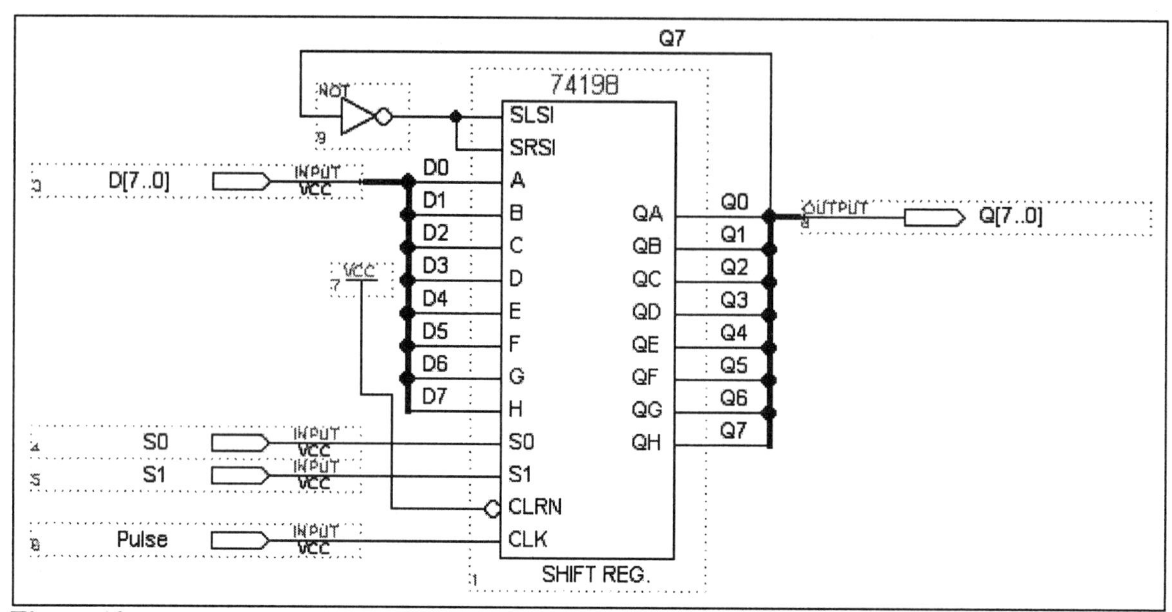

Figure 10

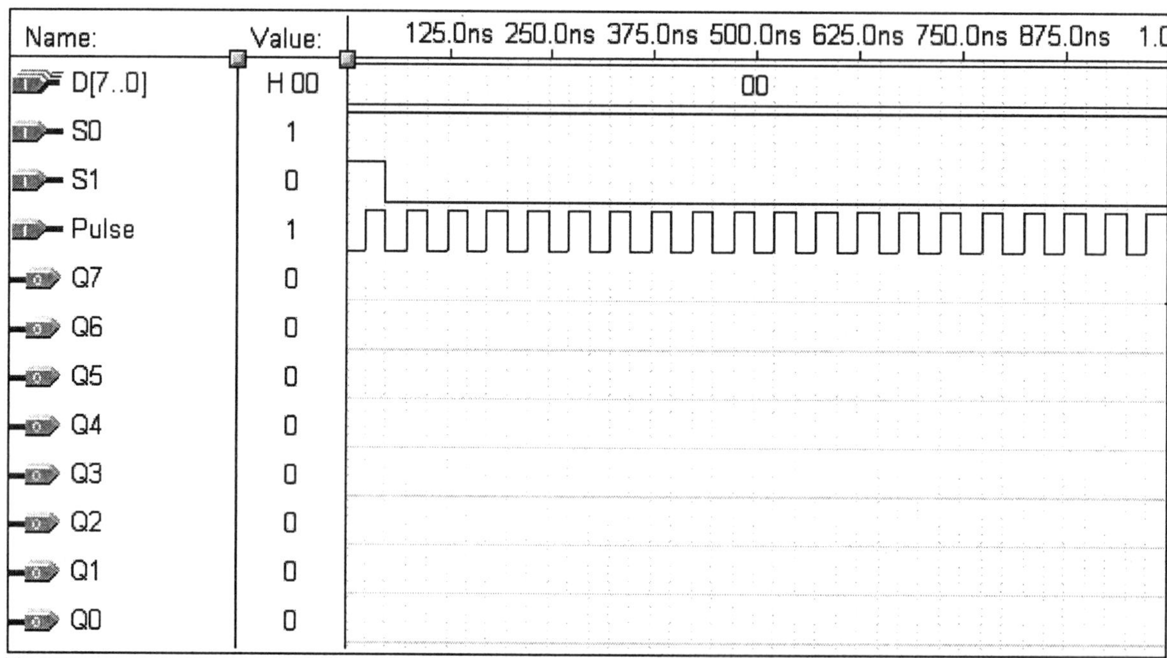

Figure 11

Clock	Q_H Q_G Q_F Q_E Q_D Q_C Q_B Q_A	Hex equiv.
1st	_ _ _ _ _ _ _ _	____H
2nd	_ _ _ _ _ _ _ _	____H
3rd	_ _ _ _ _ _ _ _	____H
4th	_ _ _ _ _ _ _ _	____H
5th	_ _ _ _ _ _ _ _	____H
6th	_ _ _ _ _ _ _ _	____H
7th	_ _ _ _ _ _ _ _	____H
8th	_ _ _ _ _ _ _ _	____H
9th	_ _ _ _ _ _ _ _	____H

Clock	Q_H Q_G Q_F Q_E Q_D Q_C Q_B Q_A	Hex equiv.
10th	_ _ _ _ _ _ _ _	____H
11th	_ _ _ _ _ _ _ _	____H
12th	_ _ _ _ _ _ _ _	____H
13th	_ _ _ _ _ _ _ _	____H
14th	_ _ _ _ _ _ _ _	____H
15th	_ _ _ _ _ _ _ _	____H
16th	_ _ _ _ _ _ _ _	____H
17th	_ _ _ _ _ _ _ _	____H

Table 6

9. Click the Compiler and Simulator buttons. Correct all errors before continuing.

10. Describe how the modifications to the circuit affected the circuit operation.

    ____

11. List the new count sequence (in hexadecimal) of the Johnson counter.

    _____H _____H _____H _____H _____H _____H _____H _____H

    _____H _____H _____H _____H _____H _____H _____H _____H

12. Demonstrate your circuit and waveforms to your professor. Obtain the signature of approval directly on the answer page for this lab.

13. Obtain a hard copy of the Graphic and Waveform Editor files. Label these hard copies **Part 4, Step 13A** and **Part 4, Step 13B**, respectively.

14. Save all files to Drive A as **johnson2**, then exit the Graphic and Waveform Editors.

15. Write a 1- to 2-page summary based on the 74198 wired as a Johnson and Ring counter. Provide a set of waveforms and discuss the results obtained after simulation.

16. Staple the following in the sequence listed to your instructor for grading.
    - Cover page
    - Completed answer page for this lab
    - Hard copy of the Graphic Editor, **Part 4, Step 13A**
    - Hard copy of the Waveform Editor, **Part 4, Step 13B**

Lab 19: Johnson and Ring Counters Answer Pages

Name: ______________________________

Part 1

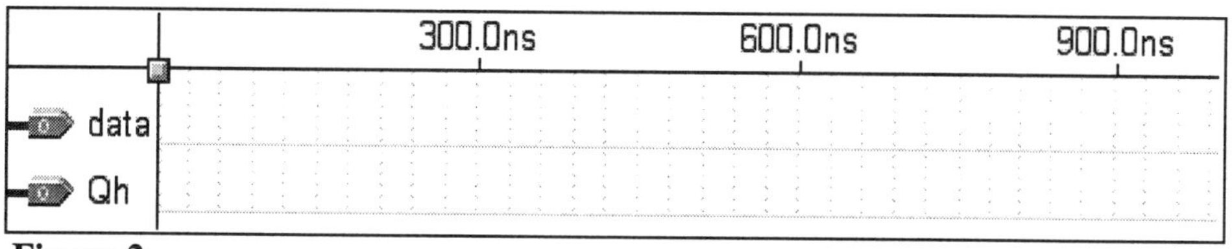

Figure 2

6. __________

7. __________

8. __________

9. A B

10. __________

11. A B

12. __________

Clock	Binary Pattern H G F E D C B A	Hex Value
Initial contents	_ _ _ _ _ _ _ _	$b7_H$
1st	_ _ _ _ _ _ _ _	______H
2nd	_ _ _ _ _ _ _ _	______H
3rd	_ _ _ _ _ _ _ _	______H
4th	_ _ _ _ _ _ _ _	______H
5th	_ _ _ _ _ _ _ _	______H
6th	_ _ _ _ _ _ _ _	______H
7th	_ _ _ _ _ _ _ _	______H
8th	_ _ _ _ _ _ _ _	______H
9th	_ _ _ _ _ _ _ _	______H

Table 1

Part 2

6. (leading/trailing)

7. __________

8. S0: ____ S1: ____

9. __________

10. __________

11. (serial shift right/ serial shift left/ parallel input)

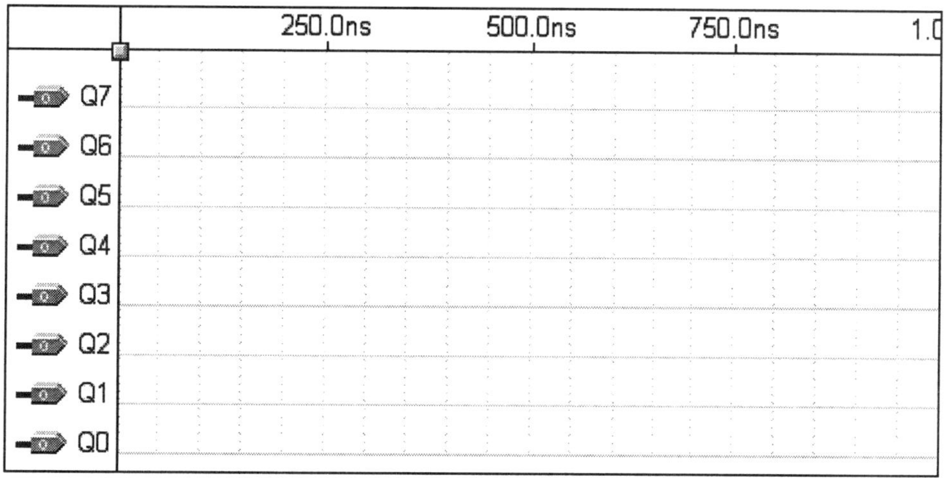

Figure 4

Clock	Binary Pattern H G F E D C B A	Hex Value
1st	_ _ _ _ _ _ _ _	01_H
2nd	_ _ _ _ _ _ _ _	______H
3rd	_ _ _ _ _ _ _ _	______H
4th	_ _ _ _ _ _ _ _	______H
5th	_ _ _ _ _ _ _ _	______H
6th	_ _ _ _ _ _ _ _	______H
7th	_ _ _ _ _ _ _ _	______H
8th	_ _ _ _ _ _ _ _	______H
9th	_ _ _ _ _ _ _ _	______H
10th	_ _ _ _ _ _ _ _	______H

Table 2

Clock	Binary Pattern H G F E D C B A	Hex Value
1st	_ _ _ _ _ _ _ _	24_H
2nd	_ _ _ _ _ _ _ _	______H
3rd	_ _ _ _ _ _ _ _	______H
4th	_ _ _ _ _ _ _ _	______H
5th	_ _ _ _ _ _ _ _	______H
6th	_ _ _ _ _ _ _ _	______H
7th	_ _ _ _ _ _ _ _	______H
8th	_ _ _ _ _ _ _ _	______H
9th	_ _ _ _ _ _ _ _	______H
10th	_ _ _ _ _ _ _ _	______H

Table 3

16. ________ 17. (Yes/No) 18. A B C

Part 3

Clock	Q_H Q_G Q_F Q_E Q_D Q_C Q_B Q_A	10101100_2
Initial	0 0 0 0 0 0 0 0	
1st	__ __ __ __ __ __ __ __	⇐ bit 7
2nd	__ __ __ __ __ __ __ __	⇐ bit 6
3rd	__ __ __ __ __ __ __ __	⇐ bit 5
4th	__ __ __ __ __ __ __ __	⇐ bit 4
5th	__ __ __ __ __ __ __ __	⇐ bit 3
6th	__ __ __ __ __ __ __ __	⇐ bit 2
7th	__ __ __ __ __ __ __ __	⇐ bit 1
8th	__ __ __ __ __ __ __ __	⇐ bit 0

Table 4

10. ______________________________________

____

____

____

Clock	Q_H Q_G Q_F Q_E Q_D Q_C Q_B Q_A	Hex Equiv.
Initial	__ __ __ __ __ __ __ __	AC_H
1st	__ __ __ __ __ __ __ __	______H
2nd	__ __ __ __ __ __ __ __	______H
3rd	__ __ __ __ __ __ __ __	______H
4th	__ __ __ __ __ __ __ __	______H
5th	__ __ __ __ __ __ __ __	______H
6th	__ __ __ __ __ __ __ __	______H
7th	__ __ __ __ __ __ __ __	______H
8th	__ __ __ __ __ __ __ __	______H
9th	__ __ __ __ __ __ __ __	______H
10th	__ __ __ __ __ __ __ __	______H
11th	__ __ __ __ __ __ __ __	______H
12th	__ __ __ __ __ __ __ __	______H
13th	__ __ __ __ __ __ __ __	______H
14th	__ __ __ __ __ __ __ __	______H
15th	__ __ __ __ __ __ __ __	______H
16th	__ __ __ __ __ __ __ __	______H

Table 5

Part 4

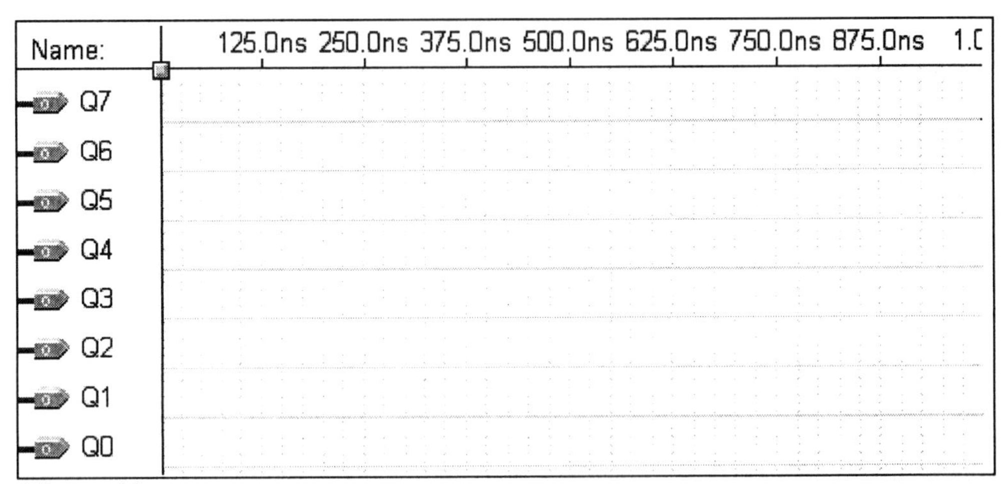

Figure 11

Clock	Q_H Q_G Q_F Q_E Q_D Q_C Q_B Q_A	Hex equiv.
1st	__ __ __ __ __ __ __ __	______H
2nd	__ __ __ __ __ __ __ __	______H
3rd	__ __ __ __ __ __ __ __	______H
4th	__ __ __ __ __ __ __ __	______H
5th	__ __ __ __ __ __ __ __	______H
6th	__ __ __ __ __ __ __ __	______H
7th	__ __ __ __ __ __ __ __	______H
8th	__ __ __ __ __ __ __ __	______H
9th	__ __ __ __ __ __ __ __	______H

Clock	Q_H Q_G Q_F Q_E Q_D Q_C Q_B Q_A	Hex equiv.
10th	__ __ __ __ __ __ __ __	______H
11th	__ __ __ __ __ __ __ __	______H
12th	__ __ __ __ __ __ __ __	______H
13th	__ __ __ __ __ __ __ __	______H
14th	__ __ __ __ __ __ __ __	______H
15th	__ __ __ __ __ __ __ __	______H
16th	__ __ __ __ __ __ __ __	______H
17th	__ __ __ __ __ __ __ __	______H

Table 6

10. ______________________________________

11. ______H ______H ______H ______H ______H ______H ______H ______H

______H ______H ______H ______H ______H ______H ______H ______H

12. Demonstrated to: ______________

Grade: __________

Date: __________

Lab 20: Tristate Logic

Objectives:

1. To examine the characteristics of the tristate logic
2. To use tristate logic on bus lines
3. To create bi-directional I/O pins

Materials List:

- Max+plus II software by Altera Corporation
- University Board with CPLD (Optional)
- Computer requirements:
  Minimum 486/66 with 8 MB RAM

Discussion:

In computers, it is often necessary for several devices to share a common data bus for transmitting or receiving data to and from various devices. Buffers, based on tristate logic, are designed to isolate one circuit from another. When the device is "ON," the output will be either a logic-HIGH or a logic-LOW, depending on the inputs to that device. When the device is "OFF," the output will exhibit a high impedance state.

The symbol of a tristate gate is distinguished from others by the addition of an input to the side of the gate. Figure 1 shows several common tristate gates that may be created inside an integrated circuit. The Max+plus II software includes a tristate buffer (D) that may be connected in series with an output to create various tristate functions. It is important that only one tristate output is active at a time when several of these devices share common signal lines (buses).

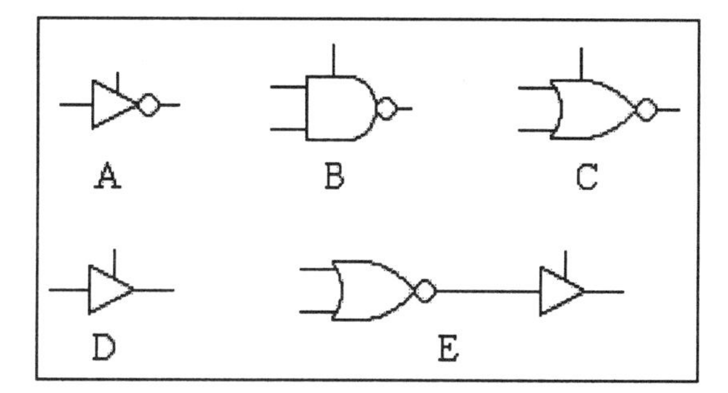

Figure 1

This lab utilizes the basic tristate buffer (Figure 1 D) to develop uni- or bidirectional bus control that may be used in microprocessor-based applications.

Part 1 Procedure

1. Open the Max+plus II software. Assign the project name **buffer1** and MAX7000S as the device family.

2. Open the Graphic Editor and construct the circuit shown in Figure 2.

3. Open the Waveform Editor, set Grid Size to 25 ns, and create the waveforms shown in Figure 3.

4. Click the Compiler and Simulator buttons. Correct all errors before continuing.

5. Sketch the resulting output waveform in the area provided at the bottom of Figure 3.

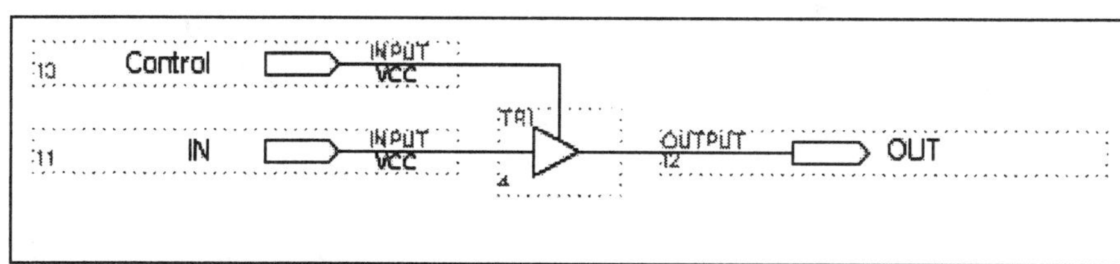

Figure 2

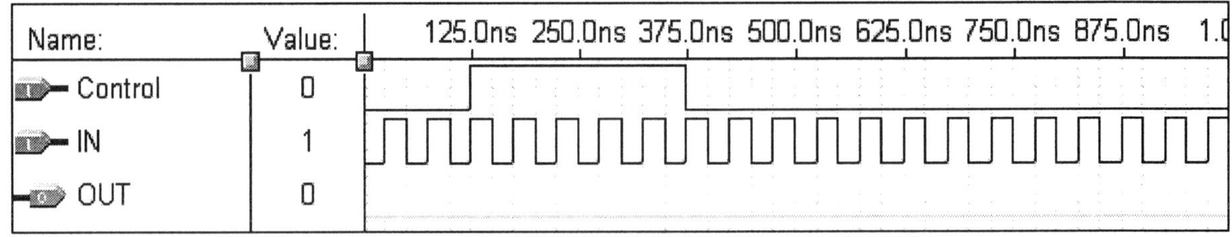

Figure 3

6. When using tristate logic, there are three logic states: a logic-LOW, a logic-HIGH, and a high impedance, Z_H, state. Which logic state is represented by the output from 0 ns to 125 ns? _____

7. Which logic state(s) is represented by the output from 125 ns to 375 ns? __________

8. Based on the recorded results, the tristate input to the buffer shown in Figure 2 is active- (LOW/HIGH)

9. Save the files to Drive A as **buffer1**, then exit the Graphic and Waveform Editors.

Part 2 Procedure

1. Open the Max+plus II software. Assign the project name **buffer2**.

2. Open the Graphic Editor and construct the circuit shown in Figure 4. Use bus lines as shown and use the BIDIR symbol for the INOUT pin designator on the right side of the top buffer.

3. Open the Waveform Editor, set Grid Size to 25 ns, and create the waveforms shown in Figure 5.

4. Click the Compiler and Simulator buttons. Correct all errors before continuing.

5. Sketch the resulting output waveform in the area provided at the bottom of Figure 5.

6. When the Control input is a logic-LOW, which buffer is turned ON? (Top/Bottom)

7. Why is INOUT a high impedance state? ______________________________

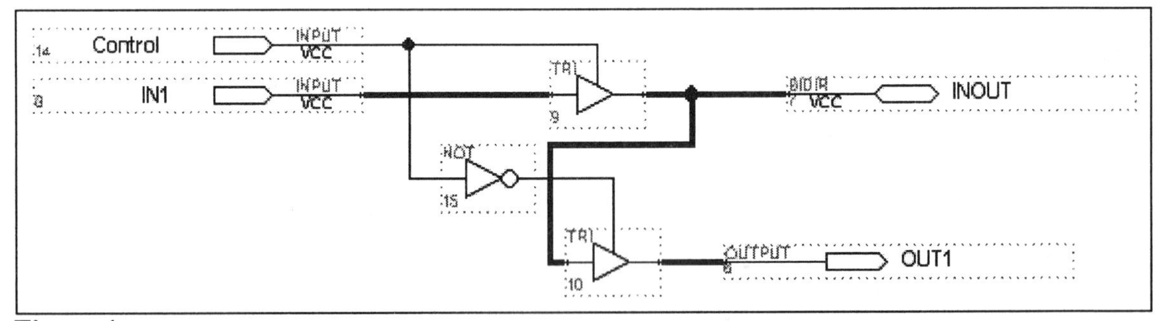

Figure 4

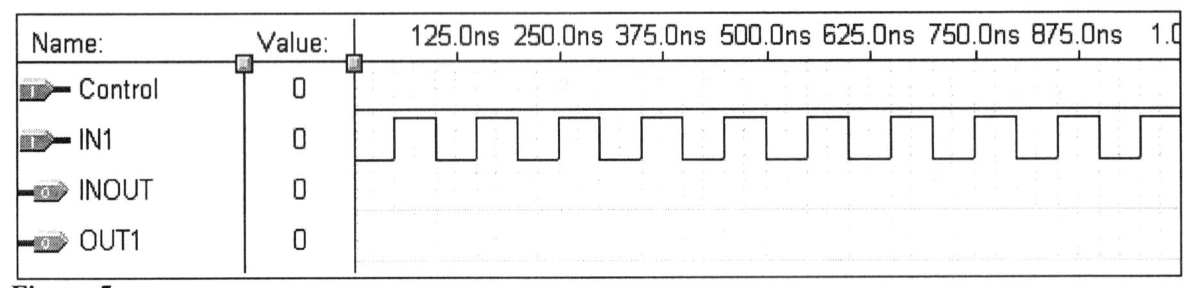

Figure 5

8. Describe the signal that appeared on the OUT1 pin connector. ______________________________

____

9. Set the Control signal to a constant logic-HIGH.

10. Click the Simulate button.

11. Which output now has valid data? (INOUT/OUT1)

12. Make the following changes to the Waveform Editor file:
    - Place the Control signal to a constant logic-LOW
    - Change the I/O type of the INOUT signal to an input pin
    - Change the grid size to 100 ns, highlight the INOUT signal, click the Overwrite Count Value button, then click OK. The INOUT signal should now be a square wave with a pulse width of 100 ns.

13. Click the Simulator button.

14. Sketch the resulting output waveforms in Figure 6.

15. Which tristate gate is turned "ON"? (Top/Bottom)

16. The OUT1 signal is from which input? (IN1/INOUT)

17. Modify your Graphic Editor file to reflect all changes as shown in Figure 7.

18. Delete the IN1, INOUT, and OUT1 waveforms from the Waveform Editor.

19. Set theGrid Size to 100 ns and add the waveforms as shown in Figure 8. (Notice Control = 1)

20. Click the Compiler and Simulator buttons. Correct all errors before continuing.

21. Sketch the resulting output waveforms in the area provided at the bottom of Figure 8.

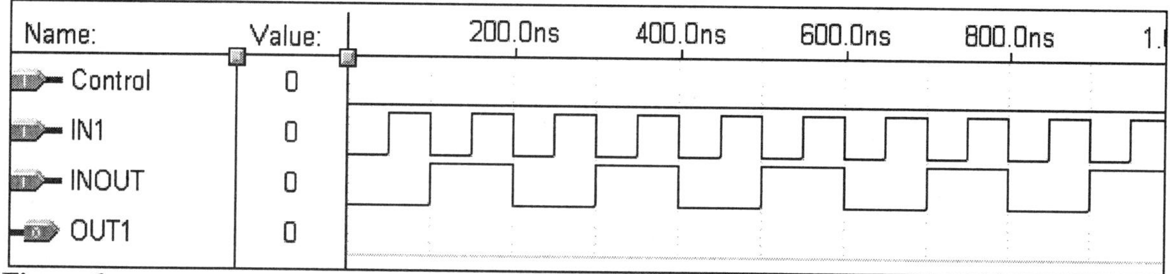

Figure 6

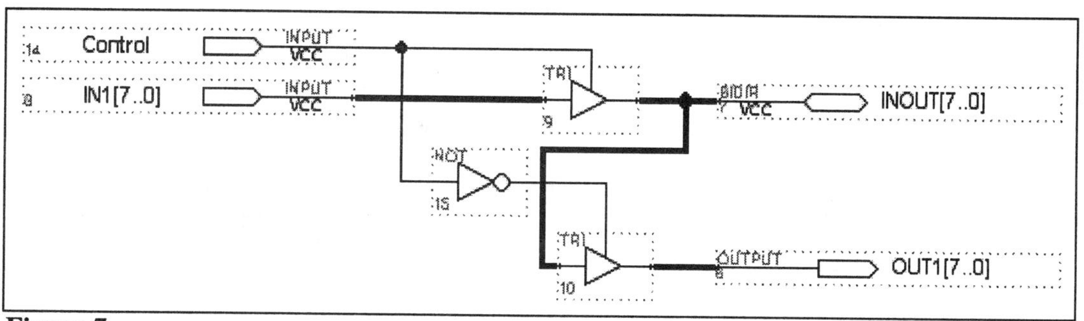

Figure 7

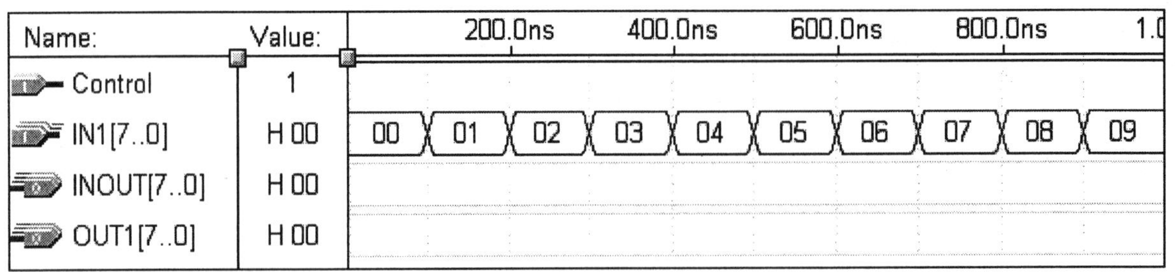

Figure 8

22. Since the tristate buffer symbol may be used with single or multiple inputs and outputs, the need for specialty buffer chips like the 74240, 74241, and 74244 are not needed for circuit designs using the Max+plus II software. However, these buffer chips may be necessary to isolate the EPM7128LC84 chip from heavy current-demanding loads.

23. Save all files to your disk in Drive A as **buffer2**, then exit the Graphic and Waveform Editors.

Part 3 Procedure

1. Open the Max+plus II software. Assign the project name **buffer3**.

2. Open the Graphic Editor and construct the circuit shown in Figure 9. Use BIDIR symbols for the INOUTA and INOUTB signals.

3. Open the Waveform Editor and create the waveforms shown in Figure 10.

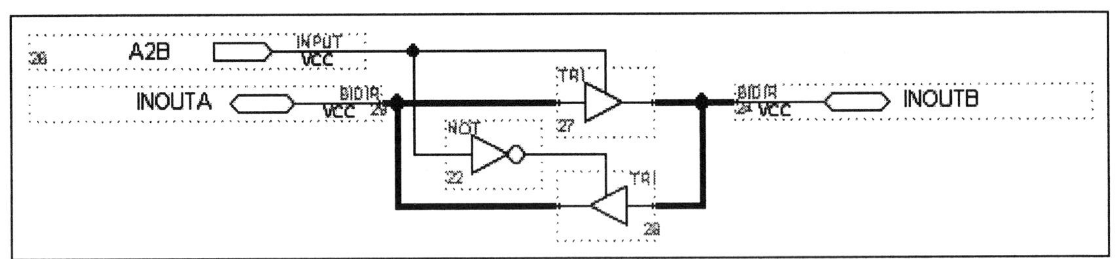

Figure 9

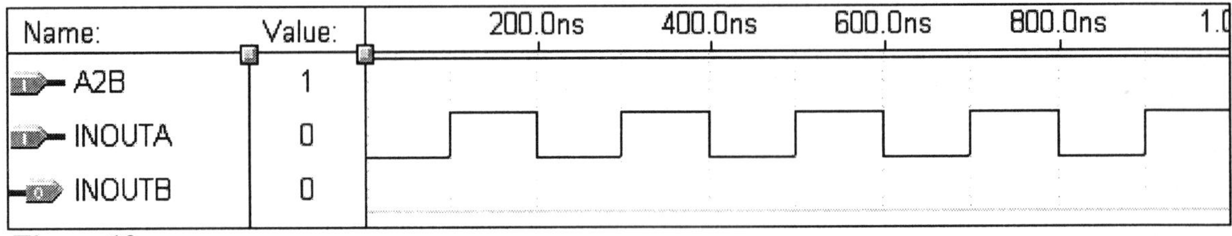

Figure 10

4. Click the Compiler and Simulator buttons. Correct all errors before continuing.

5. Sketch the resulting output waveform in the area provided at the bottom of Figure 10.

6. What error message (if any) would be displayed if the control line was a logic-LOW when the circuit was simulated?

   ____

7. Modify the waveforms as shown in Figure 11.

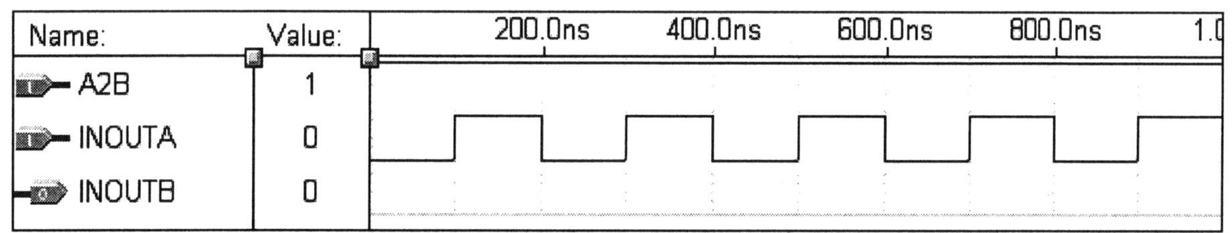

Figure 11

8. Click the simulator button, and draw the resulting output waveform in the area provided at the bottom of Figure 11.

9. When the control input A2B is a logic-LOW, data flow from input (INOUTA/INOUTB) to output (INOUTA/INOUTB).

10. When the control input A2B is a logic-HIGH, data flow from input (INOUTA/INOUTB) to output (INOUTA/INOUTB).

11. When using a TRI buffer, you must observe the following rules[1]:

    - A TRI buffer may drive only one BIDIR or BIDIRC pin. You must use a BIDIR or BIDIRC pin if feedback is included after the TRI buffer.
    - If a TRI buffer feeds logic, it must also feed a BIDIR or BIDIRC pin unless it is part of a tristate bus. If it feeds a BIDIR or BIDIRC pin, it may not feed any other outputs.

12. Obtain a hard copy of your modified circuit and simulated waveforms with the R/W control input. Label these hard copies **Part 3, Step 12A** and **Part 3, Step 12B**, respectively.

13. Demonstrate the results obtained from the circuit designed to your instructor. Obtain the signature of approval on the answer page for this lab.

14. Save both editor files to Drive A as **buffer3**, then exit the Graphic and Waveform Editors.

15. Write a 1- to 2-page technical summary pertaining to the results obtained from this lab.

16. Place all papers for this lab in the following sequence, then submit the lab to your instructor for grading.
    - Cover page
    - Typed summary
    - The completed answer page for this lab
    - Hard copy of the Graphic Editor, **Part 3, Step 12A**
    - Hard copy of the Waveform Editor, **Part 3, Step 12B**

1 From the Max+plus II Help screen for the TRI Primitive. Altera Corporation.

Lab 20: Tristate Logic Answer Pages

Name: ______________________________

Part 1

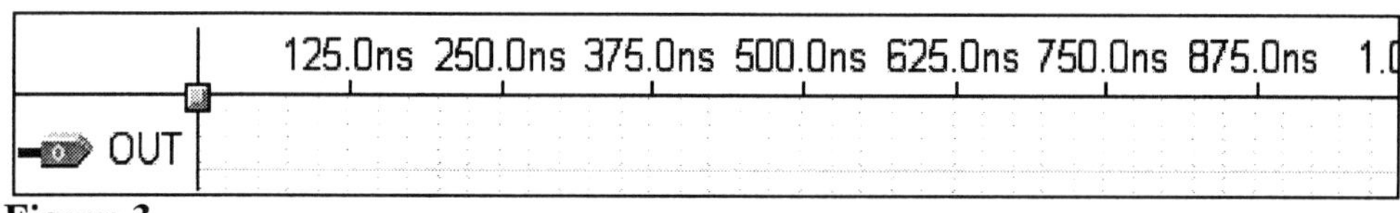

Figure 3

6. ____________ 7. ____________ 8. LOW/HIGH

Part 2

6. (Top/Bottom) 7. ____

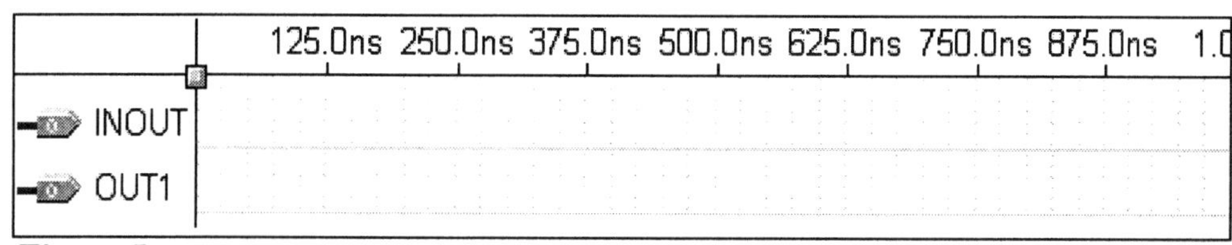

Figure 5

8. ____

11. (INOUT/OUT1) 15. (Top/Bottom) 16. (IN1/INOUT)

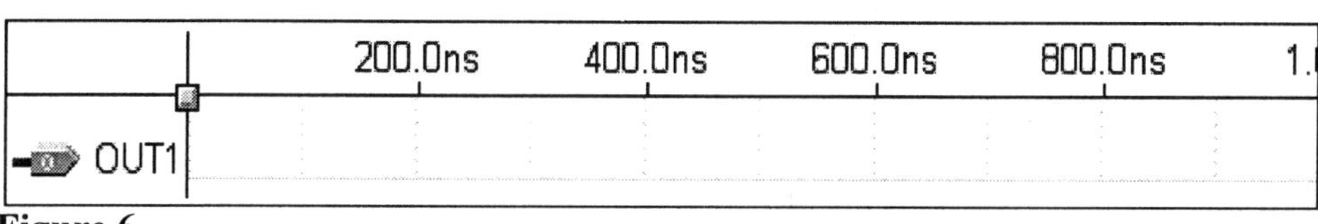

Figure 6

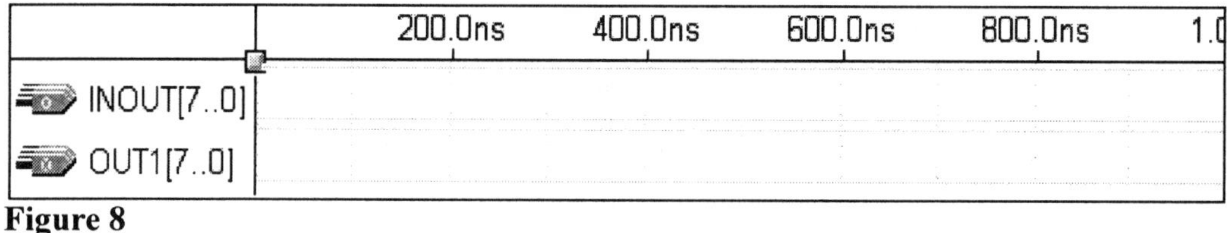

Figure 8

Part 3

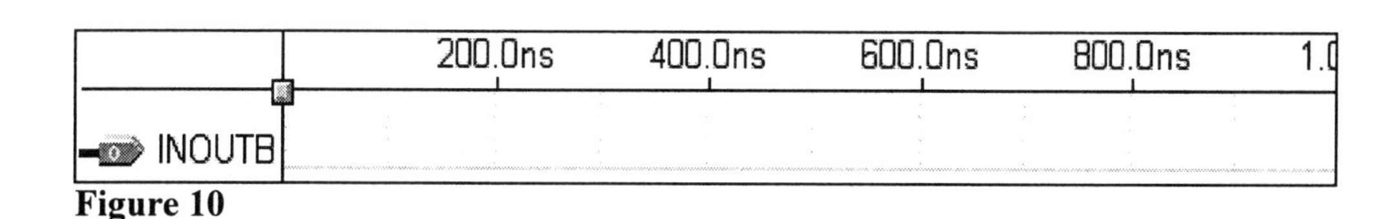

Figure 10

6. ____

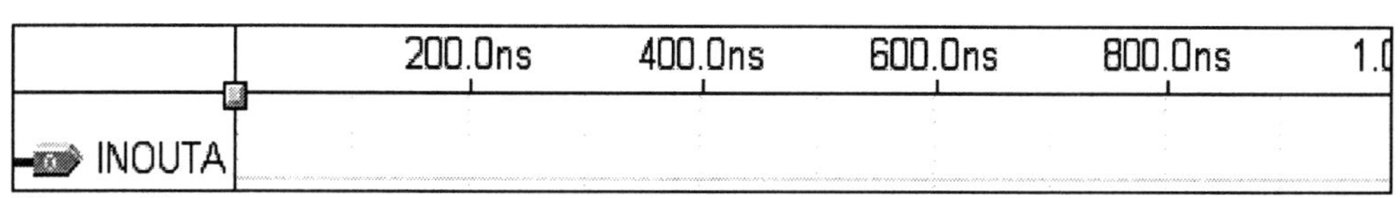

Figure 11

9. input (INOUTA/INOUTB) to output (INOUTA/INOUTB)

10. input (INOUTA/INOUTB) to output (INOUTA/INOUTB)

13. Demonstrated to: ________________________________ Date: ______________

Grade: _____________

Lab 21: The D/A Converter

Objectives:

1. Convert digital quantities to analog voltages using the DAC0808 D/A converter
2. Build a digital-controlled analog amplifier
3. Program the MAX7000S to produce different bit patterns to apply to the D/A converter for creating ramp and trapezoidal waveforms

Materials List:

- (1) 220 Ω resistor
- (4) 4.7 KΩ resistors
- (3) 10 KΩ resistors
- (5) 20 KΩ resistors
- (1) 20 μF electrolytic capacitor
- (2) Quad SPDT switches
- Power supply
- Bench generator
- (1) DAC0808 D/A converter
- (1) 741 operational amplifier
- (2) 0.1 μF capacitors
- Max+plus II software
- Computer requirements:
  Minimum 486/66 with 8 MB RAM
- Circuit card with the EPM7128SLC84
- Oscilloscope
- Digital voltmeter

Discussion:

The D/A converter converts digital quantities to an analog voltage. The simplest D/A converter with load resistor is shown in Figure 1. This resistive D/A converter is dependent on the relationship of each input resistor with respect to each other. Ideally, the most significant input, input D for the circuit shown, will have a resistance "R" and the next significant input, input C, will have a value of "2R," input B will have a resistance of "4R," and input A has a resistance of "8R."

If standard resistor values are used, R_B will probably be 39 KΩ, and R_A will be 68 KΩ or 86 KΩ. Either way, the accuracy of the converter will be affected since the relationship of the resistors is no longer R, 2R, 4R, and 8R. The use of potentiometers may increase accuracy but also increases costs. Additionally, the value of R_L should be at least ten times that of R_A so that the load doesn't affect the resistive network as various DCBA codes are applied.

Assume the DCBA code 0001_2 is applied to the inputs of Figure 1, with a logic-LOW representing 0 volts and a logic-HIGH representing 5 volts.

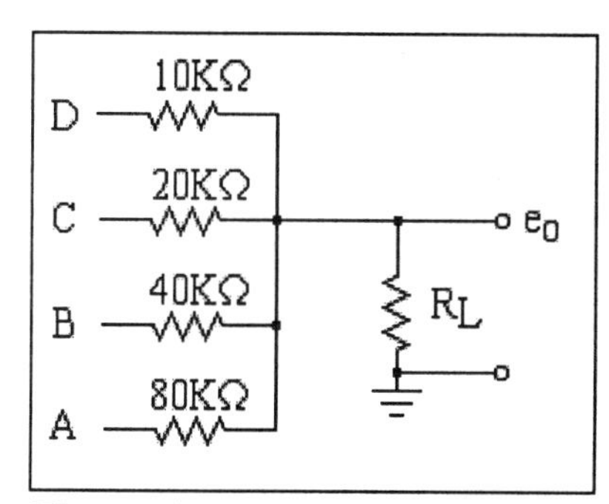

Figure 1

Figure 2 shows the equivalent circuit with input A wired to Vcc and inputs B, C, and D wired to ground. The equivalent resistance of the 10 K, 20 K, and 40 K resistors is 5714 Ω in series with the 80 KΩ resistor yields 0.333 V at the output. If 0010_2 is applied to Figure 1, the 80 KΩ and 40 KΩ resistors in Figure 2 would be swapped, giving 0.667 volts on e_o.

The ladder R/2R circuit in Figure 3 and Thevenin's equivalent in Figure 4 has an advantage over the resistive network in that only two resistor values are required. If the DCBA code 0001_2 is applied to the inputs of the R/2R network, the output voltage will be 0.3125 V. A voltage on input D will have a greater effect of the output than a voltage applied to inputs C, B, or A.

Once the step voltage is determined (0.333 V for Figure 1 or 0.3125 V for Figure 2) the output voltage for any DCBA code applied may be calculated by multiplying the decimal equivalent of the number applied by the step voltage. Otherwise, doubling the binary value doubles the output voltage. If the input binary number increases by a factor of 4, the output increases by a factor of 4.

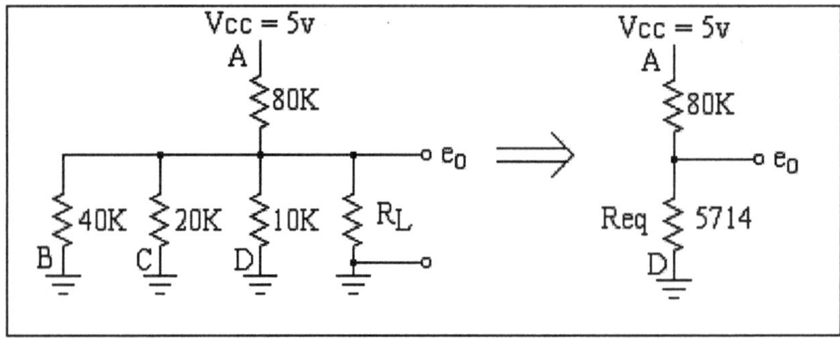

Figure 2

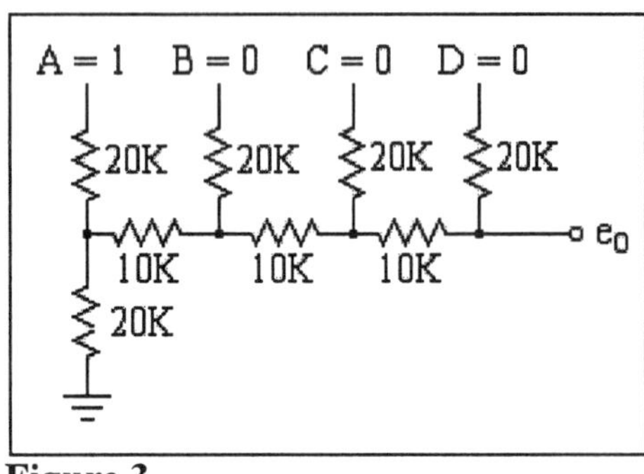

Figure 3

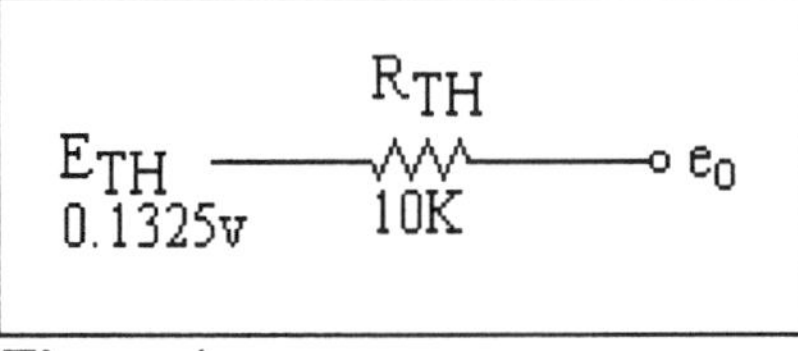

Figure 4

The bit resolution, *N*, is the total number of binary digits in the converter. For both Figure 1 and Figure 3, the bit resolution is 4.

The voltage resolution, sometimes called the step voltage, E_{STEP}, may be calculated by the following equation:

For the resistive network:

$$E_{STEP} = \frac{V_{OH} - V_{OL}}{2^N - 1} \qquad (1)$$

For the ladder R/2R network:

$$E_{STEP} = \frac{V_{OH} - V_{OL}}{2^N} \qquad (2)$$

(Source output voltage for a logic-HIGH, V_{OH}, or a logic-LOW, V_{OL}.)

For any D/A converter, the output voltage for any given input number may be determined by:

$$E_{OUT} = E_{STEP} \times \text{Number Applied (IN DECIMAL)} \qquad (3)$$

In integrated form, D/A converters come in various bit sizes from 8 to 16 bits. This lab focuses mainly on the 8-bit DAC0808 D/A converter by National Semiconductors. Product specification sheets are available online at *http://www.national.com*. Enter DAC0808 in the search engine on National's site to locate the specific data sheets for the digital-to-analog converter used in this lab.

The DAC0808 (Figure 5) has a bit resolution of 8 and operates with a maximum of +18 and − 18 VDC. The DAC0808 has an internal R/2R ladder D/A converter so equations 2 and 3 may be used to calculate the step voltage and output voltage for any binary number applied. Full scale output current (max 2 mA), I_{FS}, is determined by the selection of V_{REF} and R_{REF} attached to pin 14 of the IC. If V_{REF} is 5 V and R_{REF} is 4.7 K, I_{FS} will be 1.064 mA. The 1.064 mA current flowing through the 4.7 K feedback resistor of the inverting operational amplifier shown in Figure 5 produces +5 V on the amplifier output. The operational amplifier protects the D/A converter from load variations.

The output current, and thus output voltage of the operational amplifier, may be determined for any binary number applied to the D/A converter using the binary weighted equation shown in the lower right of Figure 5. As long as the feedback resistor, R_F, in the inverting amplifier is identical to the reference resistor at pin 14 of the D/A converter, equations 2 and 3 may also be used to determine the output voltage of Figure 5.

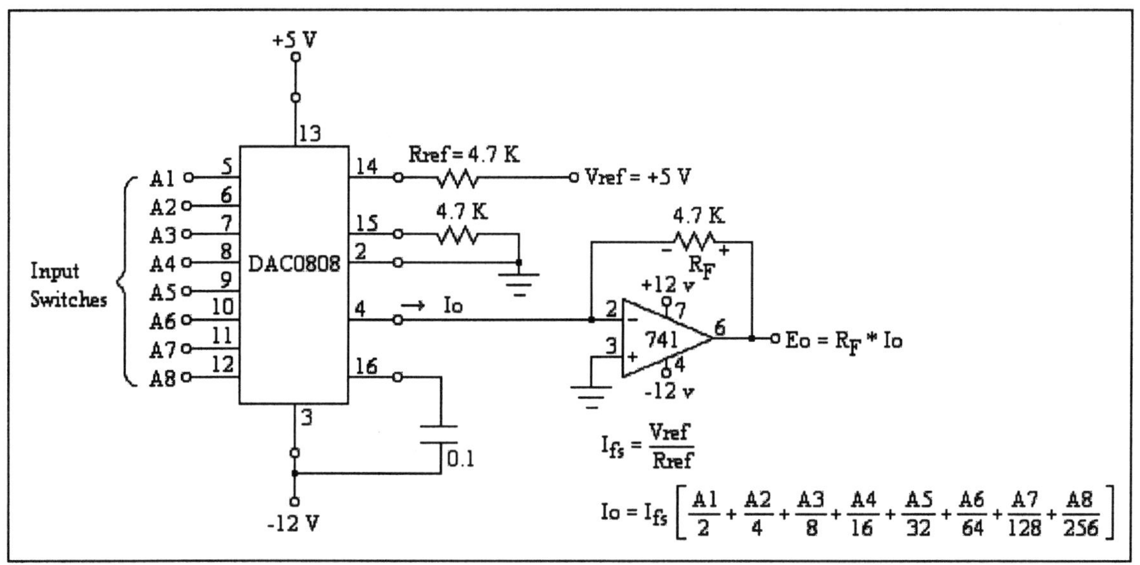

Figure 5

Part 1 Procedure

1. Construct the circuit shown in Figure 3. Wire the data inputs D, C, B, and A to switches providing 5 V for a logic-HIGH and 0 V for a logic-LOW.

2. Calculate the output voltage to the nearest millivolt of Figure 3 for the different DCBA input combinations. Record your calculated answers in Table 1.

3. Complete Table 1 by recording the measured output voltage for each DCBA input code shown in Table 1.

DCBA Code	Calculated Output	Measured Output
0 0 0 0	______	______
0 0 0 1	______	______
0 0 1 0	______	______
0 0 1 1	______	______
0 1 0 0	______	______
0 1 0 1	______	______
0 1 1 0	______	______
0 1 1 1	______	______
1 0 0 0	______	______
1 0 0 1	______	______
1 0 1 0	______	______
1 0 1 1	______	______
1 1 0 0	______	______
1 1 0 1	______	______
1 1 1 0	______	______
1 1 1 1	______	______

Table 1

Part 2 Procedure

1. Construct the D/A converter circuit shown in Figure 5.

2. Calculate and record the output voltage (to the nearest millivolt) for each HGFEDCBA bit pattern applied to Figure 5 as shown in Table 2.

3. Set the data switches of Figure 5 to each bit combination shown in Table 2 and record the measured output voltages.

D/A Inputs (MSB) A_1	A_2	A_3	A_4	A_5	A_6	A_7	A_8 (LSB)	Calculated Output, Eo	Measured Output, Eo
0	0	0	0	0	0	0	0	______	______
0	0	0	0	0	0	0	1	______	______
0	0	0	0	0	0	1	0	______	______
0	0	0	0	0	1	0	0	______	______
0	0	0	0	1	0	0	0	______	______
0	0	0	1	0	0	0	0	______	______
0	0	1	0	0	0	0	0	______	______
0	1	0	0	0	0	0	0	______	______
1	0	0	0	0	0	0	0	______	______
1	1	1	1	1	1	1	1	______	______

Table 2

4. As the magnitude represented by the binary pattern increases, the output voltage (decreases/increases).

5. Do not disassemble your circuit.

Part 3 Procedure

1. Re-wire pin 14 of your circuit as shown in Figure 6.

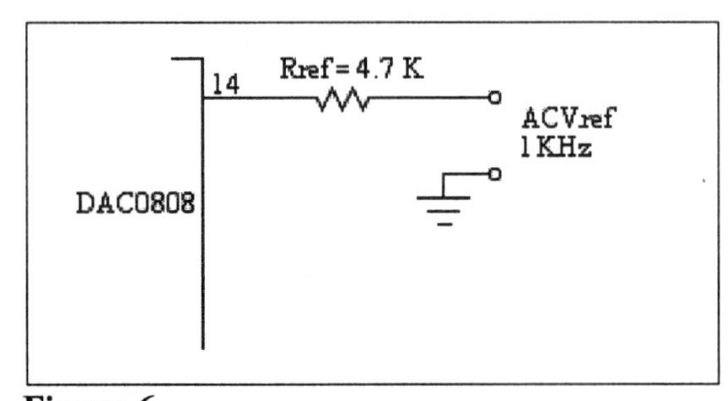

Figure 6

2. Set the bench signal generator for a 1 KHz sine wave with a low voltage amplitude.

3. Set all data switches for full scale, FF_H, output of the D/A converter.

4. Monitor the ACV at the output of the operational amplifier using an oscilloscope.

5. Vary the bench signal generator amplitude and DC offset controls until the observed sine wave at the amplifier output is at its maximum amplitude without distortion.

6. Use the oscilloscope to measure and record the bench signal generator sine wave amplitude and DC offset that yielded the maximum amplifier output amplitude without distortion. ACV: ____________ V (p-p)

   DC offset: ____________ V

7. Monitor and record (Table 3) the output of the operational amplifier with the digital voltmeter measuring DCV and the oscilloscope measuring ACV as each input switch is set to a logic-LOW. Be sure to change the meter and scope sensitivity settings for accurate readings as each switch is toggled to a logic-LOW.

8. You should notice that the amplitude of the AC signal at the output of the operational amplifier is determined by the digital bit pattern applied to the D/A converter. A computer operator inserts a CD-ROM or DVD-ROM into a player, increases the on-screen volume control, which in turn increases the digital code applied to a D/A converter, and the analog sound coming out of the speaker increases.

D/A Inputs								Measured E_O ACV	Measured E_O DCV
(MSB) A_1	A_2	A_3	A_4	A_5	A_6	A_7	A_8 (LSB)		
1	1	1	1	1	1	1	1	________	________
0	1	1	1	1	1	1	1	________	________
0	0	1	1	1	1	1	1	________	________
0	0	0	1	1	1	1	1	________	________
0	0	0	0	1	1	1	1	________	________
0	0	0	0	0	1	1	1	________	________
0	0	0	0	0	0	1	1	________	________
0	0	0	0	0	0	0	1	________	________
0	0	0	0	0	0	0	0	________	________

Table 3

Part 4 Procedure

A ramp signal, either positive stepping or negative stepping, can be created by applying the output of a binary counter to the digital inputs of a D/A converter.

1. Connect the circuit card with the EPM7128SLC84 chip to the computer's printer port. If the computer has a software key attached to LPT1, connect the card to the parallel port, LPT2. See your instructor to determine the correct connection port.

2. Open the Max+plus II software. Assign the project name **dabinctr**, MAX7000S as the device family, and the EPM7128SLC84 as the device.

3. Open the Graphic Editor and construct the circuit shown in Figure 7. Assign pins according to the user manual for the circuit card you are soon to program.

4. Open the Waveform Editor, set Grid Size to 20 ns, and create the waveforms shown in Figure 8.

5. Click the Compiler and Simulator buttons. Correct all errors before continuing.

6. Save all files as **dabinctr** to your disk in Drive A.

7. Select Programmer in the Max+plus II menu item (left of the File option).

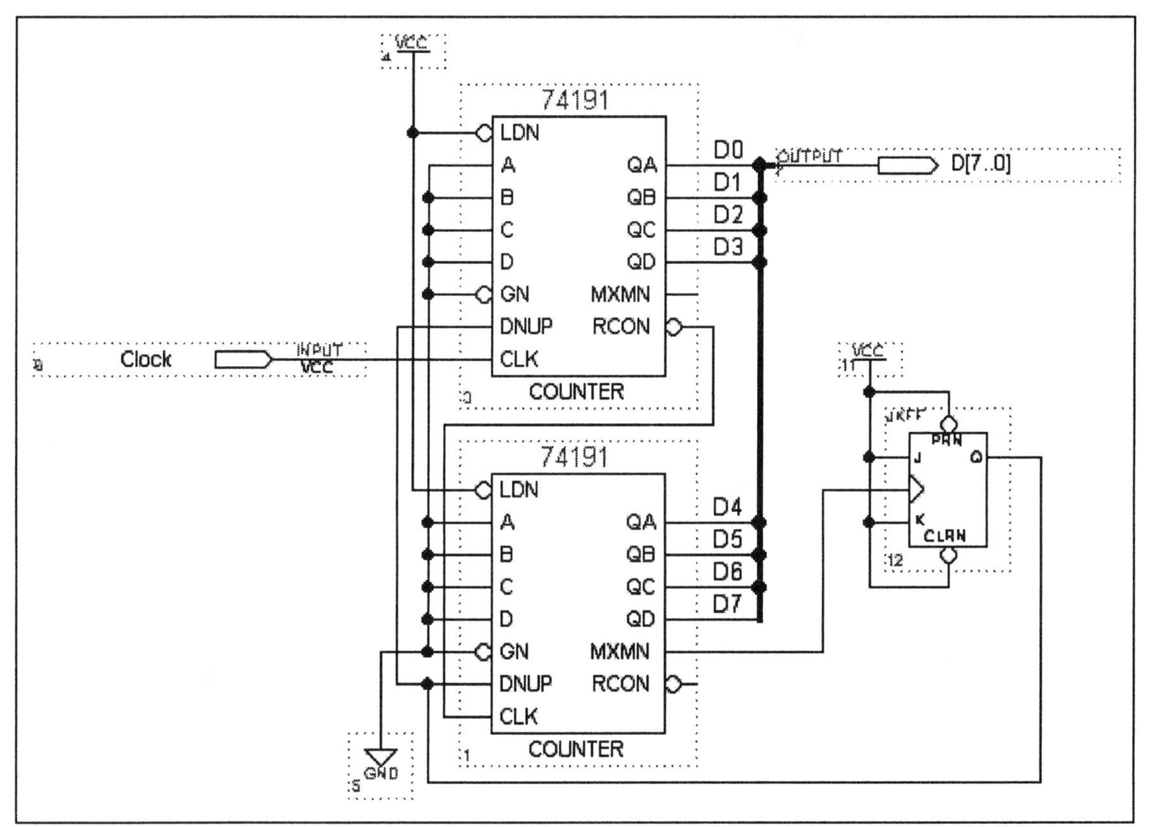

Figure 7

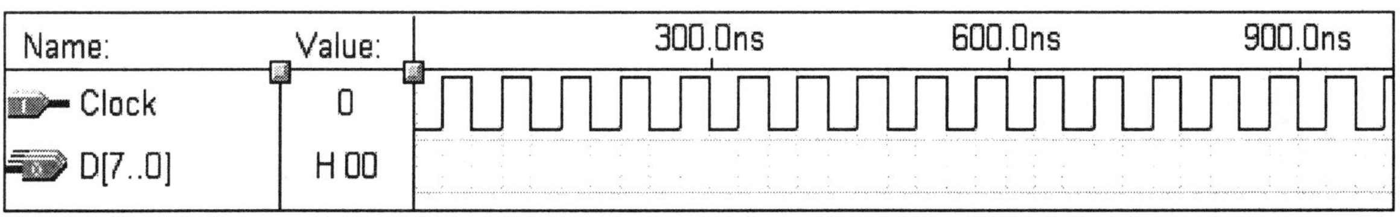

Figure 8

8. Proceed to program the EPM7128SLC84 chip following the instructions in Appendix C or refer to the circuit board manufacturer's programming instructions.

9. Save all files to your disk in Drive A as **dabinctr**, then exit the Graphic and Waveform Editors.

10. Assuming the board was successfully programmed, construct the circuit shown in Figure 9. Refer to the **dabinctr.rpt** file on your disk that was created during compilation for pin assignments of the EPM7128SLC84 IC.

11. Apply a 10 KHz square wave clock (with a zero volt DC offset) to the circuit card containing the EPM7128LC84 chip. See the board manufacturer's programming notes for switch settings when applying an external clock signal.

12. Monitor the output of the operational amplifier with an oscilloscope, then neatly and accurately sketch the resulting output waveform in Figure 10. Identify the amplitude and time base for the waveform sketched.

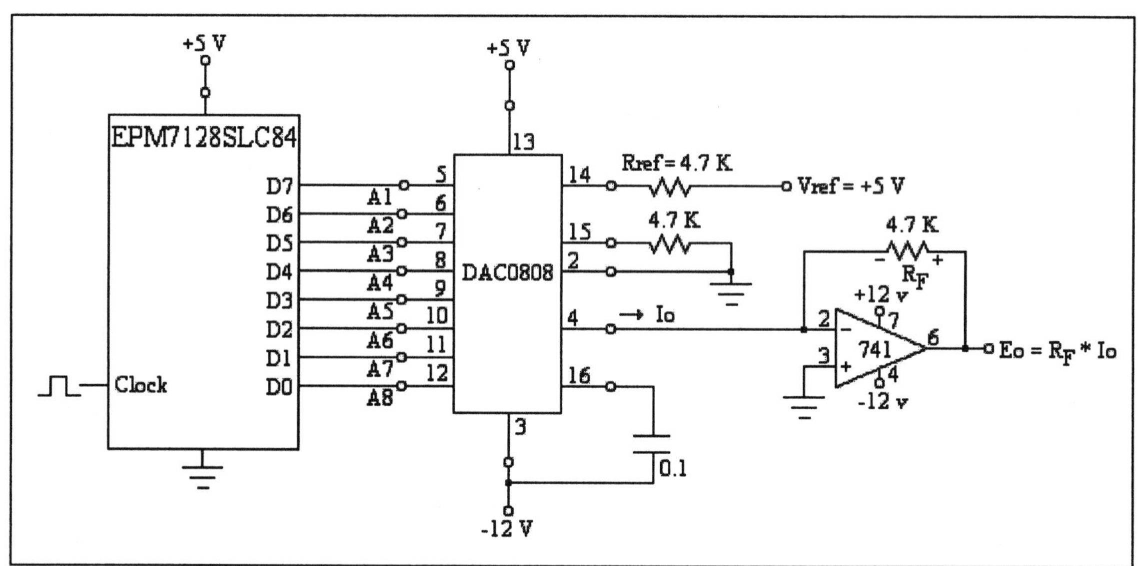

Figure 9

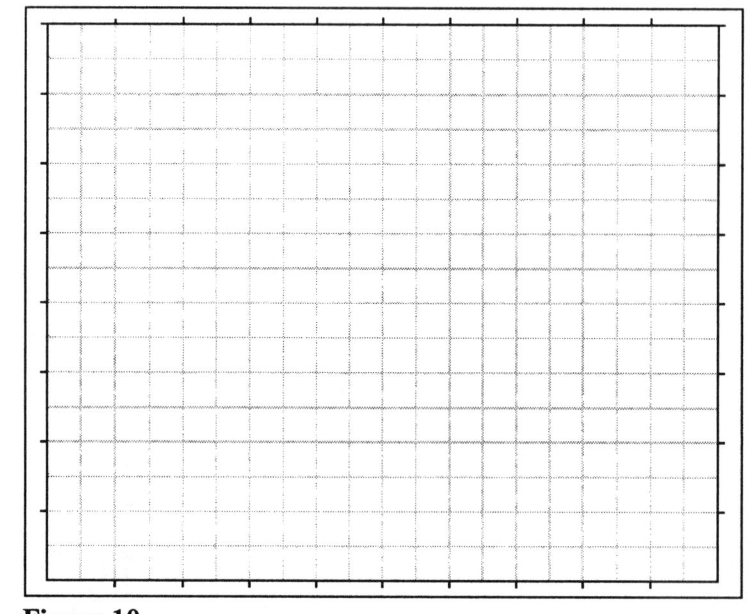

Figure 10

a. Show the ground reference

b. $+E_{PEAK}$ = ____________ V

c. $-E_{PEAK}$ = ____________ V

d. Time for 1 cycle: ____________

e. Frequency of output: ____________

Part 5 Procedure

Assume we want to create a 100% modulated waveform with an amplitude of 0 V to 5 V. Let's also take 16 samples of the modulated waveform in one complete cycle. This equates to one sample every 22.5 degrees (360 ÷ 16). From the sketch of the modulated waveform (see Figure 11) we can see that the envelope of the waveform appears to be two sine waves: one on a 1.25 VDC reference and the other on a 3.75 VDC reference. Also notice that every other sample is taken from the top envelope of the modulated waveform.

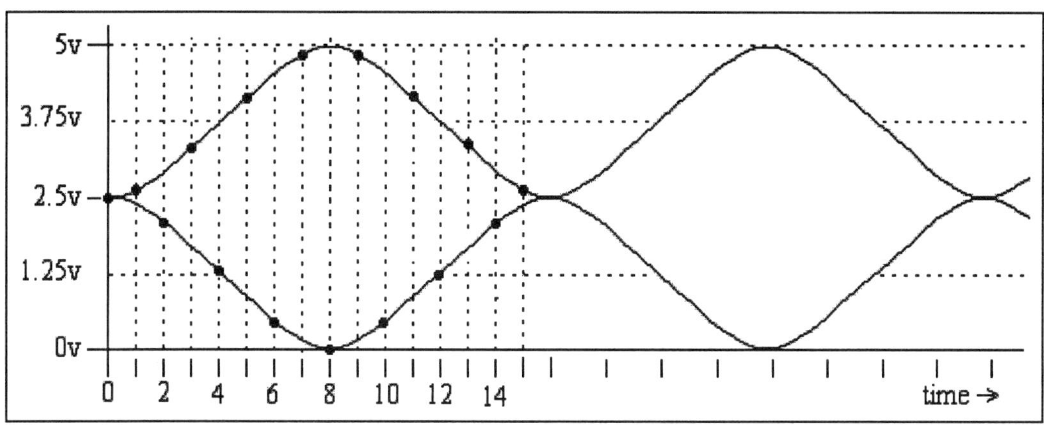

Figure 11

1. Calculate the amplitude of each point on the waveform using the following equation: $E_{INIT} = E_{PEAK} \sin \theta + E_{REF}$. The peak voltage for each sine wave is 1.25 V. The DC reference for the top sine wave is 3.75 V and the DC reference for the bottom sine wave is 1.25 V. Complete the calculations in Table 4 to determine the magnitude of each point on the waveforms.

Time	Degree	$E_{INIT} = E_{PK} \sin \theta + E_{REF}$ (Step 1)	÷ E_{STEP} (Step 2)	Hex Equivalent	Binary hgfedcba
0	90	1.25 V × sin90+1.25 = ________V	______	______	______
1	292.5	1.25 V × sin292.5+3.75 = ________V	______	______	______
2	135	1.25 V × sin135+1.25 = ________V	______	______	______
3	337.5	1.25 V × sin337.5+3.75 = ________V	______	______	______
4	180	1.25 V × sin180+1.25 = ________V	______	______	______
5	22.5	1.25 V × sin22.5+3.75 = ________V	______	______	______
6	225	1.25 V × sin225+1.25 = ________V	______	______	______
7	67.5	1.25 V × sin67.5+3.75 = ________V	______	______	______
8	270	1.25 V × sin270+1.25 = ________V	______	______	______
9	112.5	1.25 V × sin112.5+3.75 = ________V	______	______	______
10	315	1.25 V × sin315+1.25 = ________V	______	______	______
11	157.5	1.25 V × sin157.5+3.75 = ________V	______	______	______
12	0	1.25 V × sin0+1.25 = ________V	______	______	______
13	202.5	1.25 V × sin202.5+3.75 = ________V	______	______	______
14	45	1.25 V × sin45+1.25 = ________V	______	______	______
15	247.5	1.25 V × sin247.+3.75 = ________V	______	______	______

Table 4

2. Calculate E_{STEP} for the DAC0808 using equation 2, then divide each amplitude recorded for Step 1 in Table 4 by the voltage resolution of the DAC0808 D/A converter. Record the answers in the fourth column in Table 4.

3. Convert each answer recorded for Step 2 to hexadecimal and show the binary equivalents. Record these conversions in Table 4.

4. If the binary equivalents are applied to the D/A converter inputs at a given rate, the waveform shown in Figure 11, theoretically, will be created.

5. Open the Max+plus II software. Assign the project name **am-mod**, assign MAX7000S as the device family, and assign EPM7128SLC84 as the device.

6. Open the Graphic Editor and construct the circuit shown in Figure 12 (next page).

7. Assign pins for the inputs and outputs according to the user manual for the circuit board you will soon program.

8. Add Figure 13 to the bottom of Figure 12. Connect the bus line in Figure 13 to the bus line in Figure 12.

9. Open the Waveform Editor, set Grid Size to 30 ns, and create the waveforms shown in Figure 14.

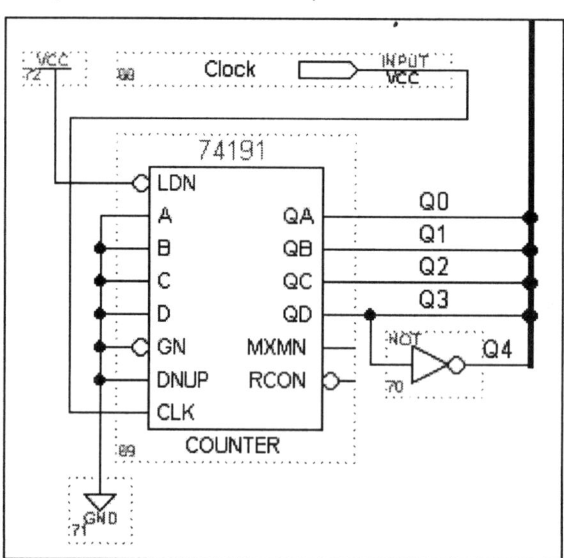

Figure 13

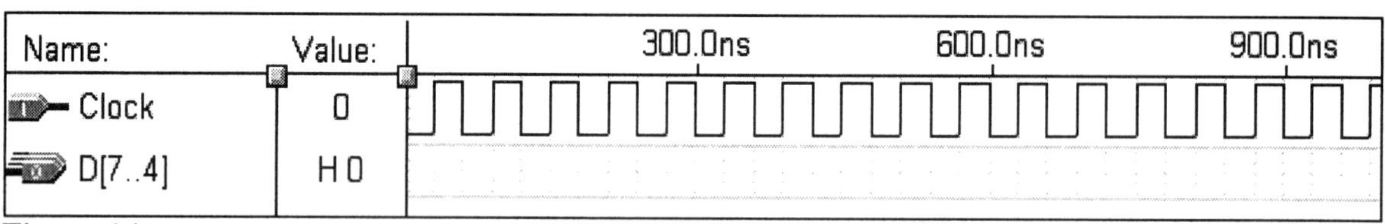

Figure 14

10. Click the Compiler and Simulator buttons. Correct all errors before continuing.

11. Neatly and accurately draw the resulting output waveform in the area provided in Figure 14.

12. Connect the circuit card with the EPM7128SLC84 chip to the computer's printer port. If the computer has a software key attached to LPT1, connect the card to the parallel port, LPT2. See your instructor to determine the correct connection port.

13. Select Programmer in the Max+plus II menu item (left of the File option).

14. Proceed to program the EPM7128SLC84 chip following the instructions in Appendix C or refer to the circuit board manufacturer's programming instructions.

15. Save all files to your disk in Drive A as **am-mod**, then exit the Graphic and Waveform Editors.

16. Assuming the board was successfully programmed, construct the circuit shown in Figure 15. Refer to the **am-mod.rpt** file on your disk that was created during compilation for pin assignments of the EPM7128SLC84 IC.

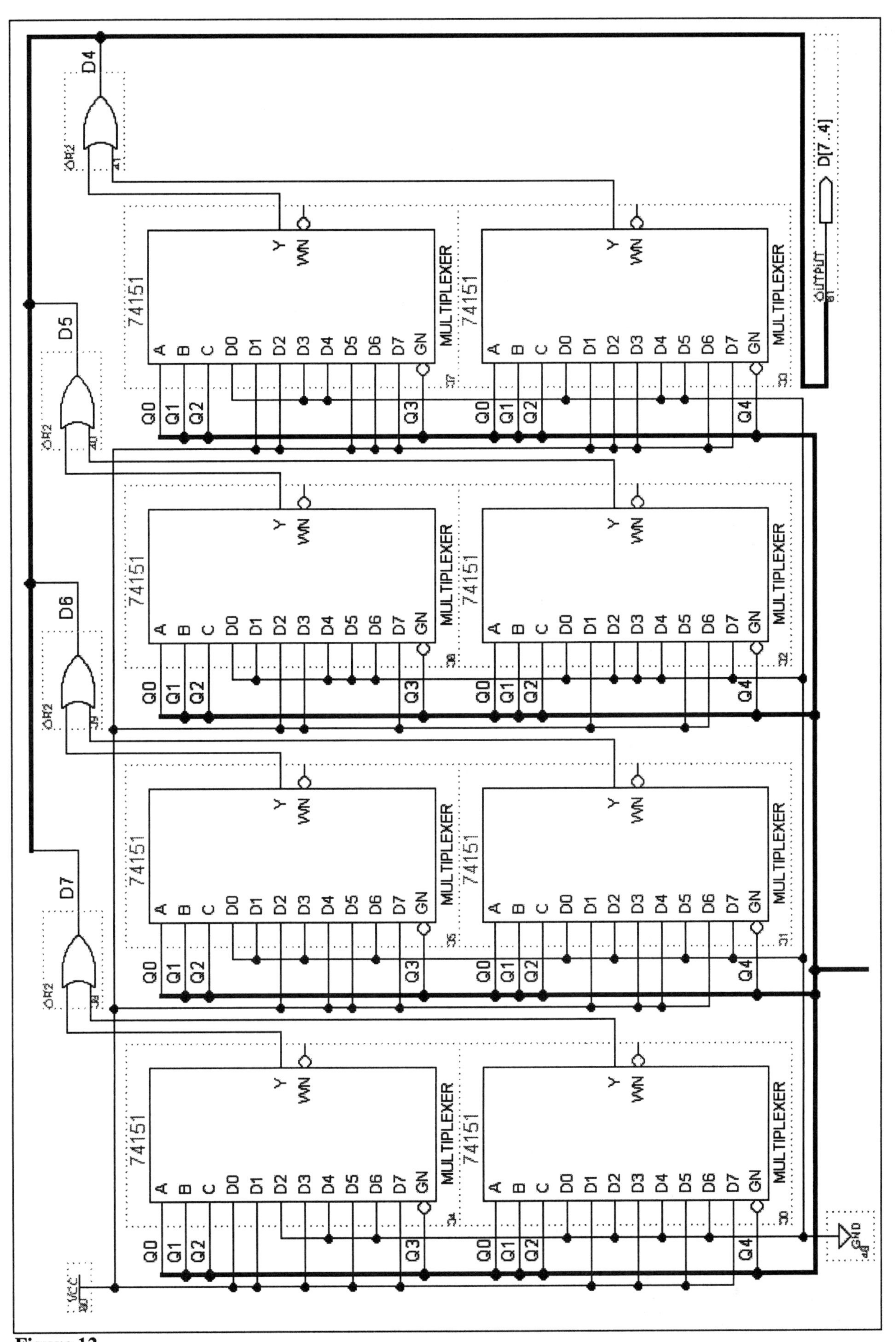
D4
D5
D6
D7
OR2
74151
MULTIPLEXER
A
B
C
D0
D1
D2
D3
D4
D5
D6
D7
GN
Y
WN
Q0
Q1
Q2
Q3
Q4
OUTPUT
D[7..4]
VCC
GND

Figure 12

17. Apply a 10 KHz square wave clock (with a zero volt DC offset) to the circuit card containing the EPM7128LC84 chip. See the board manufacturer's programming notes for switch settings when applying an external clock signal.

18. Monitor the output of the operational amplifier with an oscilloscope, then neatly and accurately sketch the resulting output waveform in Figure 16. Do identify the amplitude and time base for the waveform sketched.

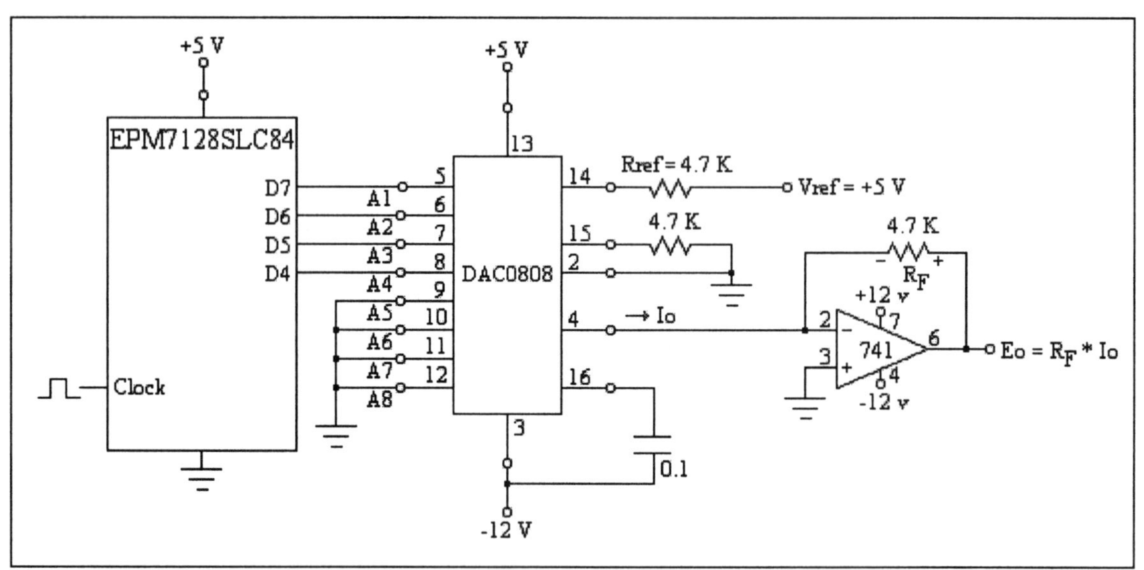

Figure 15

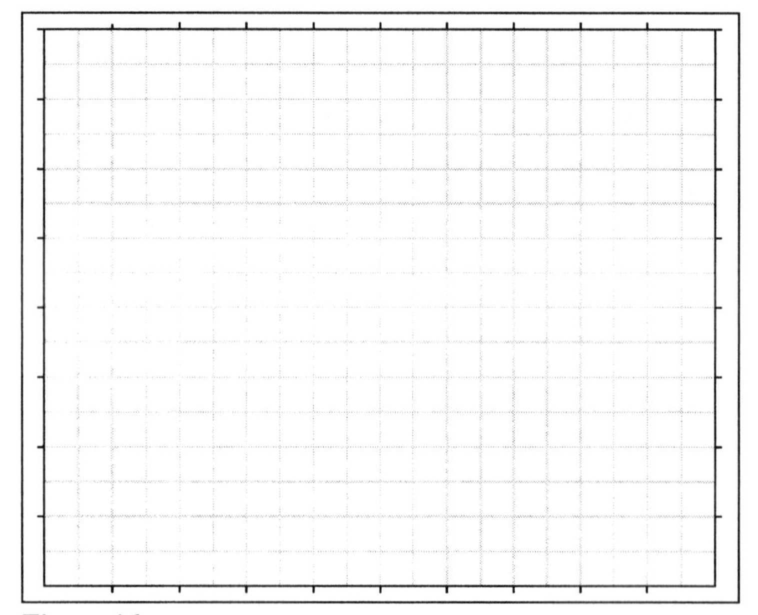

Figure 16

a. Show the ground reference

b. $+E_{PEAK}$ = ____________V

c. $-E_{PEAK}$ = ____________V

d. Time for 1 cycle: ________

e. Frequency of output: _____

19. Demonstrate your functional circuit and displayed waveforms to your instructor. Obtain the signature of approval directly on the answer page for this lab.

20. Write a 1- to 2-page technical summary of this lab. Include a paragraph based on each part of the lab and an embedded graphic of the waveforms for Figure 14.

21. Submit the following for grading:
    - Cover page
    - Typed summary
    - The completed answer page for this lab

Lab 21: The D/A Converter Answer Pages

Name: ______________________

Part 1

DCBA Code	Calculated Output	Measured Output
0 0 0 0	______	______
0 0 0 1	______	______
0 0 1 0	______	______
0 0 1 1	______	______
0 1 0 0	______	______
0 1 0 1	______	______
0 1 1 0	______	______
0 1 1 1	______	______
1 0 0 0	______	______
1 0 0 1	______	______
1 0 1 0	______	______
1 0 1 1	______	______
1 1 0 0	______	______
1 1 0 1	______	______
1 1 1 0	______	______
1 1 1 1	______	______

Table 1

Part 2

D/A Inputs								Calculated	Measured
A_1	A_2	A_3	A_4	A_5	A_6	A_7	A_8	E_O	E_O
0	0	0	0	0	0	0	0	______	______
0	0	0	0	0	0	0	1	______	______
0	0	0	0	0	0	1	0	______	______
0	0	0	0	0	1	0	0	______	______
0	0	0	0	1	0	0	0	______	______
0	0	0	1	0	0	0	0	______	______
0	0	1	0	0	0	0	0	______	______
0	1	0	0	0	0	0	0	______	______
1	0	0	0	0	0	0	0	______	______
1	1	1	1	1	1	1	1	______	______

Table 2

4. (decreases/increases)

Part 3

6. ____________

D/A Inputs								Measured E_O	Measured E_O
A_1	A_2	A_3	A_4	A_5	A_6	A_7	A_8	ACV	DCV
1	1	1	1	1	1	1	1	______	______
0	1	1	1	1	1	1	1	______	______
0	0	1	1	1	1	1	1	______	______
0	0	0	1	1	1	1	1	______	______
0	0	0	0	1	1	1	1	______	______
0	0	0	0	0	1	1	1	______	______
0	0	0	0	0	0	1	1	______	______
0	0	0	0	0	0	0	1	______	______
0	0	0	0	0	0	0	0	______	______

Table 3

Part 4

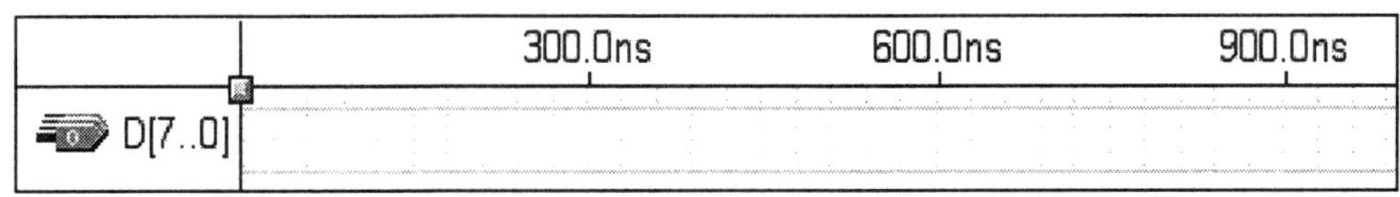

Figure 8

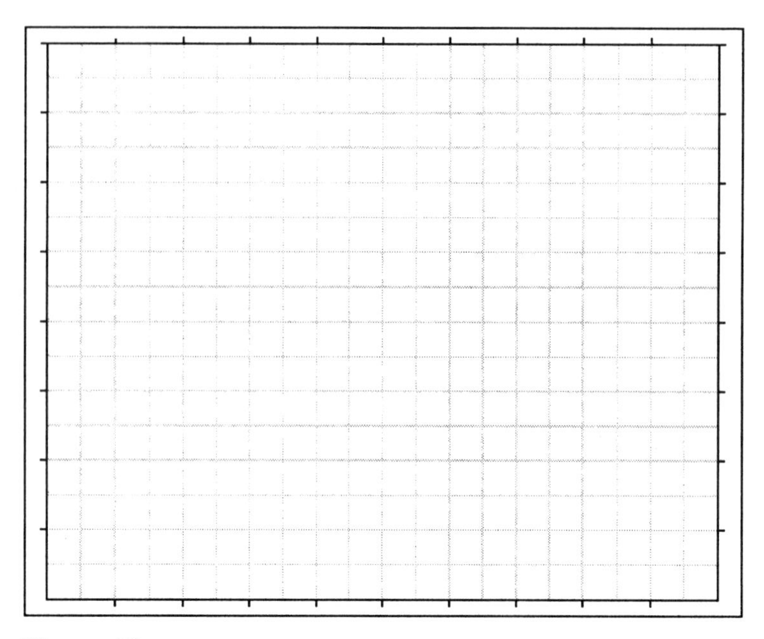

Figure 10

a. Show the ground reference

b. $+E_{PEAK}$ = ____________ V

c. $-E_{PEAK}$ = ____________ V

d. Time for 1 cycle: ____________

e. Frequency of output: __________

Time	Degree	$E_{INIT} = E_{PK}\,\mathrm{Sin}\,\theta + E_{REF}$ (Step 1)	÷ E_{STEP} (Step 2)	Hex Equivalent	Binary hgfedcba
0	90	1.25 V × sin90+1.25 = __________V	_______	_______	_______
1	292.5	1.25 V × sin292.5+3.75 = __________V	_______	_______	_______
2	135	1.25 V × sin135+1.25 = __________V	_______	_______	_______
3	337.5	1.25 V × sin337.5+3.75 = __________V	_______	_______	_______
4	180	1.25 V × sin180+1.25 = __________V	_______	_______	_______
5	22.5	1.25 V × sin22.5+3.75 = __________V	_______	_______	_______
6	225	1.25 V × sin225+1.25 = __________V	_______	_______	_______
7	67.5	1.25 V × sin67.5+3.75 = __________V	_______	_______	_______
8	270	1.25 V × sin270+1.25 = __________V	_______	_______	_______
9	112.5	1.25 V × sin112.5+3.75 = __________V	_______	_______	_______
10	315	1.25 V × sin315+1.25 = __________V	_______	_______	_______
11	157.5	1.25 V × sin157.5+3.75 = __________V	_______	_______	_______
12	0	1.25 V × sin0+1.25 = __________V	_______	_______	_______
13	202.5	1.25 V × sin202.5+3.75 = __________V	_______	_______	_______
14	45	1.25 V × sin45+1.25 = __________V	_______	_______	_______
15	247.5	1.25 V × sin247.5+3.75 = __________V	_______	_______	_______

Table 4

	300.0ns	600.0ns	900.0ns
D[7..4]			

Figure 14

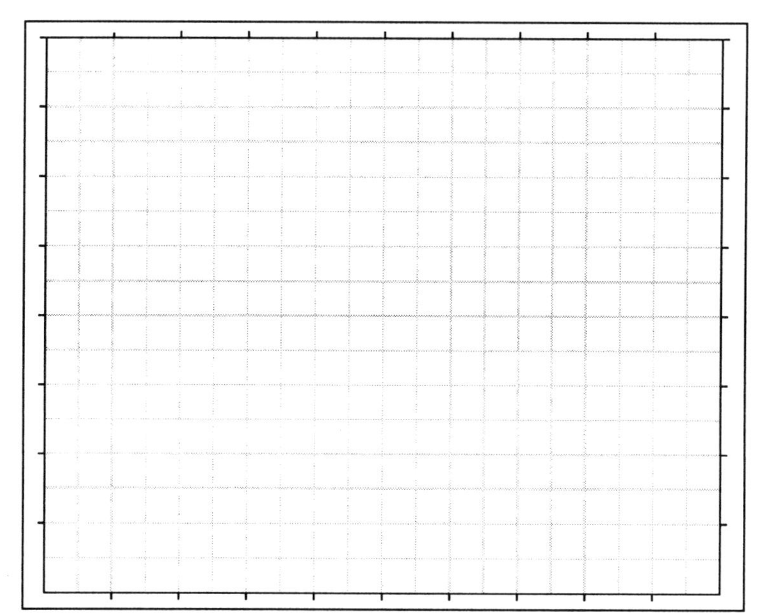

Figure 16

a. Show the ground reference

b. $+E_{PEAK}$ = ____________V

c. $-E_{PEAK}$ = ____________V

d. Time for 1 cycle: ________

e. Frequency of output: _____

19. Demonstrated to: ______________________________________ Date: ___________

Grade: _________

Lab 22: The A/D Converter

Objectives:

1. Construct and demonstrate the operation of an A/D converter in free running mode
2. Convert temperature variations to a digital pattern
3. Construct and test a network containing both D/A and A/D converters

Materials list:

- Signal generator
- Power supply
- Dual trace oscilloscope
- ADC0804 A/D converter
- (1) 10 KΩ potentiometer
- (1) Thermistor
- (1) 741 op-amp
- (8) Light emitting diodes
- (1) 0.1 μF capacitor
- (1) 0.001 μF capacitor
- (9) 1 KΩ resistors
- (3) 4.7 KΩ resistors
- (1) 10 KΩ resistor
- (5) 100 KΩ resistors
- (1) Pushbutton switch
- (8) SPDT switches

Discussion:

The analog-to-digital (A/D) converter converts analog quantities to digital values. The 8-bit converter shown in Figure 1 may be used as an I/O port on a microprocessor or as a stand-alone unit. The A/D has one active-LOW output and three active-LOW input control signals.

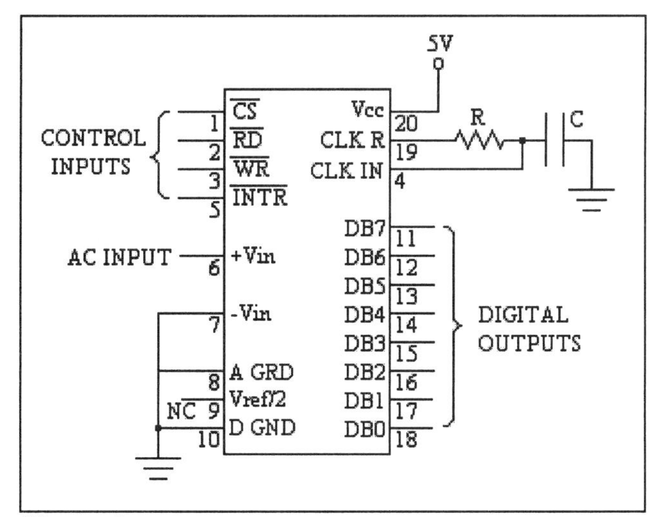

Figure 1

An internal successive approximation register (SAR) is *reset* when both the chip select, CS, and write, WR, inputs are a logic-LOW. *Start of conversion* takes place from 1 to 8 clock cycles after either CS or WR go to a logic-HIGH. Once conversion starts, the internal SAR and a comparator will convert the analog input to a digital quantity in 64 clock pulses. The contents of the SAR will then be transferred to output tristate latches and cause an interrupt, INTR, to become active. Figure 2 shows the timing diagram of the A/D converter for the write cycle.

To read the data in the tristate output latches, both CS and RD must be a logic-LOW. Figure 3 shows a typical read cycle based on an interrupt occurring on a microprocessor pin. Based on the interrupt, the microprocessor addresses the A/D converter, then sends a read command, RD. Data is valid on the output pins during the time segment while both CS and RD are logic-LOW.

The A/D converter may be operated in *free running mode* by connecting the INTR output to the WR input. Initially, WR will need to be forced to a logic-LOW to initiate the free running mode, but once started, the response of INTR

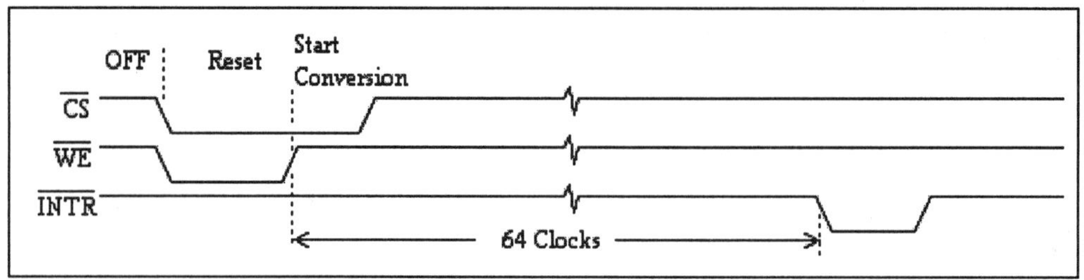

Figure 2

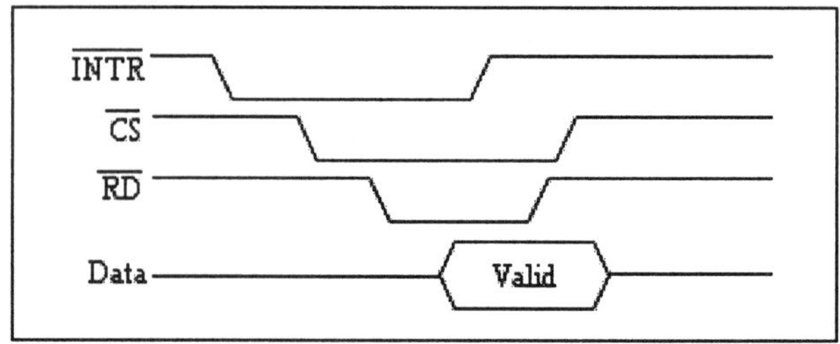

Figure 3

will cause triggering of the next conversion time. Figure 4 shows the typical response of the A/D converter in free running mode. Initially, the A/D is turned off by the WR input at a logic-HIGH. When WR drops to a logic-LOW, the A/D is reset. When WR goes to a logic-HIGH, the A/D starts its conversion of the analog input signal. After 64 clocks, the INTR goes low, forcing reset (when CS and WR are logic-LOW) and causing the output latches to latch onto the new sample. The new output data is the result of the conversion (after 64 clock cycles). Since CS and RD are tied low, the INTR is automatically returned to a logic-HIGH, hence starting the next conversion cycle.

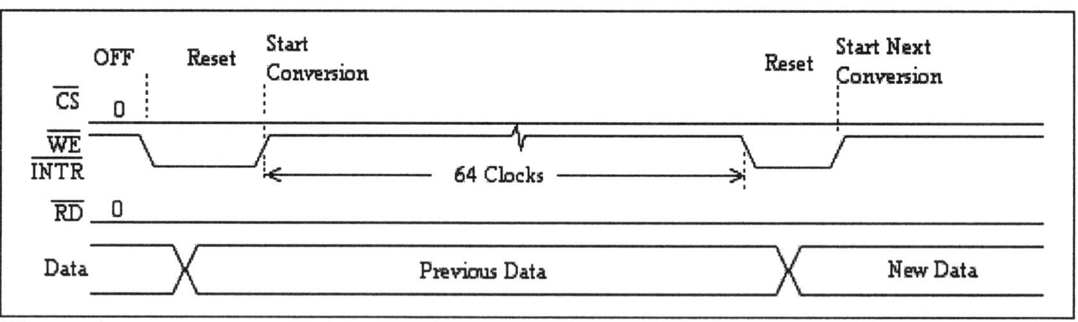

Figure 4

Part 1 Procedure

1. Construct the circuit shown in Figure 5. Use 1-KΩ resistors in series with all light emitting diodes.

2. Set the voltage at the wiper arm of the potentiometer to each value listed in Table 1, then record the binary equivalent as shown by the LEDs.

3. Convert each 8-bit binary entry in Table 1 to its decimal equivalent. Record the equivalents in the third column in Table 1.

4. Multiply the decimal equivalents of the binary numbers by the step voltage (19.5 mV) for the 8-bit A/D converter. Record your answers in the fourth column in Table 1.

5. The frequency of the internal clock may be calculated by: $f = \frac{1}{1.1RC}$ (see Figure 1).

   Based on this equation and the parts shown in Figure 5, the oscillator frequency is: _________Hz

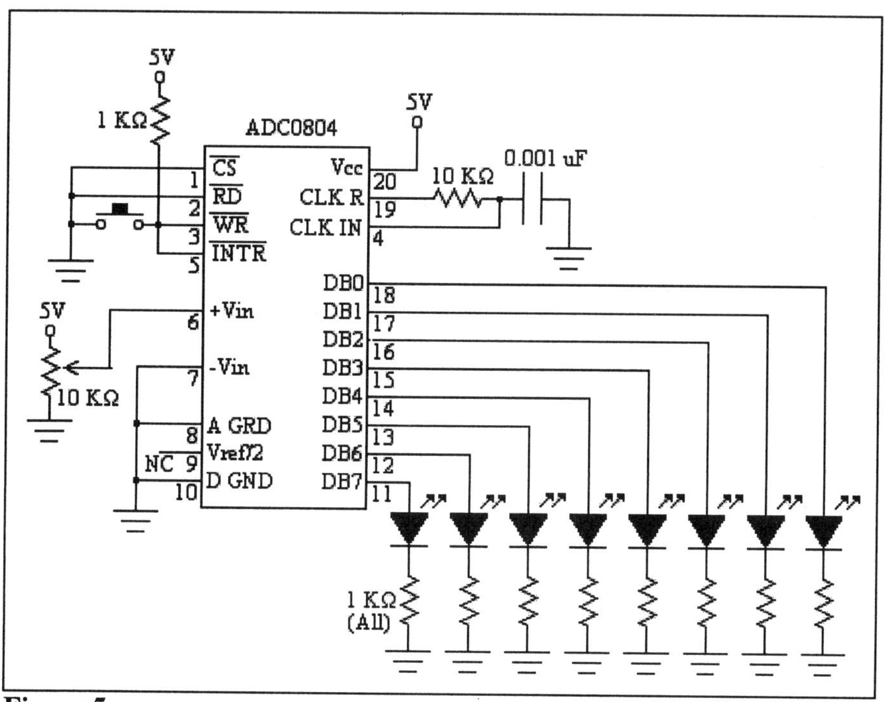

Figure 5

Input voltage (DCV)	Output H G F E D C B A	Decimal Equivalent of Binary Value	Multiply Decimal Equivalent by E_{STEP}
0 V	__ __ __ __ __ __ __ __	________10	________
0.5 V	__ __ __ __ __ __ __ __	________10	________
1.0 V	__ __ __ __ __ __ __ __	________10	________
1.5 V	__ __ __ __ __ __ __ __	________10	________
2.0 V	__ __ __ __ __ __ __ __	________10	________
2.5 V	__ __ __ __ __ __ __ __	________10	________
3 V	__ __ __ __ __ __ __ __	________10	________
3.5 V	__ __ __ __ __ __ __ __	________10	________
4 V	__ __ __ __ __ __ __ __	________10	________
4.5 V	__ __ __ __ __ __ __ __	________10	________
5 V	__ __ __ __ __ __ __ __	________10	________

Table 1

6. Replace the 10-KΩ potentiometer with the circuit shown in Figure 6. Wire the $\mathbf{E}_{ERROR}$ output of the circuit shown in Figure 6 to the $+\mathbf{V}_{IN}$ input of the A/D converter.

7. Resistance of the thermistor (out of circuit): ______________ Ω. Be careful not to touch the body of the thermistor for the resistance will change when heat is applied.

8. Using the voltage divider equation: $V_1 = \dfrac{Vcc \times R_T}{R_1 + R_T}$ where R_1 is 100 KΩ and R_T is the measured value recorded for Step 7, calculate $\mathbf{V}_1$ when **Vcc** is 5 V. $\mathbf{V}_1$ = ______________

9. Adjust the potentiometer until the output of the A/D converter is 70_H.

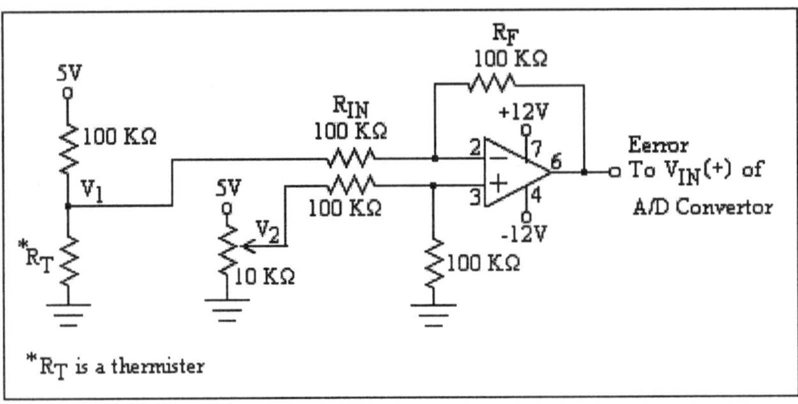

Figure 6

10. Measure and record the voltages at
    A. the potentiometer wiper arm: $\mathbf{V_2}$ = ____________ V.
    B. the junction of the thermistor and the two 100-KΩ resistors: $\mathbf{V_1}$ = ____________ V.
    C. the output of the operational amplifier: $\mathbf{E}_{ERROR}$ = ____________ V.

11. Was the measured value of V_1 recorded for Step 10 very close to the calculated value recorded for Step 8? (Yes/No)

12. The voltage at the output of the difference amplifier of Figure 6 may be calculated using the equation: $\mathbf{E}_{ERROR} = \mathbf{A}_V(\mathbf{V}_2 - \mathbf{V}_1)$, where $\mathbf{A}_V = \mathbf{R}_F / \mathbf{R}_{IN}$. Based on the recorded voltages in Step 10, calculate the error voltage when $R_F = R_{IN}$. E_{ERROR} = ____________ V

13. Was the calculated error voltage (Step 12) very close to the measured value? (Yes/No)

14. Place a voltmeter at the output of the operational amplifier to monitor the voltage.

15. Squeeze the thermistor, being careful not to short the thermistor wires to each other. Record the highest binary number seen at the output of the A/D converter and highest voltage measured at the output of the operational amplifier.

    Binary number: ________________$_2$

    Highest DCV on operational amplifier output: ____________ V

16. The circuit is not the best or most accurate temperature-measuring device but it does show how temperature may be converted to a digital value. If the digital output of the A/D converter is read by a microprocessor, the software controlling the microprocessor may equate different digital values read to an equivalent temperature. The microprocessor may take a temperature reading once an hour or more often as desired to monitor temperature, tabulate the results maybe once a week, and send the data to a central office to be included in monthly reports. Use your imagination for other applications of such a circuit.

Part 2 Procedure

1. Construct the circuit shown in Figure 7.

2. For various values of switch settings listed in Table 2, measure the voltage at the output of the operational amplifier and record the pattern displayed by the LEDs.

3. Demonstrate your operational circuit (Figure 7) to your instructor. Obtain the signature of approval directly on the answer page for this lab.

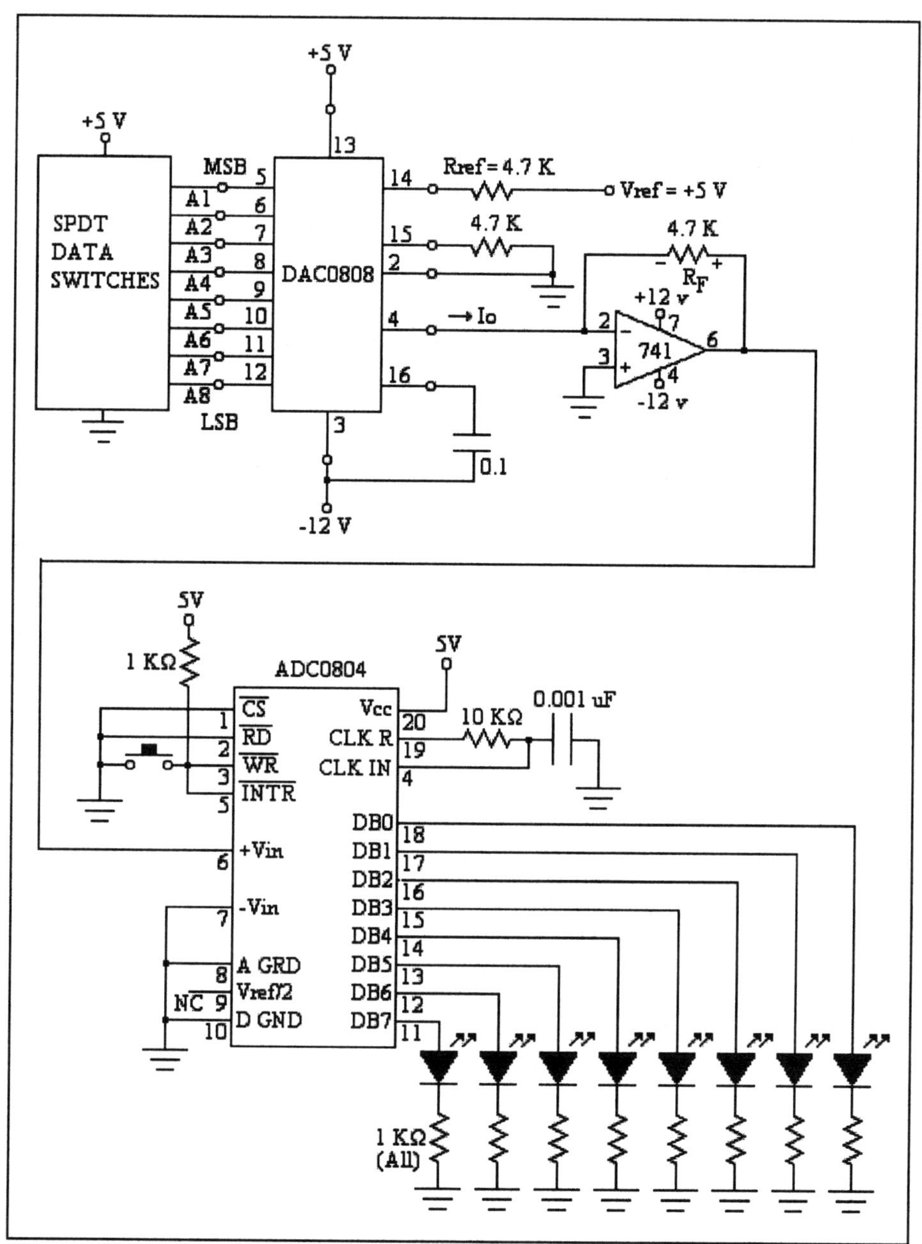

Figure 7

Set Switches A_1 A_2 A_3 A_4 A_5 A_6 A_7 A_8	Op-amp Pin 6 Voltage	Binary Code Displayed DB_7 DB_6 DB_5 DB_4 DB_3 DB_2 DB_1 DB_0
0 0 0 0 0 0 0 0	________	_ _ _ _ _ _ _ _
0 0 0 0 0 0 0 1	________	_ _ _ _ _ _ _ _
0 0 0 0 0 0 1 0	________	_ _ _ _ _ _ _ _
0 0 0 0 0 1 0 0	________	_ _ _ _ _ _ _ _
0 0 0 0 1 0 0 0	________	_ _ _ _ _ _ _ _
0 0 0 1 0 0 0 0	________	_ _ _ _ _ _ _ _
0 0 1 0 0 0 0 0	________	_ _ _ _ _ _ _ _
0 1 0 0 0 0 0 0	________	_ _ _ _ _ _ _ _
1 0 0 0 0 0 0 0	________	_ _ _ _ _ _ _ _

Table 2

4. Write a 1- to 2- page technical summary pertaining to this lab. Include embedded graphics of the schematic used for Part 2 of this lab.

5. Staple the following in the sequence listed and submit the papers to the instructor for grading:
   - Cover page
   - The completed answer page for this lab
   - Typed summary

Lab 22: The A/D Converter Answer Page

Name: ______________________

Part 1

Input voltage (DCV)	Output H G F E D C B A	Decimal Equivalent of Binary Value	Multiply Decimal Equivalent by E_{STEP}
0 V	_ _ _ _ _ _ _ _	________$_{10}$	________
0.5 V	_ _ _ _ _ _ _ _	________$_{10}$	________
1.0 V	_ _ _ _ _ _ _ _	________$_{10}$	________
1.5 V	_ _ _ _ _ _ _ _	________$_{10}$	________
2.0 V	_ _ _ _ _ _ _ _	________$_{10}$	________
2.5 V	_ _ _ _ _ _ _ _	________$_{10}$	________
3 V	_ _ _ _ _ _ _ _	________$_{10}$	________
3.5 V	_ _ _ _ _ _ _ _	________$_{10}$	________
4 V	_ _ _ _ _ _ _ _	________$_{10}$	________
4.5 V	_ _ _ _ _ _ _ _	________$_{10}$	________
5 V	_ _ _ _ _ _ _ _	________$_{10}$	________

Table 1

5. ____________ Hz 7. ____________ 8. ____________

10. A. $\mathbf{V}_2$ = ____________ V B. $\mathbf{V}_1$ = ____________ V C. $\mathbf{V}_{ERROR}$ = ____________ V

11. (Yes/No) 12. $\mathbf{V}_{ERROR}$ = ____________ V 13. (Yes/No)

15. ________________$_2$ ____________ V

Part 2

Set Switches A_1 A_2 A_3 A_4 A_5 A_6 A_7 A_8	Op-amp Pin 6 Voltage	Binary Code Displayed DB_7 DB_6 DB_5 DB_4 DB_3 DB_2 DB_1 DB_0
0 0 0 0 0 0 0 0	________	_ _ _ _ _ _ _ _
0 0 0 0 0 0 0 1	________	_ _ _ _ _ _ _ _
0 0 0 0 0 0 1 0	________	_ _ _ _ _ _ _ _
0 0 0 0 0 1 0 0	________	_ _ _ _ _ _ _ _
0 0 0 0 1 0 0 0	________	_ _ _ _ _ _ _ _
0 0 0 1 0 0 0 0	________	_ _ _ _ _ _ _ _
0 0 1 0 0 0 0 0	________	_ _ _ _ _ _ _ _
0 1 0 0 0 0 0 0	________	_ _ _ _ _ _ _ _
1 0 0 0 0 0 0 0	________	_ _ _ _ _ _ _ _

Table 2

3. Demonstrated to: ______________________________ Date: __________

Grade: ____________

Lab 23: Memory Addressing

Objectives:

1. Learn the basic operation of RAM (Random Access Memory)
2. Construct a circuit to read from or write to any address in a RAM
3. Control read/write cycles and memory addressing with the EPM7128LC84

Materials List:

- Max+plus II software by Altera Corporation
- Circuit card with the EPM7128SLC84 chip
- Computer requirements:
  Minimum 486/66 with 8 MB RAM
- (1) 6116 - 2 K × 8 memory (or equivalent)
- (8) 1 KΩ resistors
- Floppy disk
- (2) 0.1 μFd capacitors
- (2) 0.01 μFd capacitors
- Power supply
- Breadboard
- (8) LEDs
- (10) Toggle switches
- (2) Pushbutton switches
- (1) NE555
- (3) 4.7 KΩ resistors
- (1) 33 KΩ resistor
- (2) 47 KΩ resistors

Discussion:

The memory integrated circuit that you will use in this lab is the 6116 (or equivalent), which is a 2 K × 8 static RAM. 2 K × 8 implies 2048 memory addresses, 2^{11}, with 8 bits of data storage at each address. Because the 6116 is a static device, the data stored in the memory elements is stable and does not require refresh, as with DRAM (Dynamic RAM). Also, data read is nondestructive and will be the same data that was written to that location.

The pin configuration, logic symbol, and pin names are shown in Figure 1. The 11 address inputs, A_0 to A_{10}, provide the 2048 (2^{11}) addresses. The eight data input/output lines, I/O_1 to I/O_8, will be connected to a bidirectional data bus to write or read data to each address specified by the bit pattern on A_0 to A_{10}. The chip select, CS, input enables the memory when CS is a logic-LOW. The CS input must be a logic-LOW for either the read or write operation.

Data are written into memory when the Write Enable, WE, input is a logic-LOW. Data are read when Write Enable is a logic-HIGH and the Output Enable, OE, is a logic-LOW. The data out are tristate logic, allowing for more than one memory IC to be connected to the data bus. The memory chip operation is illustrated in Table 1 with respect to the control line logic levels.

Mode	CS	WE	OE	I/O_0 to I/O_8
Off	H	X	X	Z_H
Write	L	L	X	Data inputs
Read	L	H	L	Data outputs
Read	L	H	H	Z_H

Table 1

The general configuration of the 16348 (2 K × 8) bits in the 6116 is illustrated in Figure 2. Each of the squares represents a memory location for one bit. Thus, the eight squares under the "0" column indicate that these eight bits are accessed simultaneously (for read or write) when binary address 00000000000_2 is applied to address lines A_0 to A_{10}. Also, the eight squares under the "1" column are simultaneously accessed when the binary address 00000000001_2 is applied. The remaining locations, 2 through 2047, are accessed by binary addresses

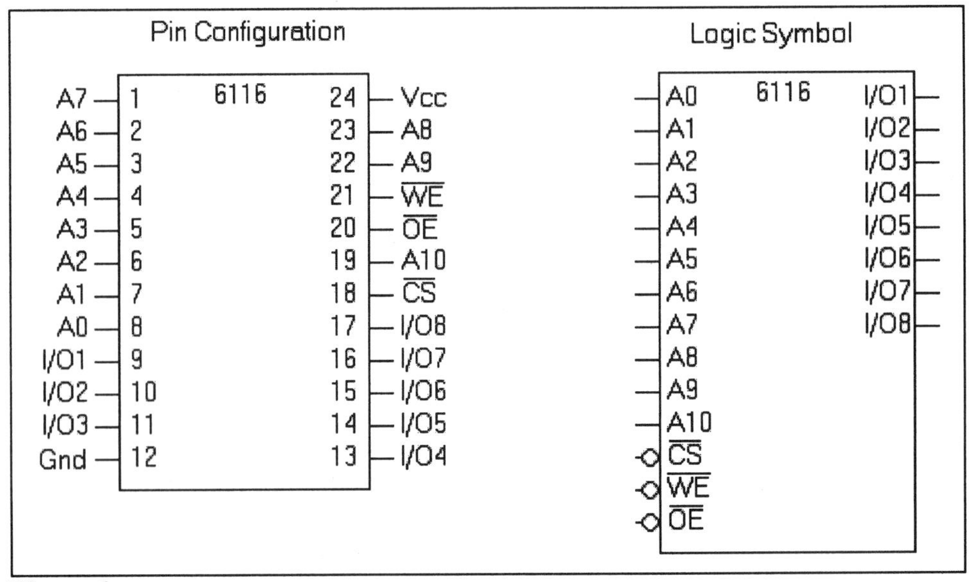

Figure 1

00000000010_2 to 11111111111_2 (0002_H to $7FFF_H$). The address applied to A_0 to A_{10} identifies where the data contents at that location can be found.

The eight I/O lines in Figure 2 are data lines used for writing data to memory or reading data from memory. Other RAM memory may have separate data input and data output lines. How memory is structured depends solely on the internal design of the device.

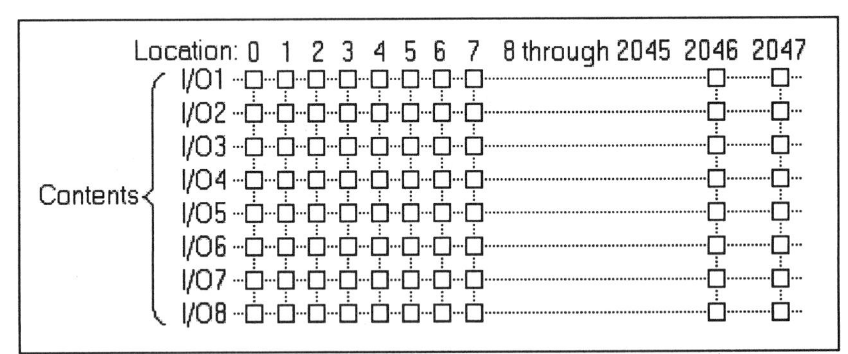

Figure 2

To prevent this lab from becoming tedious and repetitious, only a portion of the 2048 addresses will be accessed for read and write for Part 1. For Part 2, you will build a mod 2048 counter to access all memory addresses and use an 8-bit Johnson counter to supply data to these addresses. To prevent contention in Part 1, where more than one active output share a common load, a 74244 (octal tristate buffer) will be used to isolate the "data" switches from the memory chip during the read mode. In Part 2, tristate buffers will be used to prevent contention during read and write operations.

Finally, in Part 3 and Part 4, you will program the EPM7128SLC84 IC and interface the circuit card containing the chip to a 6116 memory chip. Part 3 is based on **memory.gdf** and **memory.scf** created in Part 1 of this lab. Part 4 is based on **memory2.gdf** and **memory2.scf** as constructed in Part 2.

Part 1 Procedure

1. Open the Max+plus II software. Assign the project name **memory**.

2. Open the Graphic Editor and create the circuit shown in Figure 3.

3. Open the Waveform Editor, turn off the Snap to Grid feature in the Options menu, and create the waveforms shown in Figure 4 following these setup instructions.

- Set Grid Size to 6 ns
- Set the clear waveform to a constant logic-HIGH
- Highlight 0 to 6 ns, set this section of the clear waveform to a logic-LOW
- Highlight 498 to 504 ns, set this section of the clear waveform to a logic-LOW
- Set Grid Size to 20 ns
- Highlight the Clock waveform
- Click the Clock Overwrite button in the Draw tool bar, then click the OK button
- Set Grid Size to 500 ns
- Highlight the R/W waveform
- Click the Clock Overwrite button in the Draw tool bar, then click the OK button
- Set Grid Size to 40 ns
- Highlight the first 40 ns of the S[7..0] waveform
- Click the Overwrite Group Value button in the Draw tool bar
- Enter 33 as the group value, then click the OK button
- Complete the S[7..0] values as shown in Figure 4

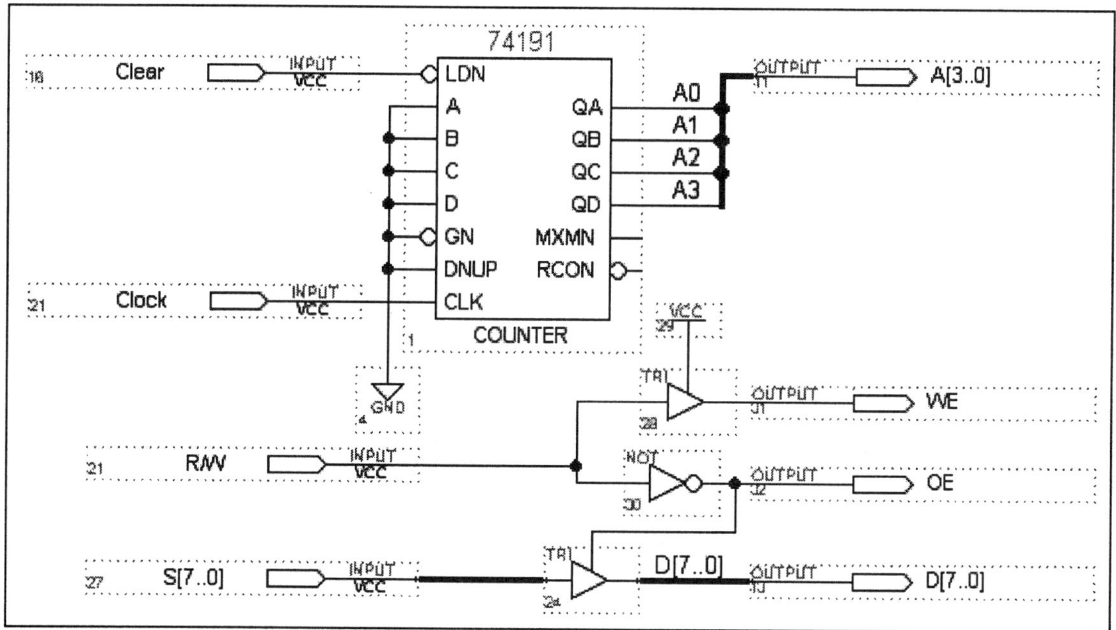

Figure 3

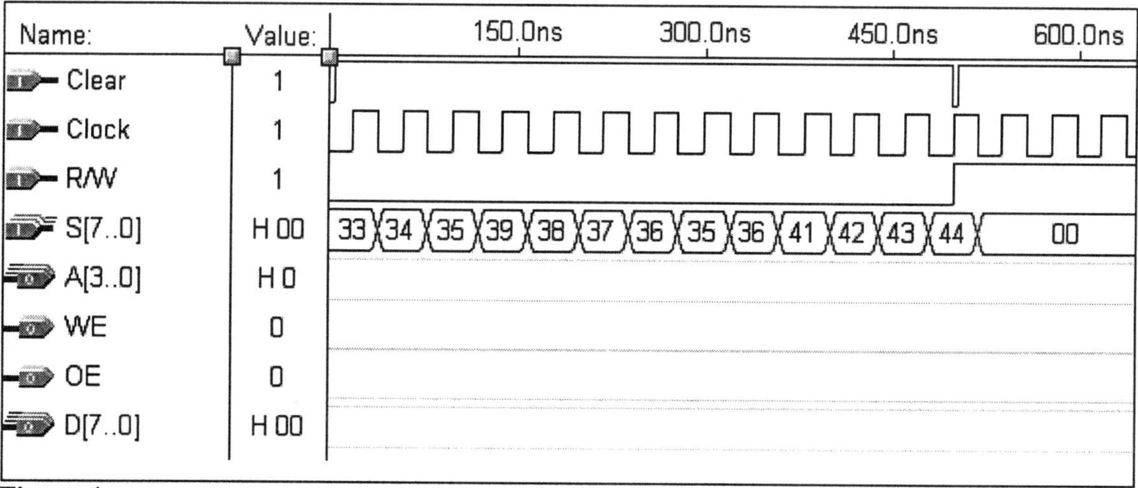

Figure 4

4. Run the Compiler and Simulator. Assuming zero errors, sketch the resulting output waves (0 ns to 650 ns) in the area provided at the bottom of Figure 4.

5. What is the counter count sequence as displayed in the Graphic Editor? ______________________

6. What prevented the counter from reaching the "F" count?

____

7. Data appears on the data bus during the (HIGH/ LOW) level of the R/W control signal.

8. Describe the D[7..0] waveform from 500 ns to 1 μs.

____

9. The data was given in standard ASCII code. Identify the character of each ASCII code shown in Table 2. Refer to your textbook or research the library for the ASCII table.

ASCII	Character
33H	______
34H	______
35H	______
39H	______
38H	______
37H	______
36H	______

ASCII	Character
35H	______
36H	______
41H	______
42H	______
43H	______
44H	______

Table 2

10. WE and OE will be connected to the active-LOW WE and OE control inputs of a 6116 memory chip. During the time segment 0 to 500 ns, as shown in the Waveform Editor, data will be (written to/read from) memory.

11. During the time segment 500 ns to 1 μs, as shown in the Waveform Editor, data will be (written to/read from) memory.

12. Obtain a hard copy of the Graphic and Waveform Editor files. Label these hard copies **Part 1, Step 12A** and **Part 1, Step 12B**, respectively.

13. Save all files to Drive A as **memory**, exit the Graphic and Waveform Editors.

Part 2 Procedure

1. Open the Max+plus II software. Assign the project name **memory1**.

2. Open the Graphic Editor and create the circuit shown in Figure 5.

3. Open the Waveform Editor and create the waveforms shown in Figure 6 following these setup instructions.

    - Set Grid Size to 10 ns
    - Create the Clear waveform as shown in Figure 6
    - Change the End Time to 81.98 μs
    - Highlight the Clock waveform
    - Click the Clock Overwrite button in the Draw tool bar, then click the OK button

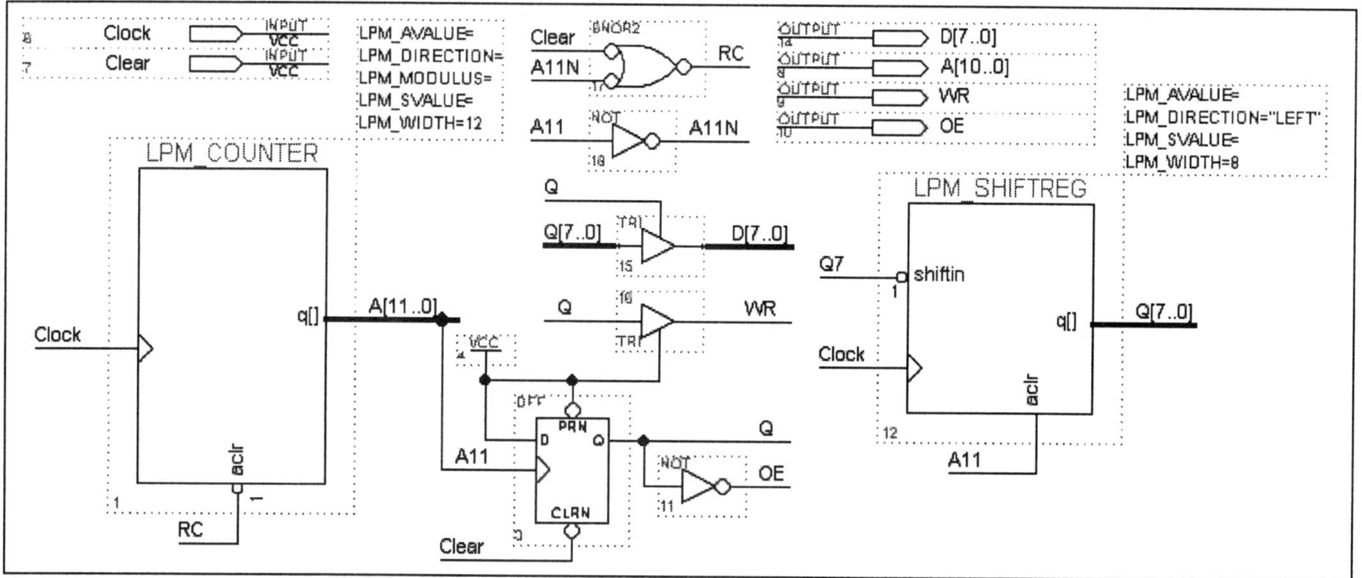

Figure 5

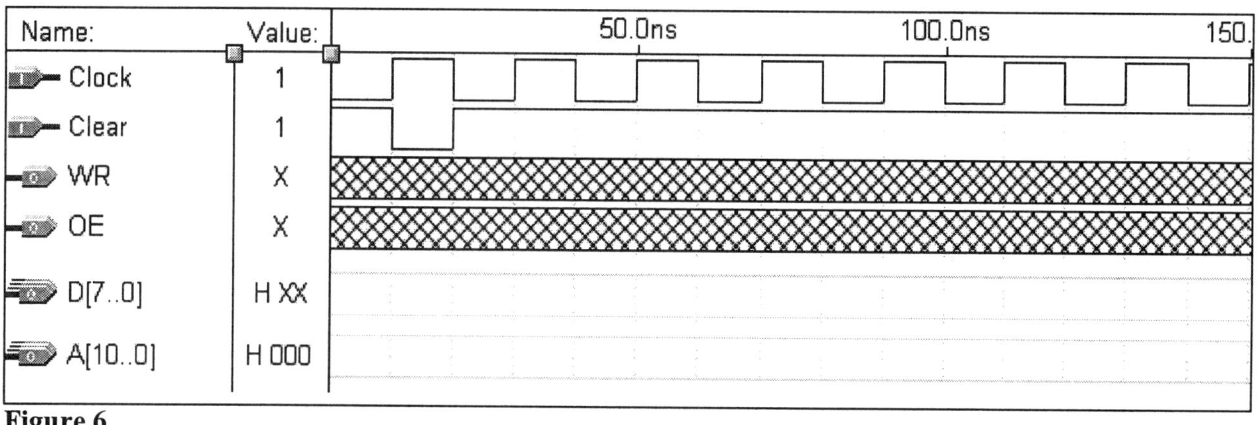

Figure 6

4. Run the Compiler and Simulator. Assuming zero errors, sketch the resulting output waves (0 ns to 150 ns) in the area provided at the bottom of Figure 6.

5. What is the count sequence of the 12-bit LPM_COUNTER megafunction? ______________________

6. What is the count sequence of the LPM_SHIFTREG megafunction as wired in Figure 5? (all 16 counts)

   ____

7. The clear input to the circuit of Figure 5 is (synchronous/asynchronous) and is active- (HIGH/LOW).

8. As long as the counter counts and data from the 74198 starts to repeat, you may proceed with the next step, otherwise correct all mistakes and repeat Steps 2 through 9 of Part 2.

9. Obtain a hard copy of the Graphic and Waveform Editor files. Label these hard copies **Part 2, Step 9A** and **Part 2, Step 9B**, respectively.

10. Save all files to Drive A as **memory1**, then exit the Graphic and Waveform Editors.

Part 3 Procedure

1. Connect the circuit card with the EPM7128SLC84 chip to the computer's printer port. If the computer has a software key attached to LPT1, connect the card to the parallel port, LPT2. See your instructor to determine the correct connection port.

2. Open the Max+plus II software. Assign the project name **memory**, assign the MAX7000S device family, and select the EPM7128SLC84 device.

3. Open the **memory.gdf** file created in Part 1 of this lab.

4. Open the **memory.scf** file created in Part 1 of this lab.

5. Run the Compiler and Simulator. Correct all errors before continuing.

6. The software will automatically assign pin 83 as the clock and pin 1 as clear since these are detected as global. Assign all other pins to match the input/output (I/O) configuration of your circuit card containing the EPM7128SLC84 chip. Refer to the user manual for the circuit card I/O pin reference.

7. Re-compile the circuit so the software will use your pin assignments.

8. Select Programmer in the Max+plus II menu item (left of the File option).

9. The software should automatically detect the board, providing the switches on the board are in the correct settings. It is best to follow the board manufacturer's notes on programming the 7128 chip since the procedure will be different for each board. Appendix C at the end of this lab contains typical instructions for programming the EPM7128SLC84. Program your chip following the procedure outlined in Appendix C. Potential problems that inhibit programming the IC are:

   - Not all required wires in the interface cable are connected
   - The speed of transmission between the computer and board is too fast
   - The cable from the computer to the board is too long
   - You need the patch program from Altera
   - Your computer misidentified the board, assuming the board is a printer
   - Lack of power to the EPM7128SLC84
   - Poor ground
   - For problems not mentioned above, see your professor for suggestions.

10. Assuming the board was successfully programmed, construct the circuit shown in Figure 7. Refer to the **memory.rpt** file on your disk that was created during compilation for pin assignments of the EPM7128SLC84 IC. You may also want to place LEDs at the A3, A2, A1, and A0 address lines to monitor the address as you proceed to program the 6116.

11. Pin 2 of the NE555 on-shot is an active-LOW trigger input. When clicked, the Q output (pin 3) becomes unstable for a period of time before returning to a logic-LOW. Calculate the time the one-shot is unstable. Unstable time: ____________

12. The CS input of the 6116 is active-LOW. Grounding CS keeps the 6116 memory chip turned on for the following read and write process.

13. The clear input (see Figure 3) of the 74191 is active-LOW. Set the clear switch to the GND side, then place this switch to the Vcc side. This action loads the counter with zero (DCBA inputs), and provides address 00000000000_2 to the 6116.

14. Set the R/W switch to the GND position for writing data to memory.

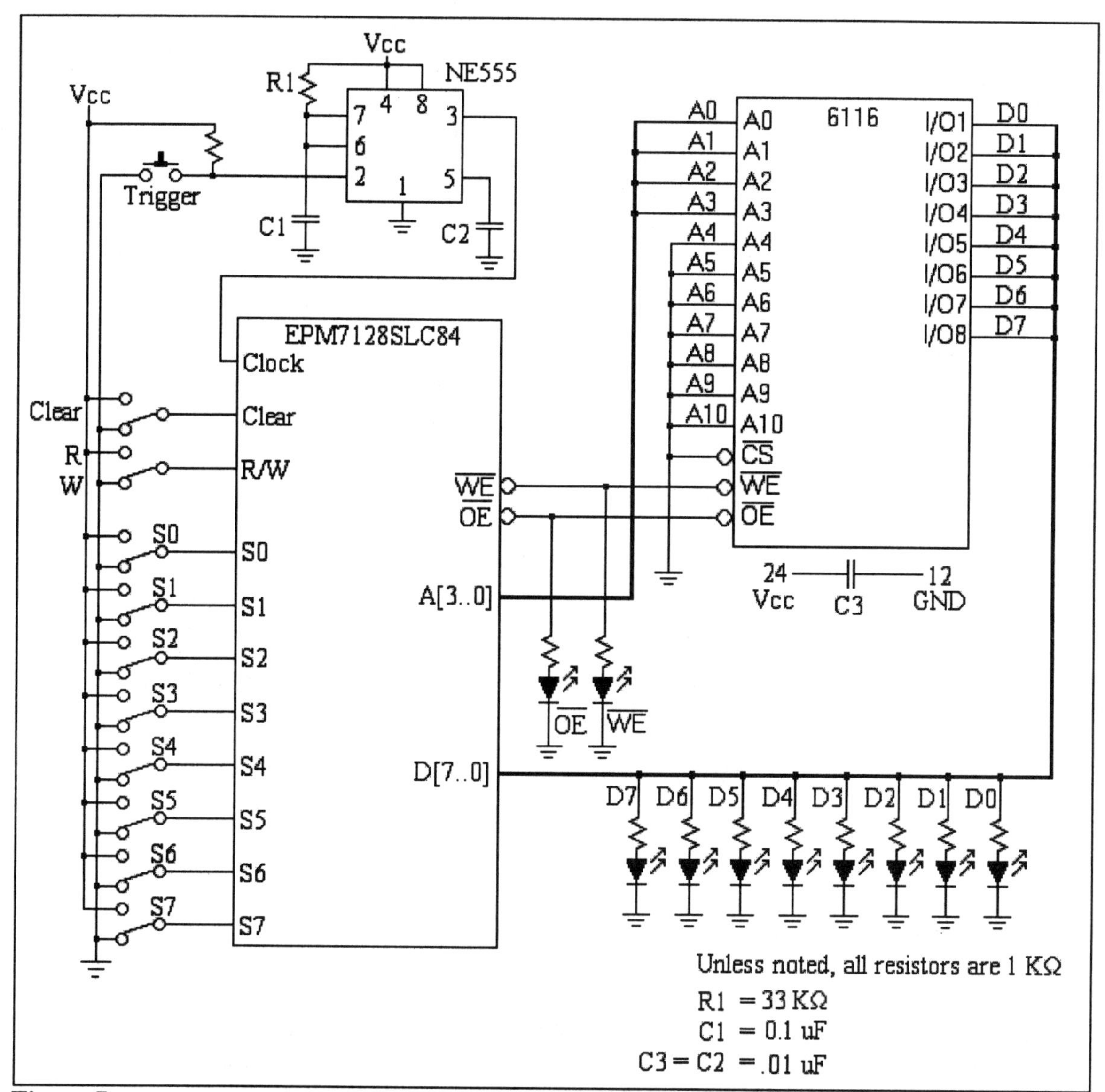

Figure 7

15. Set switches S7 through S0 for 33_H. Note, the LEDs D7 through D0 should show 33_H; 33_H was loaded into the 6116 at address 00000000000_2.

16. Click and release the trigger switch. The counter advances to 1_H, creating address 00000000001_2.

17. Set switches S7 through S0 for 34_H. Note, the LEDs D7 through D0 should show 34_H; 34_H was loaded into the 6116 at address 00000000001_2.

18. Click and release the trigger switch. The counter advances to 2_H, creating address 00000000010_2.

19. Repeat Steps 17 and 18 to program the data shown in Table 3 to the addresses shown.

20. Place the R/W switch to a logic-HIGH. This places the 6116 in read mode. Note that switches S7 through S0 no longer have an effect on the LEDs.

21. Place the clear switch to its GND side, then place it back to the Vcc side to clear the counter for address zero.

22. Verify that the contents of the 6116 matches Table 3 clicking the trigger switch to advance to the next address.

Address	Data
00000000010_2	35_H
00000000011_2	39_H
00000000100_2	38_H
00000000101_2	37_H
00000000110_2	36_H
00000000111_2	35_H
00000001000_2	36_H

Address	Data
00000001001_2	20_H
00000001010_2	41_H
00000001011_2	42_H
00000001100_2	43_H
00000001101_2	44_H
00000001110_2	45_H
00000001111_2	46_H

Table 3

23. Demonstrate the programmed EPM7128SLC84 and 6116 to your instructor. Obtain the signature of approval directly on the answer page.

24. Exit the Graphic and Waveform Editors.

Part 4 Procedure

1. Connect the circuit card with the EPM7128SLC84 chip to the computer's printer port. If the computer has a software key attached to LPT1, connect the card to the parallel port, LPT2. See your instructor to determine the correct connection port.

2. Open the Max+plus II software. Assign the project name **memory1**, assign the MAX7000S device family, and select the EPM7128SLC84 device.

3. Open the **memory1.gdf** and **memory1.scf** files created in Part 2 of this lab.

4. Run the Compiler and Simulator.

5. The software will automatically assign pin 83 as the clock and pin 1 as clear since these are detected as global. Assign all other pins to match the input/output (I/O) configuration of your circuit card containing the EPM7128SLC84 chip. Refer to the user manual for the circuit card I/O pin reference.

6. Re-compile the circuit so the software will use your pin assignments.

7. Select Programmer in the Max+plus II menu items (left of the File option).

8. Proceed to program the EPM7128SLC84 chip following the instructions in Appendix C or refer to the circuit board manufacturer's programming instructions.

9. Assuming the board was successfully programmed, construct the circuit shown in Figure 8. Refer to the **memory1.rpt** file on your disk that was created during compilation for pin assignments of the EPM7128SLC84 IC. You may want to place LEDs at the A10 through A0 address lines to monitor the address since the 6116 is programmed with the data generated by the 74194.

10. The 555 is wired as a(n) (bistable, astable, monostable) multivibrator.

11. Calculate the frequency of the oscillator. **f** = ____________________

12. Calculate the time it takes to store data at all memory addresses within the 6116. _____________

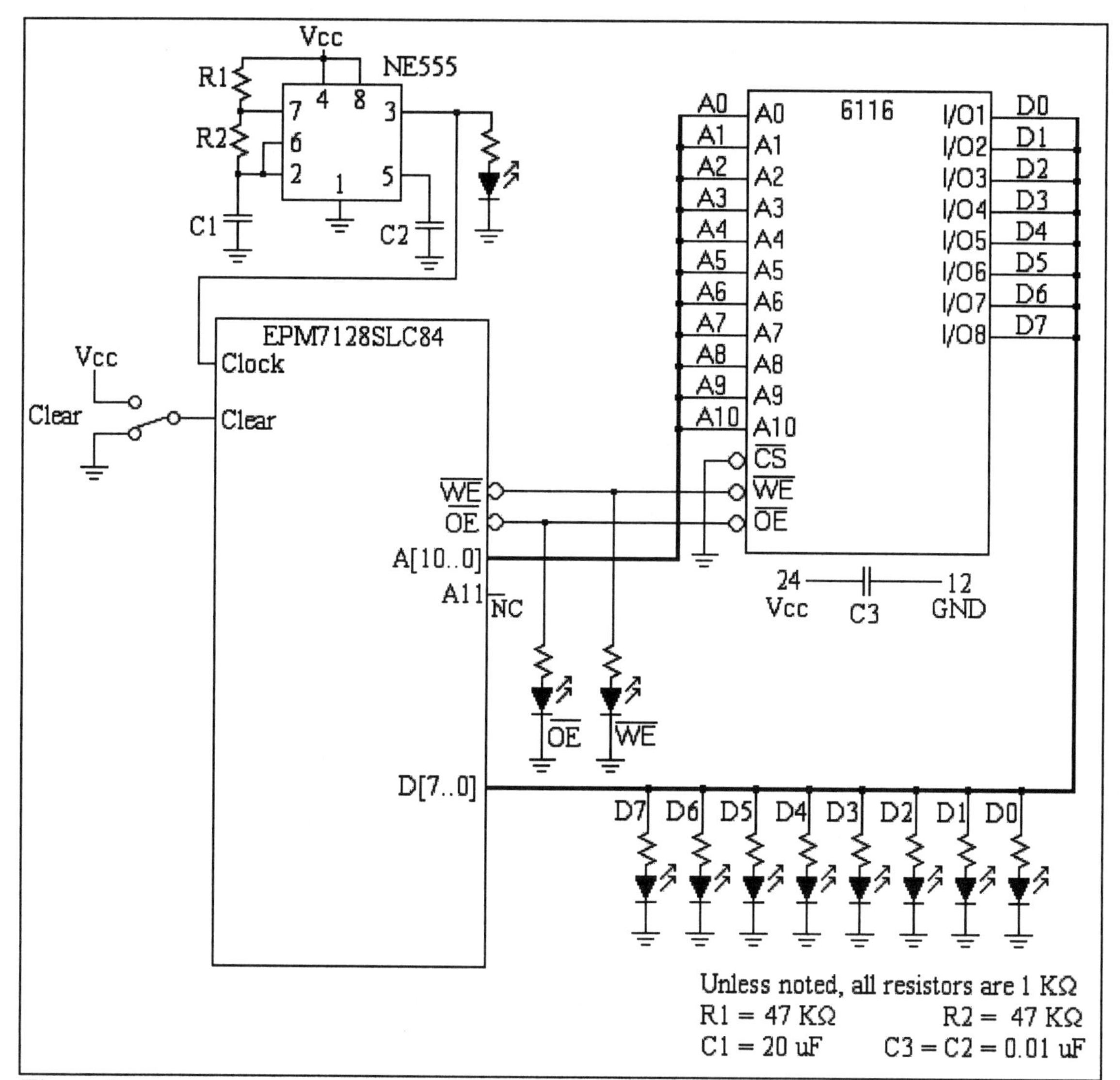

Figure 8

13. In order for data to be loaded into various memory addresses, the clear switch must be set to a logic-(HIGH/LOW).

14. Demonstrate the programmed EPM7128SLC84 and 6116 to your instructor. Obtain the signature of approval directly on the answer page.

15. Exit the Graphic and Waveform Editors.

16. Write a 1- to 2-page summary containing charts and/or graphics pertaining to the results obtained from this lab.

17. Place all papers for this lab in the following sequence, then submit the lab to your instructor for grading.

    - Cover page
    - Typed summary
    - The completed answer page for this lab
    - Hard copy of the Graphic Editor, **Part 1, Step 12A**
    - Hard copy of the Waveform Editor, **Part 1, Step 12B**
    - Hard copy of the Graphic Editor, **Part 2, Step 9A**
    - Hard copy of the Waveform Editor, **Part 2, Step 9B**

Lab 23: Memory Addressing Answer Pages

Name: ______________________________

Part 1

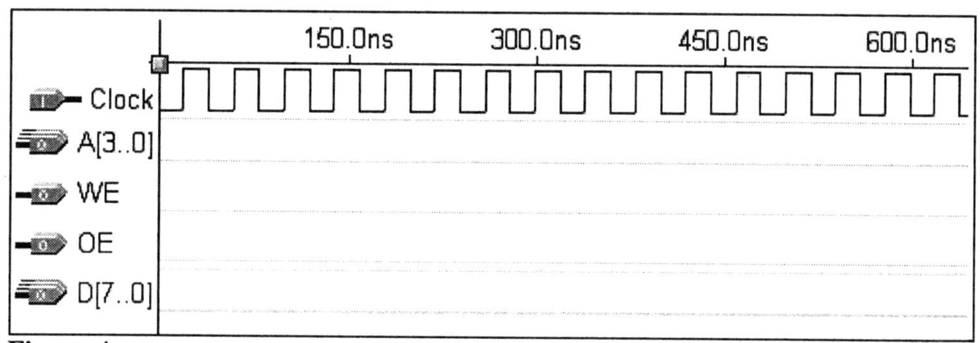

Figure 4

5. ____________

6. ______________________________

____

____

7. (HIGH/LOW)

ASCII	Character
33H	_____
34H	_____
35H	_____
39H	_____
38H	_____
37H	_____
36H	_____

ASCII	Character
35H	_____
36H	_____
41H	_____
42H	_____
43H	_____
44H	_____

Table 3

8. ____

10. (written to/read from)

11. (written to/read from)

Part 2

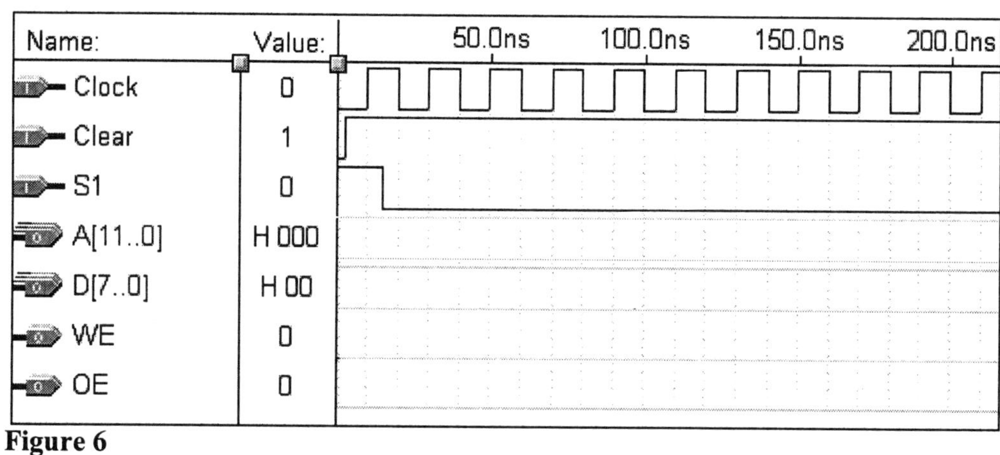

Figure 6

5. ______________________________

6. ____

7. (synchronous/asynchronous)

Part 3

11. ______________

23. Demonstrated to: ____ Date: ___________

Part 4

10. (bistable, astable, monostable)

11. **f** = ____________________

12. ______________

13. (HIGH/LOW)

14. Demonstrated to: ____ Date: ___________

Grade: _____________

Lab 24: Analog Storage

Objective:

1. Using the Max+plus II software, design digital control circuitry to control the sampling of an analog signal by an A/D converter and the read/write cycles of a memory chip shown in Figure 1 according to the following criteria.

   A. The 7128S will contain a 12-bit binary counter with outputs A0 to A10 used to address memory from 000_H to $7FF_H$.

   B. Upon reset, the 7128S will turn ON the A/D Converter, turn ON memory, and write the first analog sample converted to digital at address 000_H in memory.

   C. As the counter counts up to $7FF_H$, the analog samples converted to digital will be stored in sequential memory addresses.

   D. Once the counter overflows (goes beyond $7FF_H$), the A/D converter shuts off and the memory switches to a continuous read mode until the next manual reset pulse.

   E. The analog output of the D/A converter is continuous.

2. Construct and demonstrate the circuit illustrated in Figure 1.

Materials required:

- Max+plus II software
- Computer requirements:
  Minimum 486/66 with 8 MB Ram
- Circuit card with the EPM7128SLC84
- Signal generator
- Power supply
- Dual trace oscilloscope
- Parts to successfully construct Figure 1

Discussion:

This lab is loosely defined so that not to stifle your creativity. Many years ago, analog oscilloscopes were called oscilloscopes because there was only one type ... analog. Once digital scopes hit the market, the terms *analog* and *digital* were necessary to describe the type of oscilloscope in use. As digital oscilloscopes advanced, someone decided that the digital information can be stored to memory and later recalled.

Figure 1 shows the basic concept for a digital storage oscilloscope; an input signal is applied, is converted to digital, and then stored to memory in digital form. If/when necessary, the digital pattern in memory can reproduce the analog signal. The storage capacity is dependent on the size of installed memory.

This circuit works well to convert low-frequency sine waves, pulse waves, or triangular waves to digital.

Part 1 Procedure

1. Open the Max+plus II software. Assign the project name **lab24** and MAX7000S as the device family.

2. Open the Graphic Editor and design control circuitry and a 12-bit UP counter that will satisfy the criteria stated in the objectives.

3. Assign input and output pins according to the user manual for the University board.

4. Open the Waveform Editor and create a set of waveforms to demonstrate that your design produces the desired output control signals. You may want to use Bit 3 of the counter instead of Bit 11 to trigger your control circuitry into a memory read mode. Once the circuit is functional, reassign the proper trigger bit before programming the device.

5. Connect the circuit card with the EPM7128SLC84 chip to the computer's printer port. If the computer has a software key attached to LPT1, connect the card to the parallel port, LPT2. See your instructor to determine the correct connection port.

6. Select **Programmer** in the **Max+plus II** menu item (left of the File option).

7. Proceed to program the EPM7128SLC84 chip following the instructions in Appendix C or refer to the circuit board manufacture's programming instructions.

8. Save all files to your disk in Drive A as **lab24** then exit the Graphic and Waveform Editors.

9. Assume the board was successfully programmed, construct the circuit shown in Figure 1. Refer to the **lab24.rpt** file on your disk that was created during compilation for pin assignments of the EPM7128SLC84.

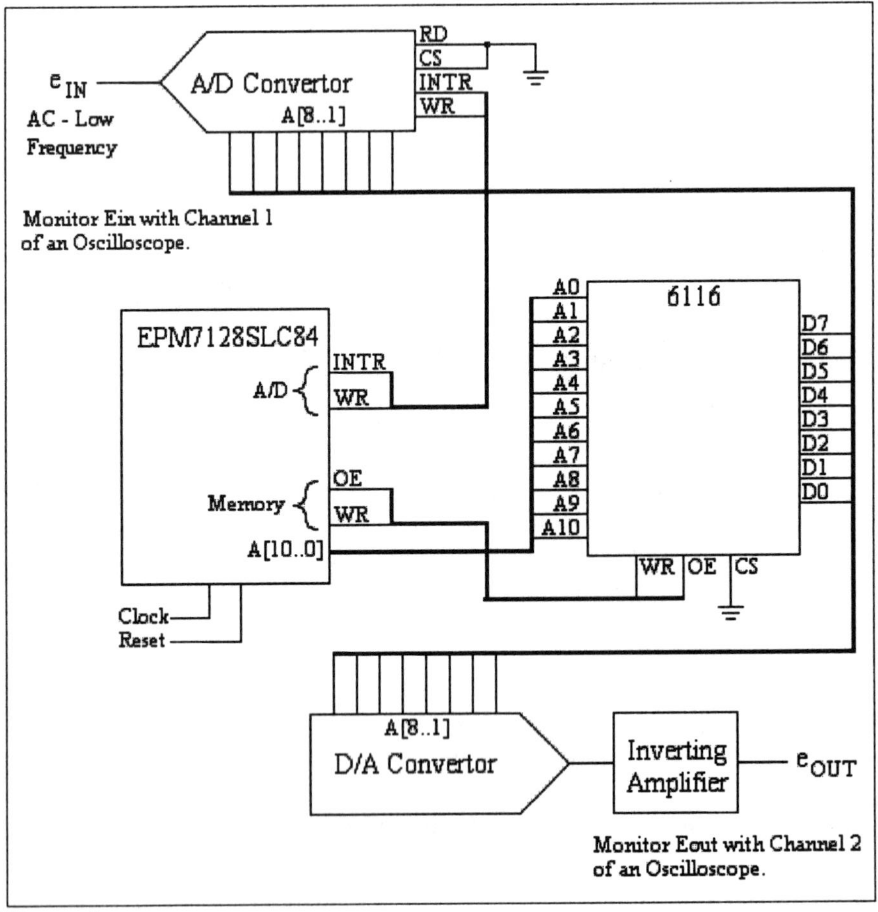

Figure 1

10. Connect LEDs with series resistors on the address/control/data lines for visual signs that the circuit indeed is writing and reading data to/from memory and that the A/D converter was shut off during memory read cycle time.

11. Demonstrate the functional circuit to your instructor by disconnecting the signal source once your circuit is in the memory read cycle. The output should then be a continuous wave without the input. Obtain the signature of approval directly on your cover page for this lab.

12. Obtain a hard copy of the Graphic and Waveform Editor files of your final design.

13. Write a 1- to 2- page summary explaining your circuit design. Provide an embedded schematic and waveforms obtained from the Graphic and Waveform Editor files.

14. Submit the following

    - Cover page
    - Typed summary
    - Hard copy of your control circuit (Graphic Editor file)
    - Hard copy of the Waveform Editor file

Lab 25: Synchronous Data Transceiver

Objectives:

1. Create an operational synchronous data transceiver
2. Program the CPLD on a circuit card with the data transceiver network
3. Use an A/D converter for the data source
4. Interface two data transceivers and demonstrate bidirectional communications

Materials list:

- Max+plus II software by Altera Corporation
- Circuit card containing the EPM7128SLC84 chip
- Computer requirements:
  Minimum 486/66 with 8 MB RAM
- Floppy disk
- (1) Pushbutton switch
- (8) SPDT switches
- (14) 1 K resistors
- (9) LEDs
- 8 STDP switches
- 5-V DC power supply
- ADC0804
- (2) .001 μF Capacitors

Discussion:

The transmitter section of a synchronous data transceiver (Figure 1) consists of two stages; a parallel-in serial-out shift register and a control circuit. The receiver contains a serial-in parallel-out shift register with latched outputs and a control circuit. When the Transmit input is pulsed, the control circuit in the transmitter sends eight clock pulses from clock-in to the register to serial output data D[7..0] and sends nine clocks to the receiver. The initial logic states of outputs, Transmitted Data and Transmitted Clock, are logic-LOW.

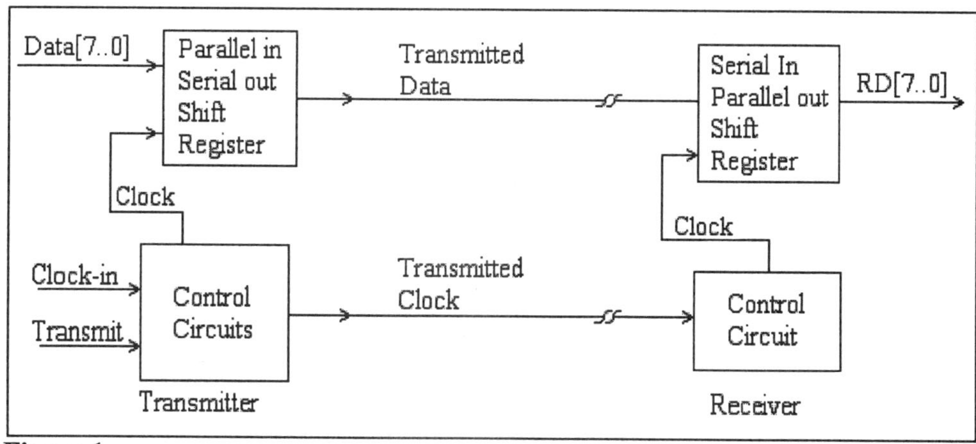

Figure 1

For the sake of discussion, each component in the transmitter control circuit of Figure 2 has been numbered. This circuit relies on the LPM_COUNTER megafunction. When selecting this megafunction, disable all inputs and outputs except for the ones used in the schematic. Be sure to select "All" for the inversion status of the aclr input.

When an active-low Transmit pulse is applied, the Q output of the D-type flip-flop (1) goes to a logic-HIGH. The enabled AND gate (1) will pass the next two clock pulses from clock-in to the first T-type flip-flop (2), creating a negative pulse on the TN line, the output of inverter 2. TN asynchronously parallel loads the 74165 Pi/So shift register. On the second clock-in signal, Clk-in-enable goes high, enabling AND gate 2 and inhibiting AND gate 1.

The first Clk-in pulse through AND gate 2 sets the Q output of D-type flip-flop #4, which in turn enables both AND gate 3 and the LPM_Counter. The next eight clock signals serial outputs data through the 74265 register and the counter counts. On the eighth clock pulse, T3 of the counter goes high, forcing reset to the flip-flops and counter. The circuit remains idle until the next transmit signal.

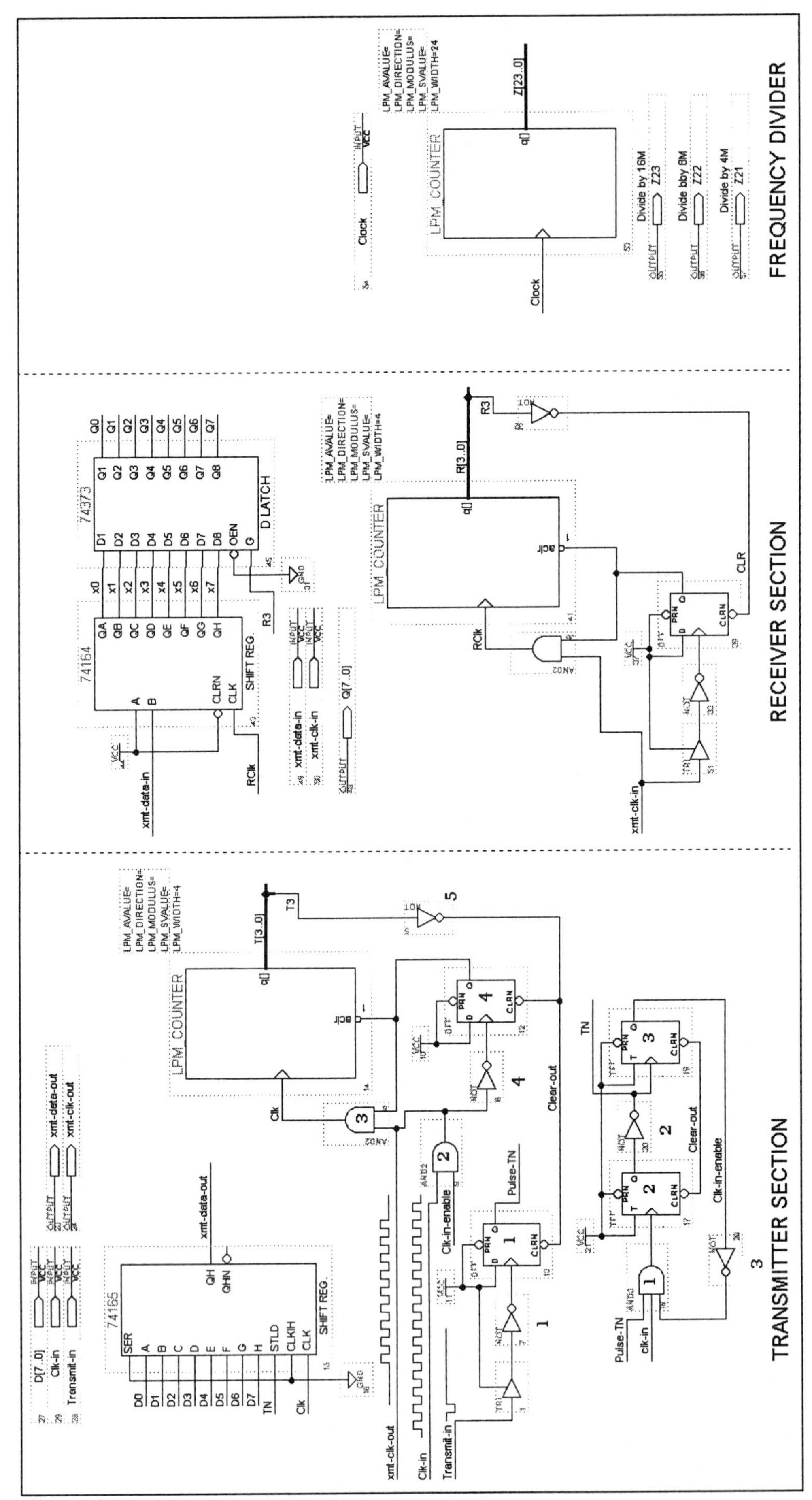
TRANSMITTER SECTION
RECEIVER SECTION
FREQUENCY DIVIDER
LPM_COUNTER
74165
SHIFT REG.
74164
74373
D LATCH
Divide by 16M
Divide bby 8M
Divide by 4M
xmt-data-out
xmt-clk-out
xmt-data-in
xmt-clk-in
Transmit-in
Clk-in
Clear-out
Pulse-TN
Clk-in-enable

Figure 2

The receiver section is not as complex as the transmitter. The 74164 converts the serially transmitted data to parallel data that is latched by the 74373 when the receiver counter counts to eight. The first of nine pulses on xmt-clk-in enables the receiver LPM_Counter to count the next eight pulses.

The first clock pulse passing through AND gate 4 in the transmitter that initiates the transmitter counter to count also initiates the receiver counter to count. Therefore both counters are operating synchronously and the eight data bits from the transmitter register will appear on the latched outputs of the receiver register.

Even though the two tri-state logic gates are not necessary for the circuit to operate in theory, in practice they are used to prevent the software from recognizing and assigning transmit and xmt_clk_in as global clock inputs.

Since most university boards containing a CPLD also contain an on-board crystal, the frequency divider circuit provides three output frequencies; Z23 divides the clock by 16 Meg, Z22 divides the clock by 8 Meg, and Z21 divides the clock by 4 Meg. Select and wire the output that gives a near 1 Hz pulse to the Clk-in pin of the programmed CPLD. If the frequency on Clk-in is too high, you may not be able to see the circuit respond to changes on the input data bus.

Part 1 Procedure

1. Open the Max+plus II software and assign the project name **s-d-xcvr**. Assign the MAX7000S as the device family and EPM7128SLC84 as the device.

2. Open the Graphic Editor and construct the circuit shown in Figure 2.

3. Assign the clock input to pin 83 (see Appendix A). Assign all remaining input and output pins according to the user manual for the university board you will be programming later. Record all pin numbers in Table 1 assigned to the input and output pins.

Pin name	Pin number
D0	
D1	
D2	
D3	
D4	
D5	
D6	
D7	
Transmit-in	
Clock-in	
Clock	83
xmt-data-out	
xmt-clk-out	

Pin name	Pin number
xmt-data-in	
xmt-clk-in	
Q0	
Q1	
Q2	
Q3	
Q4	
Q5	
Q6	
Q7	
Z23	
Z22	
Z21	

Table 1

4. Open the **s-d-xcvr.rpt** file located on your disk. Highlight and print only the pin diagram with assigned names. Verify that all pins have correctly been assigned in accordance to the circuit cards user manual. Label this hard copy **Step 1, Part 4**.

5. Open a new Waveform Editor and create the waveforms as described below.
   - Change the end time (**File – End Time**) to 1.5 μs.
   - Create a waveform called "D[7..0]." Set the group value to B5. (Use the Overwrite Group Value button.)
   - Change Grid Size to 10 ns.
   - Create a waveform called "transmit-in."
   - Change Grid Size to 40 ns.

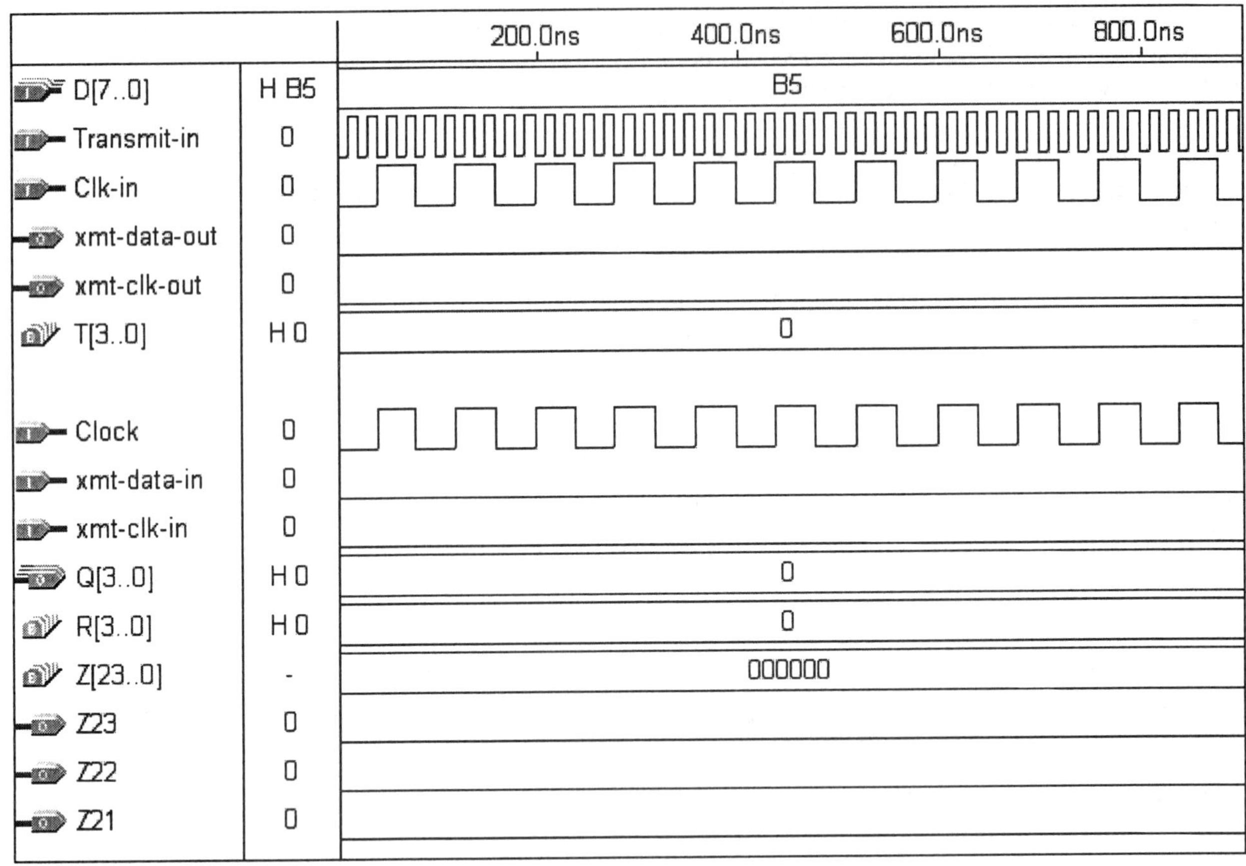

Figure 3

- Create a waveform called "clk-in."
- Create the rest of the waveforms as shown in Figure 3, including buried nodes T[4..3], R[3..0], and Z[22..0].

6. Initially xmt-data-in and xmt-clk-in will be left at a logic-LOW. Click the Simulate button. Correct all errors before continuing.

7. At this time, xmt-data-out and xmt-clk-out will show some activity. Zoom in as necessary to answer these questions.
   A. At what time does the xmt-data-out output first change logic states? ______________ ns
   B. At what time does the xmt-clk-out output first change logic states? ______________ ns

8. To simulate the receiver end of the transceiver, we need to create waveforms for the xmt-data-in and xmt-clk-in inputs in the Waveform Editor. Click and drag on the "value" logic-level indicator (current values are logic-LOWs) for the xmt-data-out and xmt-clk-out waveforms. Both waveforms and values (not including the waveform names) should be highlighted (see Figure 4). Press CTRL-C to copy the waveform to the Windows clipboard.

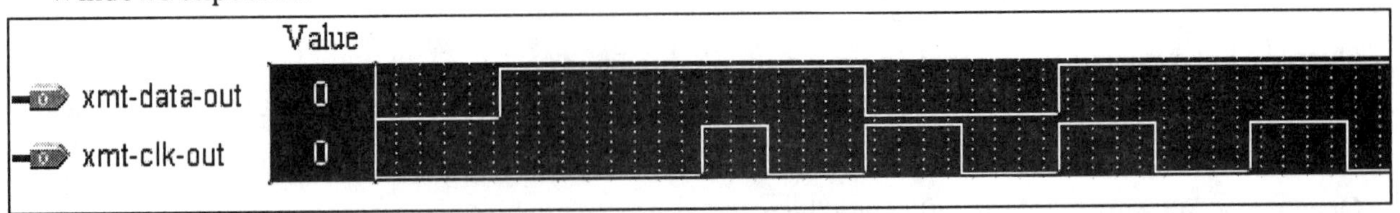

Figure 4

9. Click and drag on the "value" logic level of the xmt-data-in and xmt-clk-in waveforms. Both waveforms and values (not including the waveform name) should be highlighted. Press CTRL-V to paste the waveform from the Windows clipboard. When finished, both input waveforms should be identical to their respective output waveforms.

10. Click the simulate button. You should now see data activity on the Q[7..0] bus. At what time does B7 first appear on the Q[7..0] bus? __________ ns

11. Obtain a hard copy of the Graphic and Waveform Editor files. Label these hard copies **Part 1, Step 11A** and **Part 1, Step 11B**, respectively.

12. Proceed to Part 2.

Part 2 Procedure

1. Connect the university board containing the EPM7128LC84 CPLD to the computer printer port (or port as assigned by your professor).

2. Assign the project name **s-d-xcvr**.

3. If you have taken a break since completing Part 1, you should re-compile the project and verify you have the pins properly assigned according to the university board's user manual. This step is crucial in order to avoid damage to the CPLD.

4. Open the programmer. (Select **Max+plus** in the main menu then select **Programmer** in the list of features.)

5. Verify that the file name on the right side of the programming window is s-d-xcvr.pof and the device is EPM7128SLC84 (see Figure 5). If the file name is not correct, then repeat Step 2 and do Step 3.

6. If the Program button is black, the software has found your hardware and you only need to click the Program button to start programming your CPLD. If the Program button is a gray shade, the software did not establish a communication link between the computer and circuit card. In this case close the Programmer window, re-seat the cable on the circuit card on the printer port, connect power to the circuit card, and repeat Step 4 and Step 5.

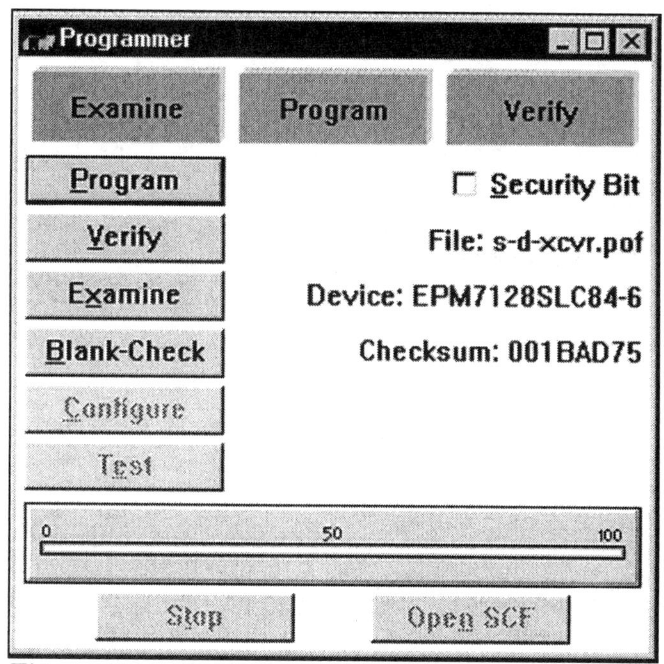

Figure 5

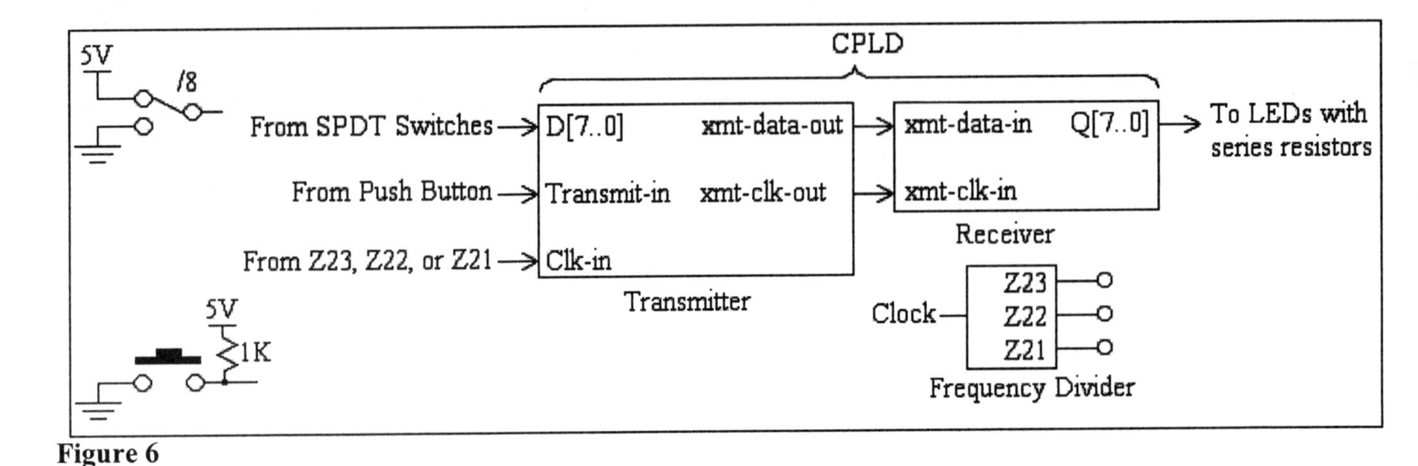

Figure 6

7. Wire your CPLD as illustrated in Figure 6. Wire eight input switches to D[7..0], and a pushbutton switch to Transmit-in. Connect LEDs with series 1 K resistors to outputs Q[7..0]. Connect a jumper wire between the xmt-data-out pin and the xmt-data-in pin. Connect a jumper between the xmt-clk-out pin and the xmt-clk-in pin. Select the frequency desired by connecting either Z23, Z22, Z21, or a low frequency (0.1 to 1 Hz pulse) from a signal generator to the clk-in pin of the CPLD.

8. Demonstrate that your data transmitter and data receiver operates. Have the instructor sign the answer page for this step.

9. Find another student who has demonstrated Step 7 operational.
   - Connect the xmt-data-out pin to the partner's xmt-data-in pin.
   - Connect the xmt-clk-out pin to the partner's xmt-clk-in pin.
   - Connect a jumper wire between the two circuit board ground pins.
   - Connect separate SPDT switches to the D[7..0] pins on both boards.
   - Connect separate LEDs with series resistors to Q[7..0] pins on both boards.

10. Demonstrate the bidirectional communications of Step 8 and Step 9 to your professor. Have the instructor sign the answer page for this step.

11. Obtain hard copies of the Graphic and Waveform Editor files. Label these hard copies **Part 2, Step 11A** and **Part 2, Step 11B**, respectively.

12. Exit the Graphic and Waveform Editors.

Part 3 Procedure

1. Disconnect all SPDT and pushbutton switches connected to D[7..0] and Transmit-in of your programmed CPLD from Part 2.

2. Construct the circuit shown in Figure 7. Refer to the data specification sheets (search and download from the Internet) for the ADC0804 analog-to-digital converter to wire the ADC0804 for free-running mode. Start with a jumper wire between test point TP1 and the input of the A/D converter. Be sure to have a common ground between the CPLD power supply and the analog circuit power supply.

3. Complete Table 2 for each test point voltage applied to the A/D converter analog input pin.

4. Demonstrate the network of Figure 7 to your instructor. Obtain the signature of approval directly on the answer page for this lab.

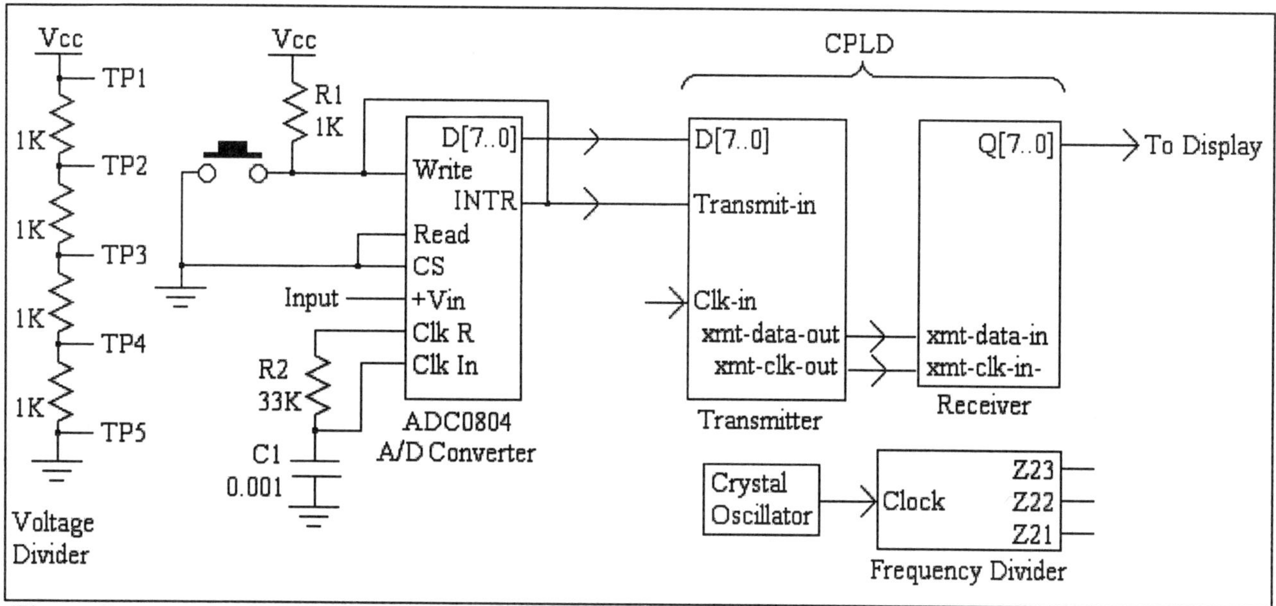

Figure 7

Test Point to A/D ACV input	Voltage Measured on Test Point	Bit Pattern on A/D Outputs	Bit Pattern on CPLD Output Bus
1	_______ VDC	$D_7\ D_6\ D_5\ D_4\ D_3\ D_2\ D_1\ D_0$ _ _ _ _ _ _ _ _	$Q_7\ Q_6\ Q_5\ Q_4\ Q_3\ Q_2\ Q_1\ Q_0$ _ _ _ _ _ _ _ _
2	_______ VDC	$D_7\ D_6\ D_5\ D_4\ D_3\ D_2\ D_1\ D_0$ _ _ _ _ _ _ _ _	$Q_7\ Q_6\ Q_5\ Q_4\ Q_3\ Q_2\ Q_1\ Q_0$ _ _ _ _ _ _ _ _
3	_______ VDC	$D_7\ D_6\ D_5\ D_4\ D_3\ D_2\ D_1\ D_0$ _ _ _ _ _ _ _ _	$Q_7\ Q_6\ Q_5\ Q_4\ Q_3\ Q_2\ Q_1\ Q_0$ _ _ _ _ _ _ _ _
4	_______ VDC	$D_7\ D_6\ D_5\ D_4\ D_3\ D_2\ D_1\ D_0$ _ _ _ _ _ _ _ _	$Q_7\ Q_6\ Q_5\ Q_4\ Q_3\ Q_2\ Q_1\ Q_0$ _ _ _ _ _ _ _ _
5	_______ VDC	$D_7\ D_6\ D_5\ D_4\ D_3\ D_2\ D_1\ D_0$ _ _ _ _ _ _ _ _	$Q_7\ Q_6\ Q_5\ Q_4\ Q_3\ Q_2\ Q_1\ Q_0$ _ _ _ _ _ _ _ _

Table 2

5. Write a technical report pertaining to the results of this lab. Your summary should include embedded graphics of each stage: the data register, the multiplexer, the demultiplexer, the control circuit for the transmitter, and the control circuit for the receiver. Include a paragraph or two explaining how each stage operates.

6. Staple and submit the following to your instructor for grading.
   - Cover page
   - Typed report
   - Completed answer page for this lab
   - Hard copy of pin diagram, **Part 1, Step 4**
   - Hard copies of Graphic and Waveform Editor files, **Part 1, Step 11A** and **Part 1, Step 11B**
   - Hard copies of Graphic and Waveform Editor files, **Part 2, Step 11A** and **Part 2, Step 11B**
   - Completed answer pages

Lab 25: Synchronous Data Transceiver Answer Page

Name: ____________________

Part 1

Pin name	Pin number
D0	
D1	
D2	
D3	
D4	
D5	
D6	
D7	
Transmit-in	
Clock-in	
Clock	83
xmt-data-out	
xmt-clk-out	

Table 1

Pin name	Pin number
xmt-data-in	
xmt-clk-in	
Q0	
Q1	
Q2	
Q3	
Q4	
Q5	
Q6	
Q7	
Z23	
Z22	
Z21	

7A. ____________ 7B. ____________ 10. ________ ns

Part 2

Step 8. Demonstrated to: ______________________________ Date: __________

Step 10. Demonstrated to: ______________________________ Date: __________

Part 3

Test Point to A/D ACV input	Voltage Measured on Test Point	Bit Pattern on A/D Outputs	Bit Pattern on CPLD Output Bus
1	_______ VDC	D_7 D_6 D_5 D_4 D_3 D_2 D_1 D_0 _ _ _ _ _ _ _ _	Q_7 Q_6 Q_5 Q_4 Q_3 Q_2 Q_1 Q_0 _ _ _ _ _ _ _ _
2	_______ VDC	D_7 D_6 D_5 D_4 D_3 D_2 D_1 D_0 _ _ _ _ _ _ _ _	Q_7 Q_6 Q_5 Q_4 Q_3 Q_2 Q_1 Q_0 _ _ _ _ _ _ _ _
3	_______ VDC	D_7 D_6 D_5 D_4 D_3 D_2 D_1 D_0 _ _ _ _ _ _ _ _	Q_7 Q_6 Q_5 Q_4 Q_3 Q_2 Q_1 Q_0 _ _ _ _ _ _ _ _
4	_______ VDC	D_7 D_6 D_5 D_4 D_3 D_2 D_1 D_0 _ _ _ _ _ _ _ _	Q_7 Q_6 Q_5 Q_4 Q_3 Q_2 Q_1 Q_0 _ _ _ _ _ _ _ _
5	_______ VDC	D_7 D_6 D_5 D_4 D_3 D_2 D_1 D_0 _ _ _ _ _ _ _ _	Q_7 Q_6 Q_5 Q_4 Q_3 Q_2 Q_1 Q_0 _ _ _ _ _ _ _ _

Table 2

Step 5. Demonstrated to: ______________________________ Date: __________

Grade: ________

Lab 26: LPM_AND

Objectives:

1. To learn how to group multiple buses into one bus
2. To create and demonstrate a circuit using the LPM_AND megafunction.

Materials list:

- Max+plus II software
- Computer requirements:
  Minimum 486/66 with 8 MB RAM
- University Board (optional)
- Floppy disk

Discussion:

For most applications, you should use the AND*n* function (*n* represents the number of inputs) for circuits. For system designs that are parameterized, one might select the LPM megafunction. A parameterized function is a function whose behavior is controlled by one or more parameters. For instance, if one needs to AND two single-bit variables, use the AND2 parameter. However, if multiple buses are to be ANDed, you could use the LPM_AND megafunction.

Altera recommends instantiating this function as described in Creating a Custom Megafunction Variation with the MegaWizard Plug-In Manager (see Apendix A). However, to give you exposure to using basic logic gate megafunctions, the LPM_AND megafunction will be discussed.

The Edit Ports/Parameters box for the LPM_ADD megafunction is shown in Figure 1. The Edit Ports/Parameter box is divided into two major parts, the ports and parameters.

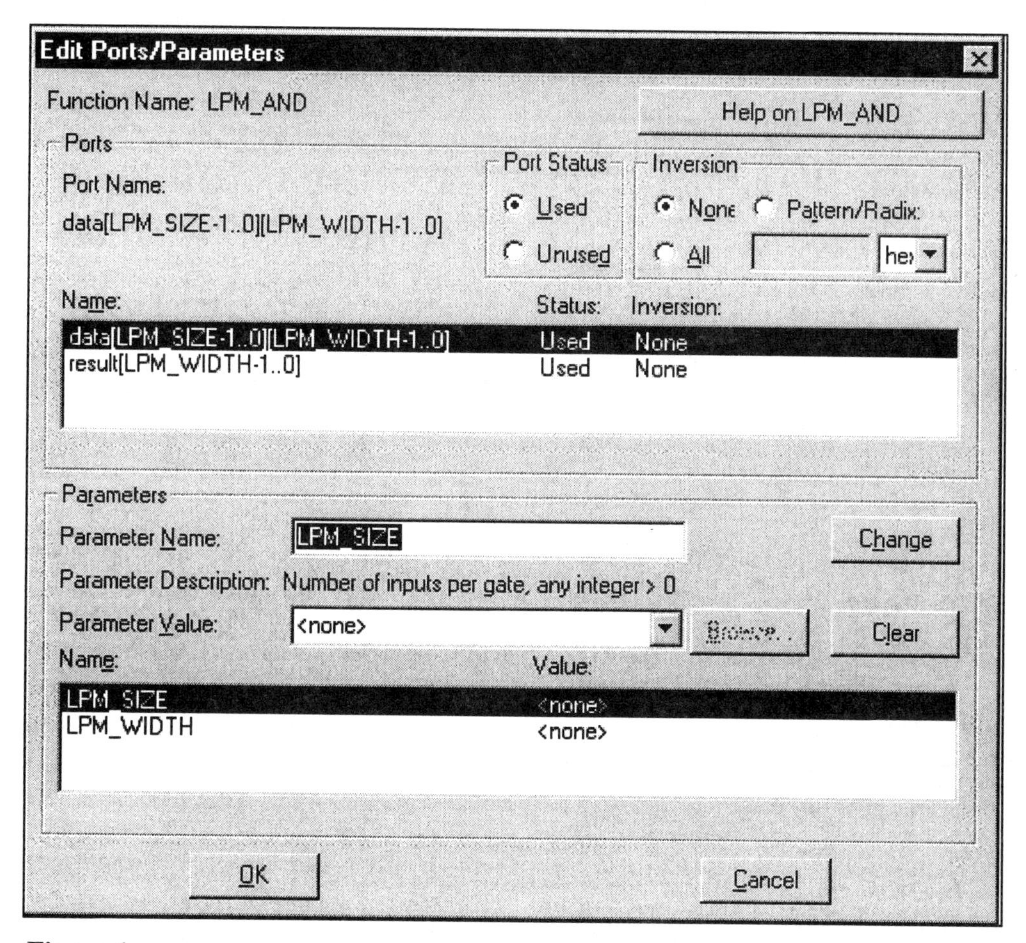

Figure 1

A port by definition is a symbolic name that represents an input or output of a design file. The ports available for the LPM_AND megafunction are data[LPM_SIZE-1..0][LPM_WIDTH-1..0] and result[LPM_WIDTH-1..0]. The LPM_SIZE is the number of input buses to be ANDed whereas the LPM_WIDTH is the bit size of each bus.

The status of a port is either "Used" or "Unused" and by default is active-HIGH (Inversion set to "none"). To make the port active-LOW and to insert a bubble on the port in the symbol, select "All" in the Inversion box (upper right).

The LPM parameter is an attribute of a megafunction or macrofunction that determines the logic created or used to implement the function; that is, characteristics that determine the size, behavior, or silicon implementation of a function. The parameters that need to be defined for the LPM_ADD megafunction are LPM_WIDTH and LPM_SIZE. To define each data input as a 4-bit bus, the LPM_WIDTH must be set to 4. To set the LPM_WIDTH value to 4, highlight LPM_WIDTH in the name field of the parameter section, place the cursor in the Parameter Value window (just above the name fields) and type 4. The LPM_SIZE may be defined in a similar matter.

Assume four 8-bit buses are to be ANDed together. If the buses are labeled A[7..0], B[7..0], C[7..0], and D[7..0], then one might construct the circuit shown in Figure 2.

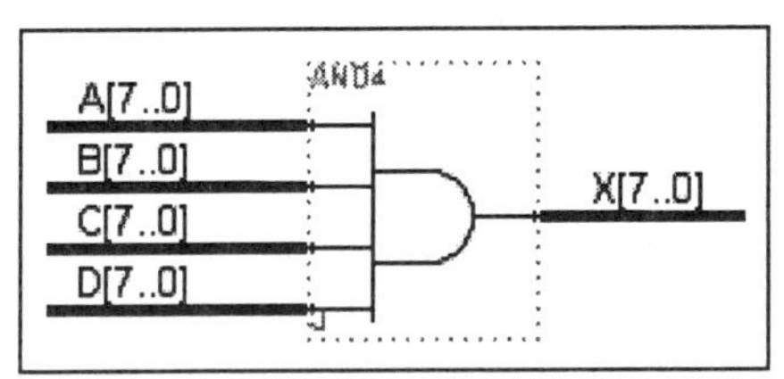

Figure 2

However, to use the LPM_AND megafunction, the inputs buses must share the same name; that is, A0_[7..0], A1_[7..0], A2_[7..0], and A3_[7..0], creating a 4-input (size) bus A[3..0][7..0] with 8 bits (width) for each input (see Figure 3).

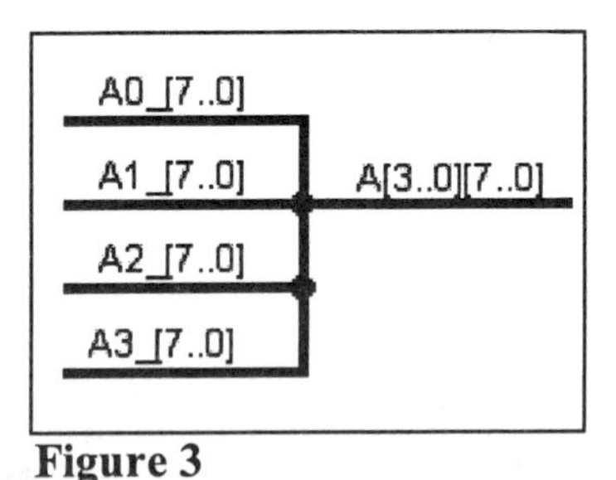

Figure 3

Read the Help-on-LPM_AND message for additional information regarding the features available for the LPM_AND megafunction.

Part 1 Procedure

1. Open the Max+plus II software. Assign the project name **4x8AND**. Assign the 7000S family but select "AUTO" as the device.

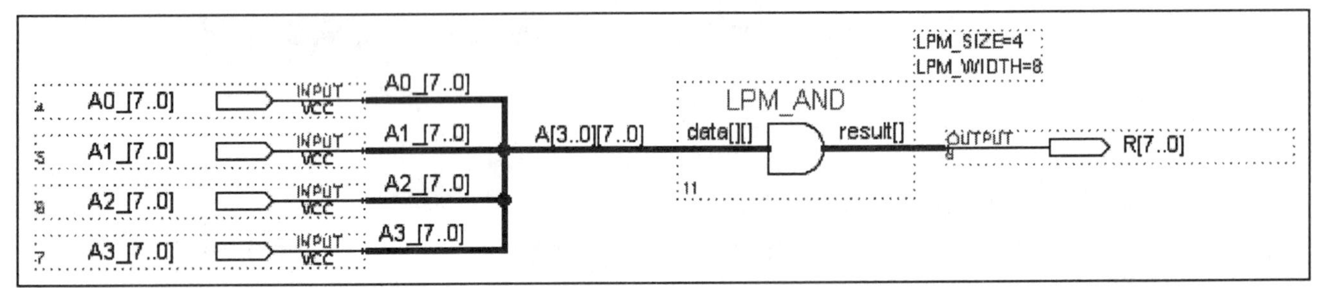

Figure 4

2. Open a new Graphic Editor file and construct the circuit shown in Figure 4.

3. Open a new Waveform Editor and create a set of four 8-bit input buses and one 8-bit output bus that share the names assigned in the Graphic Editor file.

4. Set Grid Size to 500 ns and define each input bus pattern as identified below.

Bus	Starting Value:	Increment by:
A0_[7..0]	56	67
A1_[7..0]	75	138
A2_[7..0]	C4	236
A3_[7..0]	b7	195

5. Click the Compile and Simulate buttons. Correct all errors before continuing.

6. Sketch the input and output waveform buses in Figure 5 after successfully simulating the circuit in Figure 4.

Name:	Value:	500.0ns 1.0
A0_[7..0]	H 00	
A1_[7..0]	H 00	
A2_[7..0]	H 00	
A3_[7..0]	H 00	
R[7..0]	H XX	

Figure 5

7. Based on the report file for this schematic, answer the following questions.

   A. What integrated circuit did the software assign? ________________

   B. What percentage of the chip was utilized for this design?__________

   C. How many input pins were assigned to the chip? __________

   D. For Step 4, the software assumes the numbers entered for the **Increment By:** values were (binary/decimal/hexadecimal/octal).

Assume two 2-bit numbers, A and B, are ANDed. First covert each number to binary, A_1A_0 and B_1B_0, then logical AND A_0 with B_0 and A_1 with B_1. For multiple numbers, AND all units bits together, $A_0 \bullet B_0$, then AND all 2s bits together, $A_1 \bullet B_1$, then AND all 4s bits together, $A_2 \bullet B_2$, and so on.

8. Convert each hex number in Table 1 for the inputs from 0 to 500 ns to binary, then logic-AND respective bits, A0[0] and A1[0] and A2[0] and A3[0], and so on. Equate the result back to hexadecimal.

Hex Value	(AND)	Binary Equivalent	
56		_ _ _ _ _ _ _ _ $_2$	
75	•	_ _ _ _ _ _ _ _ $_2$	
C4	•	_ _ _ _ _ _ _ _ $_2$	
B7	•	_ _ _ _ _ _ _ _ $_2$	
ANDed Result:		_ _ _ _ _ _ _ _ $_2$	= ______ $_H$

Table 1

9. Obtain a hard copy of the Graphic and Waveform Editor files. Label these **Lab 26, Step 9A** and **Lab 26, Step 9B**, respectively.

10. Demonstrate the circuit operation to your professor and obtain the signature of approval on the answer page.

11. Exit the Graphic and Waveform Editors. Save all files as **4x8AND**.

12. Write a 1- to 2- page technical summary relating to the LPM_ADD megafunction. Your summary should include the Edit Ports/Parameters dialog box and a description of each I/O port.

13. Submit the following in the sequence listed:
    - Cover page
    - Typed technical summary with embedded graphic
    - Completed answer page
    - Hard copy of the Graphic Editor, **Lab 26, Step 9A**
    - Hard copy of the Waveform Editor, **Lab 26, Step 9B**

Lab 26: LPM_ADD Answer Page

Name: ______________________

Part 1

Name:	Value:	500.0ns	1.0
A0_[7..0]	H 00		
A1_[7..0]	H 00		
A2_[7..0]	H 00		
A3_[7..0]	H 00		
R[7..0]	H XX		

Figure 5

7A. ____________________

7B. ____________________

7C. ____________________

7D. (binary/decimal/hexadecimal/octal)

Hex Value	(AND)	Binary Equivalent	
56		_ _ _ _ _ _ _ _$_2$	
75	•	_ _ _ _ _ _ _ _$_2$	
C4	•	_ _ _ _ _ _ _ _$_2$	
B7	•	_ _ _ _ _ _ _ _$_2$	
	ANDed Result:	_ _ _ _ _ _ _ _$_2$	= _____$_H$

Table 1

10. Demonstrated to: ________________________________ Date: __________

Grade: ________

Lab 27: LPM_ADD_SUB

Objectives:

1. Use the LPM _ADD_SUB symbol in a functional circuit
2. Write a technical report with embedded graphics

Materials required:

- Max+plus II software by Altera Corporation
- University Board with CPLD (optional)
- Computer requirements:
  Minimum 486/66 with 8 MB RAM
- Floppy diskette

Discussion:

The LPM_ADD_SUB symbol shown in Figure 1 may be used for signed or unsigned addition or subtraction. Synchronous operation may be implemented by using the clock input.

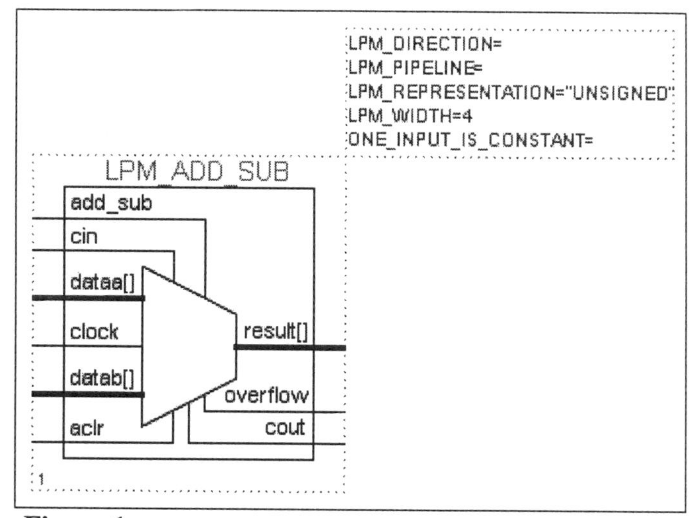

Figure 1

The dataa[ ], datab[ ], and results[ ] ports are required and LPM_WIDTH must be defined for basic summation. All other input and output ports are optional. Read the Help screen available in the Edit Ports/Parameters dialog box for pin usage and assignments. To open the Edit screen, double click on the Parameter box, upper-right corner of the LPM_ADD_SUB symbol. Click on the **Help on LPM_ADD_SUB** button for detailed descriptions of each port.

Part 1 Procedure

1. Create an asynchronous unsigned 4-bit adder with sum and carry outputs using the LPM_ADD_SUB symbol.

2. Create Waveform A[3..0], starting at 0, incrementing by 6 at a frequency of 10 MHz.

3. Create Waveform B[3..0], starting at 5, incrementing by 11 at a frequency of 10 MHz.

4. Using a word processor, create a lab cover page and write a technical summary explaining your circuit and accompanying waveforms. Embed the LPM_ADD_SUB symbol in your report and write a descriptive paragraph for each of these ports: add_sub, cin, dataa[ ], clock, datab[ ] aclr, result[ ], overflow, and count.

The waveforms, circuit, and pin assignments are to be embedded in (not attached to) your summary. All pages of your report must include a footer showing LPM_ADD_SUB and the page number as shown in the footer at the bottom of this page.

5. Modify your circuit for synchronous operation with a latency of 1. Set the clock frequency to 20 MHz.

6. Obtain hard copies of the circuit modification and waveforms verifying that the circuit operates in synchronous mode.

7. Compare the waveforms for asynchronous and synchronous modes of operation. Update your summary to include advantages of using synchronous mode on the LPM_ADD_SUB symbol.

8. Submit your summary and hard copies from Step 6 to your professor for review.

Lab 28: LPM_COMPARE

Objective:

Create and use the LPM _COMPARE symbol in a functional circuit.

Materials required:

- Max+plus II software by Altera Corporation
- University board with CPLD (optional)
- Computer requirements:
  - Minimum 486/66 with 8 MB RAM
- Floppy diskette

Discussion:

Standard TTL integrated circuits have physical limitations due to their fixed data widths and circuit functions. The OR gate in Figure 1 combines two output functions of the 7485 4-bit Magnitude Comparator to give a logic-HIGH output when Input A is greater than or equal to Input B. Additional logic gates are required to obtain functions such as less than, or equal to and not equal to. Using Very High Level Descriptive Language, VHDL, engineers can create their own logic circuit symbols with added features such as the LPM_COMPARE symbol (Figure 2).

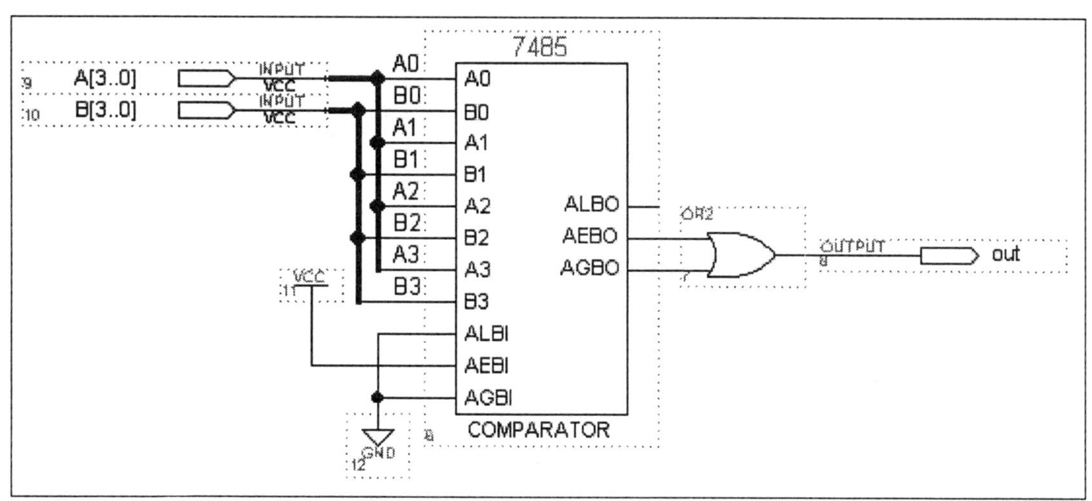

Figure 1

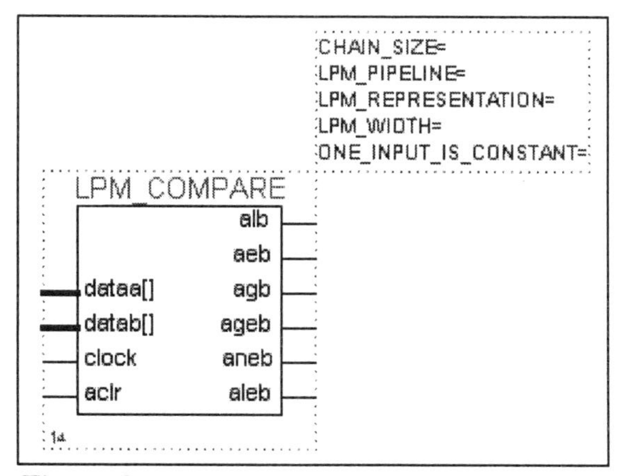

Figure 2

Added control lines such as clock and aclr make the comparator more functional. Also notice that cascade inputs are no longer necessary since the data width become software controlled, LPM_WIDTH, not hardware based.

The LPM_COMPARE megafunction is selected from the ../maxplus2/max2lib/mega_lpm directory. Complete the Edit Ports/Parameters dialog box (Figure 3) to identify ports and to define the required parameters for the symbol.

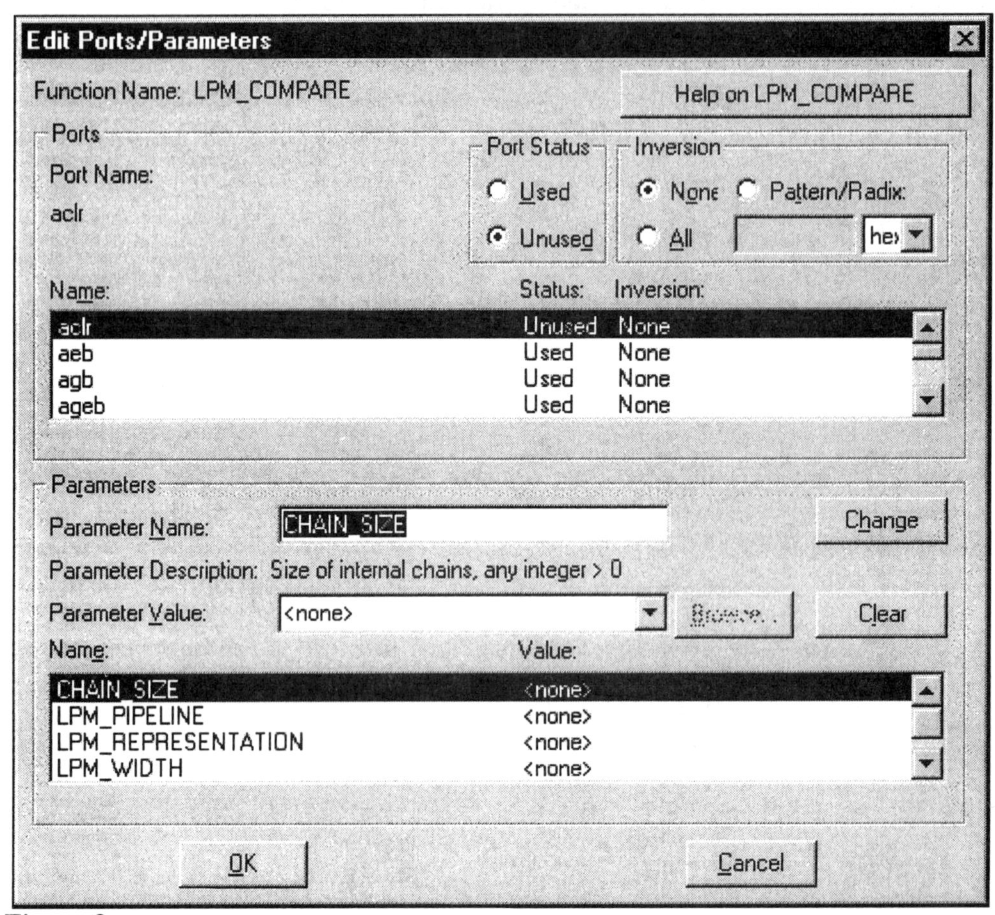

Figure 3

Several ports are available to select from; some are required whereas others are optional. Ports available for the LPM_Compare megafunction are aclr, aeb, agb, ageb, alb, aneb, clock, dataa[LPM_WIDTH-1..0] (required), and datab[LPM_WIDTH-1..0] (required). At least one output port must be selected. Select the **Help on LPM_Compare** button shown in the upper right of Figure 1 for a detailed description of each port and parameter usage.

The status of a port is either "Used" or "Unused" and by default is active-HIGH (Inversion set to "none"). To make the port active-LOW and to insert a bubble on the port in the symbol, select "All" in the Inversion box (upper right).

The LPM parameter is an attribute of a megafunction or macrofunction that determines the logic created or used to implement the function; that is, characteristics that determine the size, behavior, or silicon implementation of a function. The parameter that needs to be defined is LPM_WIDTH. To define each data input, A[3..0] or B[3..0], as a 4-bit bus, the LPM_WIDTH must be set to 4. To set the LPM_WIDTH value to 4, highlight LPM_WIDTH in the name field of the parameter section, place the cursor in the Parameter Value window (just above the name fields), and type 4.

Read the Help-on-LPM_DECODE message for additional information regarding the features available for the LPM_DECODE megafunction.

Part 1 Procedure

1. Create an asynchronous circuit using the LPM_COMPARE symbol to compare a 4-bit number (variable) to number 5 (constant). The output will be a logic-HIGH if the variable is equal to or greater than the constant, otherwise the output will be a logic-LOW.

2. Create a set of waveforms showing all possible combinations of the variable, demonstrating that the output satisfies the conditions as stated above.

3. Using a word processor, create a lab cover page and write a technical summary explaining your circuit and accompanying waveforms. Identify abnormal conditions that may occur in the output signal, time delays, parameter settings, and pin assignments. Include the following acronyms and identify their proper name and function: aeb, agb, ageb, alb, aleb, and aneb. The waveforms, circuit, and pin assignments are to be embedded in (not attached to) your summary. All pages of your report must include the footer, LPM_COMPARE, and page number.

4. Modify your circuit for synchronous operation with a latency of 1. The rising edge of the clock should occur at the same rate as the changing variable. Be sure the clock starts at a logic-LOW.

5. Obtain hard copies of the circuit modification and waveforms verifying the circuit operates in synchronous mode. Label these hard copies **Part 1, Step 5A** and **Part 1, Step 5B**, respectively.

6. Compare the waveforms for asynchronous and synchronous modes of operation. Update your summary to include advantages of using synchronous mode on the LPM_COMPARE symbol.

7. Construct a circuit that compares the output of a 4-bit binary counter to the output of a 4-bit Johnson counter. Both counters are to be asynchronously cleared in the first 5 ns. As the counters count, determine when each counter is greater than, equal to, or less than the other counter. You may use LPM functions for all parts.

8. Obtain hard copies of the Graphics and Waveform Editor files for the circuit you designed. Label these hard copies **Part 1, Step 8A** and **Part 1, Step 8B**, respectively.

9. Demonstrate the circuit designed to the instructor. Obtain the signature of approval directly on the hard copy of your .gdf file.

10. Update your technical summary comparing and contrasting the use of LPM_COMPARE megafunctions to using the 4-bit magnitude comparator IC, 7485. Provide embedded schematics to support your discussion.

11. Create a cover page for the lab and submit your summary and hard copies from Step 5 and Step 8 to your professor for grading.

12. Staple the following pages together in the sequence listed, then submit the packet to your instructor for grading.
    - Cover page
    - Typed summary
    - Hard copies of the Graphic and Waveform Editors, **Part 1, Step 5A** and **Part 1, Step 5B**
    - Hard copies of the Graphic and Waveform Editors, **Part 1, Step 8A** and **Part 1, Step 8B**

LAB 29: LPM_DECODE

Objectives:

1. Create an n-bit decoder using the LPM_DECODE megafunction.
2. Observe the affects of the clock input with a specific latency

Materials List:

- Max+plus II software
- Computer requirements:
  Minimum 486/66 with 8 MB RAM
- University board (optional)
- Floppy disk

Discussion:

Integrated circuits are nice if the application is well suited for the design of the chip; however, several chips need to be cascaded to create larger circuit functions. The megafunctions are best suited for large-scale circuits that are not readily available in a standard chip form. Megafunctions are easily expanded for n-bits by specifying the LPM_WIDTH parameter when selecting the megafunction. Written in VHDL, megafunctions have numerous asynchronous and synchronous inputs and outputs to choose from. Unlike standard integrated circuits, inputs and outputs (I/Os) of megafunctions can be selected or deselected. Only selected I/Os will appear on the symbol when inserted into the Graphic Editor work area.

Assume it is necessary to create a 1 of 32-decoder network. Either two 74154 4-line to 16-line or four 74138 3-line to 8-line macrofunction decoder chips (Lab 10) can be cascaded to compete the task. However by selecting the LPM_DECODE megafunction, it is necessary only to specify an LPM_WIDTH of 5 ($2^{\text{LPM\_WIDTH}} = 2^5 = 32$).

The notorious glitch common to asynchronous macrofunctions can be eliminated by using synchronous (clock dependent) inputs to the megafunction. When using a clock input, the eq[] output of the megafunction is registered and the number of clock pulses (latency) associated with the eq[] output is specified by the LPM_PIPELINE value.

Read the Help-on-LPM_DECODE message for additional information regarding the features available for the LPM_DECODE megafunction.

Part 1 Procedure

1. Open the Max+plus II software and assign the project name **lpm-dcdr**. Assign the MAX7000S family and EPM7128SLC84 as the device.

2. Open a new Graphic Editor. Open the Enter Symbol dialog box and select the LPM_DECODE megafunction in the **..\maxplus2\max2lib\mega_lpm** directory.

3. Set the status of the following I/Os.

Name	Status	Inversion
Aclr	Unused	None
Clock	Unused	None
Data[LPM_WIDTH-1..0]	Used	None
Enable	Unused	None
Eq[LPM_DECODES-1..0]	Used	None

4. Highlight LPM_WIDTH in the Parameters window, then enter 3 for the Parameter Value.

5. Click OK.

6. Construct the circuit shown in Figure 1.

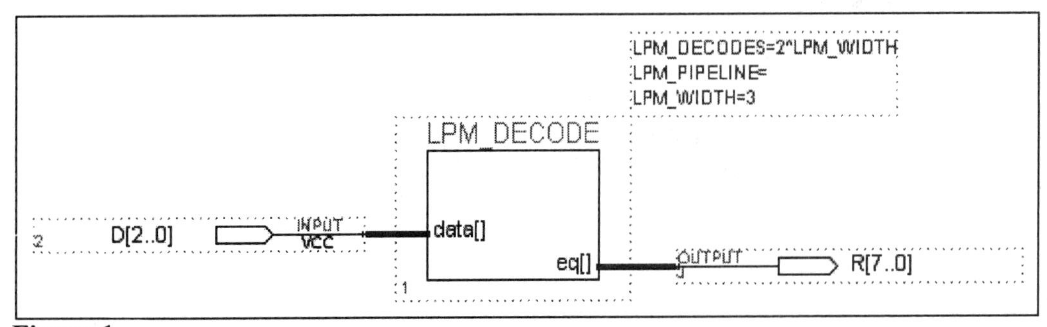

Figure 1

7. Open a new Waveform Editor file, set Grid Size to 100 ns, then create the waveforms shown in Figure 2.

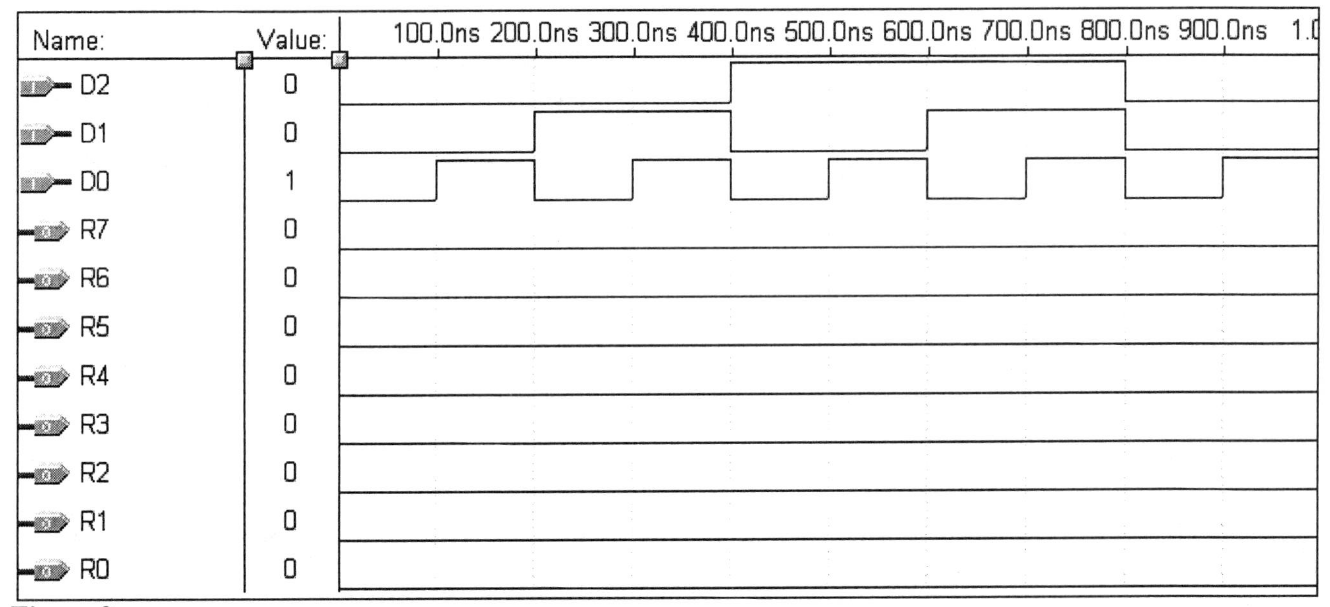

Figure 2

8. Compile and simulate the editor files. Correct all errors before continuing.

9. Open the Timing Analyzer and click Start in the Delay Matrix. The propagation delay for each output with respect to the changing data inputs is ______ ns.

10. Exit the Timing Analyzer and sketch the output waveform for R[7..0] in Figure 3.

Name:	Value:	100.0ns 200.0ns 300.0ns 400.0ns 500.0ns 600.0ns 700.0ns 800.0ns 900.0ns
R[7..0]	H 00	

Figure 3

11. Identify which output is active during the time intervals in Table 1.

Time	Active output (circle one)	Time	Active output (circle one)
0 – 100 ns	R_0 R_1 R_2 R_3 R_4 R_5 R_6 R_7	500 – 600 ns	R_0 R_1 R_2 R_3 R_4 R_5 R_6 R_7
100 – 200 ns	R_0 R_1 R_2 R_3 R_4 R_5 R_6 R_7	600 – 700 ns	R_0 R_1 R_2 R_3 R_4 R_5 R_6 R_7
200 – 300 ns	R_0 R_1 R_2 R_3 R_4 R_5 R_6 R_7	700 – 800 ns	R_0 R_1 R_2 R_3 R_4 R_5 R_6 R_7
300 – 400 ns	R_0 R_1 R_2 R_3 R_4 R_5 R_6 R_7	800 – 900 ns	R_0 R_1 R_2 R_3 R_4 R_5 R_6 R_7
400 – 500 ns	R_0 R_1 R_2 R_3 R_4 R_5 R_6 R_7	900 ns – 1 µs	R_0 R_1 R_2 R_3 R_4 R_5 R_6 R_7

Table 1

Part 2 Procedure

1. Make the necessary circuit modifications to the circuit of Figure 1, then add a clock input with a latency of 3 to the .gdf file.

2. Add a clock input to the Waveform Editor, change Grid Size to 10 ns, and then create the clock waveform using the Over Write Count button.

3. Compile and simulate the editor files. Correct all errors before continuing.

4. Zoom out until the Waveform Editor displays the first 200 to 300 ns of time, starting at 0 ns.

5. The value of the R[7..0] bus is initially zero. At what time does the R0 output change to a logic-HIGH? ________ ns

6. Count the number of leading edges of the clock until the R0 output changes from a logic-LOW to a logic-HIGH. The number of leading edges counted is ____________.

7. At 100 ns, the count of D[2..0] changes from 0 to 1. Starting at 100 ns, count the number of leading edges of the clock until the R1 output changes. The number of clock pulses should match the latency set in Part 2, Step 1.

8. Open the Timing Analyzer and click Start in the Delay Matrix. The propagation delay for each output with respect to the clock input is ______ ns.

9. Change the latency to 9.

10. Compile and simulate the editor files. Correct all errors before continuing.

11. The value of the R[7..0] bus is initially zero. At what time does the R0 output change to a logic-HIGH? ________ ns

12. Starting at 0 ns, count the number of leading edges of the clock until the R0 output changes from a logic-LOW to a logic-HIGH. The number of leading edges counted is ____________.

13. Zoom out until the entire 1 μs of time is visible in the Waveform Editor. Sketch the output bus in the area provided in Figure 4.

Name:	Value:	100.0ns 200.0ns 300.0ns 400.0ns 500.0ns 600.0ns 700.0ns 800.0ns 900.0ns
R[7..0]	H 00	

Figure 4

14. Change the latency to 15.

15. Compile and simulate the editor files. Correct all errors before continuing.

16. Zoom out until the entire 1 μs of time is visible in the Waveform Editor. Compare the results obtained to the recorded results in Figure 4. Explain what happens if the latency is set too high with respect to the rate of data change.

    ____

17. Obtain hard copies of the Graphic and Waveform Editors. Label these hard copies **Part 2, Step 17A** and **Part 2, Step 17B**, respectively.

18. Demonstrate the results obtained for a latency of 15 to the instructor. Obtain the signature of approval directly on the answer page for this lab.

19. Type a 1- to 2-page summary explaining how to configure the Edit Ports/Parameters screen to create a megafunction with a latency of 5. Identify the status for all inputs and outputs to meet the minimum requirements. Include the configured Edit Ports/Parameters screen as an embedded graphic in your summary.

20. Submit the following in the sequence listed to the instructor for grading.

- Cover page
- Typed summary
- Hard copy of the Graphic Editor, **Part 2, Step 17A**
- Hard copy of the Waveform Editor, **Part 2, Step 17B**
- Completed answer page for this lab

Lab 29: LPM_DECODE Answer Page

Name: ______________________________

Part 1

9. ________ ns

Name:	Value:	100.0ns	200.0ns	300.0ns	400.0ns	500.0ns	600.0ns	700.0ns	800.0ns	900.0ns
R[7..0]	H 00									

Figure 3

Time	Active output (circle one)		Time	Active output (circle one)
0 – 100 ns	R_0 R_1 R_2 R_3 R_4 R_5 R_6 R_7		500 – 600 ns	R_0 R_1 R_2 R_3 R_4 R_5 R_6 R_7
100 – 200 ns	R_0 R_1 R_2 R_3 R_4 R_5 R_6 R_7		600 – 700 ns	R_0 R_1 R_2 R_3 R_4 R_5 R_6 R_7
200 – 300 ns	R_0 R_1 R_2 R_3 R_4 R_5 R_6 R_7		700 – 800 ns	R_0 R_1 R_2 R_3 R_4 R_5 R_6 R_7
300 – 400 ns	R_0 R_1 R_2 R_3 R_4 R_5 R_6 R_7		800 – 900 ns	R_0 R_1 R_2 R_3 R_4 R_5 R_6 R_7
400 – 500 ns	R_0 R_1 R_2 R_3 R_4 R_5 R_6 R_7		900 ns – 1 µs	R_0 R_1 R_2 R_3 R_4 R_5 R_6 R_7

Table 1

Part 2

5. ________ ns   6. ____________   8. ________ ns

11. ________ ns   12. ____________

Name:	Value:	100.0ns	200.0ns	300.0ns	400.0ns	500.0ns	600.0ns	700.0ns	800.0ns	900.0ns
R[7..0]	H 00									

Figure 4

16. ____

____

18. Demonstrated to: ______________________________ Date: ______________

Grade: __________

Lab 30: LPM_ MUX

Objectives:

1. To configure the LPM_MUX megafunction for time division multiplexing
2. To configure the LPM_MUX megafunction for multiplexing buses

Materials list:

- Max+plus II software
- Computer requirements:
  Minimum 486/66 with 8 MB RAM
- University board (optional)
- Floppy diskette

Discussion:

Macrofunction chips like the 74150, 75151, 74153, or 74157 are designed for unique multiplexer functions with fixed number of inputs and configuration. The megafunction is unique in that the LPM_MUX can be programmed for any of the multiplexer configurations. It is recommended that the LPM_MUX be used for all multiplexer functions. This lab will explore circuit configurations for using the LPM_MUX for data bus multiplexing and time division multiplexing.

As with all megafunctions, the size represents the number of inputs or buses, and width represents the number of bits for each input. For time division multiplexing, the size is 1. For instance, the parallel data from a computer bus may be converted to a string of binary bits to be serially transmitted over long distances. For data selection of buses, the size is the number of buses to be multiplexed and the width is the number of bits in each bus. If two computers are to share one printer, the size would be 2 and the width of 8.

The symbol for the LPM_MUX is shown in Figure 1.

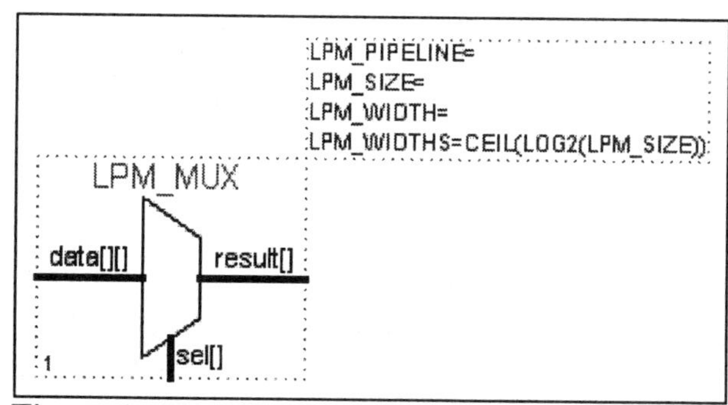

Figure 1

The port names for the LPM_MUX are identified in Table 1. Data[ ][ ], Result[ ], and Sel[ ] are required inputs for any multiplexer. Aclr (asynchronous clear) and clock may be used for synchronous control of the megafunction.

Port	Type	Required
Aclr	Input	No
Clock	Input	No
Data[ ][ ]	Input	Yes
Result[ ]	Output	Yes
Sel[ ]	Input	Yes

Table 1

Read the **Help-on-LPM_MUX** message for additional information regarding the features available for and how to label the input bus for the LPM_MUX megafunction.

Part 1 Procedure

1. Open the Max+plus II software. Assign the project name **time-div-mux**. Assign the 7000S family and the EPM7128SLC84 as the device.

2. Open a new Graphic Editor file and create a circuit illustrated in Figure 2, using the LPM_MUX megafunction.

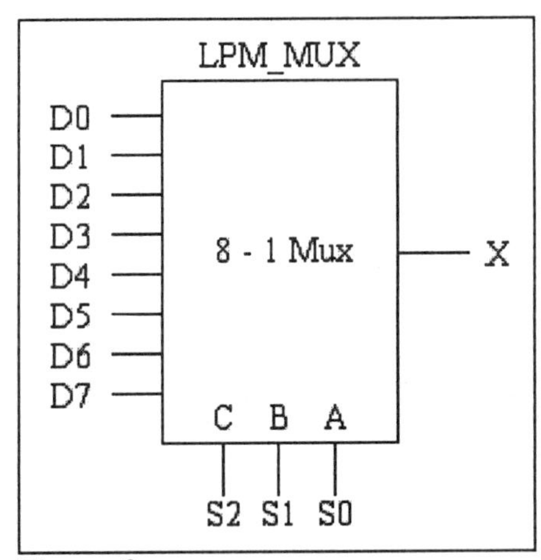

Figure 2

3. Open a new Waveform Editor file and create waveforms to demonstrate time division multiplexing. S2, S1, and S0 should change at a faster rate than the data inputs.

4. Obtain hard copies of the Graphic and Waveform Editor files. Label these hard copies **Part 1, Step 4A** and **Part 1, Step 4B**, respectively.

5. Save the files to drive A as **time-div-mux**, then exit the Graphic and Waveform Editor files.

Part 2 Procedure

1. Open the Max+plus II software. Assign the project name **bus-select**. Assign the 7000S family and the EPM7128SLC84 as the device.

2. Open a new Graphic Editor file and create a circuit illustrated in Figure 3 using the LPM_MUX megafunction.

3. Open a new Waveform Editor file and create waveforms to demonstrate the data selector of Figure 3.

4. Obtain hard copies of the Graphic and Waveform Editor files. Label these hard copies **Part 2, Step 4A** and **Part 2, Step 4B**, respectively.

5. Save the files to drive A as **bus-select**, then exit the Graphic and Waveform Editor files.

6. Demonstrate the bus multiplexer to the instructor. Obtain the signature of approval directly on the hard copy of the graphic design file.

7. Type a technical summary including embedded graphics detailing the procedure for creating a data selector network that directs data from one of sixteen 8-bit computers to a common printer.

8. Staple the pages in the sequence listed below and submit the packet to the instructor for grading.

- Cover page
- Typed summary
- Hard copy of the Graphic and Waveform Editor files, **Part 1, Step 4A** and **Part 1, Step 4B**
- Hard copy of the Graphic and Waveform Editor files, **Part 2, Step 4A** and **Part 2, Step 4B**

Lab 31: LPM_COUNTER

Objective:

This lab will step you through the process of creating a 4-bit binary up/down counter. There are many program options, but this lab will focus on the clock, counter enable, and the clear outputs.

Materials list:

- Max+plus II software by Altera Corporation
- University Board with CPLD (optional)
- Computer requirements:
  - Minimum 486/66 with 8 MB Ram
- Floppy disk

Discussion:

If you need a 4-bit binary or BCD counter, select a chip and use it in your circuit design. However, at times, larger counters are required that can be constructed using basic integrated circuits, but having a megafunction available simplifies the design considerably. The LPM_COUNTER function has all the standard synchronous and asynchronous inputs available that may be selected to customize a counter design. A dialog box appears when the LPM_COUNTER is selected, allowing one to select from a wide range of inputs by marking them as used or unused. The designer may assign either active-HIGH or active-LOW status to any one or more input or output.

It is assumed that students have a firm understanding of counters and know the differences between, advantages of, disadvantages of, and purpose of each asynchronous and synchronous input.

Part 1 Procedure

1. Open the Graphic and Waveform Editors. Assign the project name **binupctr**.

2. Select: **Symbol**, then **Enter Symbol**. Double click on the **\Max2lib\mega_lpm** in the Symbol Library. Scroll down Symbol Files and select the **lpm_counter**, then click **OK**.

3. An **Edit Ports/Parameters** dialog box should appear on the screen. Set the ports according to Table 1.

Name	Status
clock	used
cnt_en	used
q[LPM_WIDTH 1..0]	used
all other ports	unused

Table 1

4. Set **Parameters** as shown in Table 2, then click the OK button.

Name	Value
LPM_AVALUE	<none>
LPM_DIRECTION	up
LPM_MODULUS	16
LPM_SVALUE	<none>
LPM_WIDTH	4

Table 2

LPM_AVALUE is a constant value that is loaded when aset is active. The upper limit for LPM_AVALUE is 32 bits and is otherwise undefined (X) when the LPM_AVALUE exceeds the modulus of the counter as determined by LPM_MODULUS or 2^{LPM_WIDTH}.

The direction can be specified as UP, DOWN, or UNUSED by highlighting LPM_DIRECTION in the Parameters Name window and selecting the direction in the Parameter Value window using the drop-down menu. The default direction is UP.

The natural modulus for the counter is 2^{LPM_WIDTH}, or 16 for the 4-bit counter. The modulus can be changed to the maximum count plus 1 by specifying the value for the LPM_MODULUS. If not specified, the LPM_MODULUS defaults to 2^{LPM_WIDTH}.

LPM_SVALUE is a constant value loaded on the active edge of the clock when either sset or sconst is active. LPM_SVALUE must be used if sconst is used.

The number of bits for the counter is determined by the value assigned for LPM_WIDTH. The LPM_WIDTH must be specified for the counter.

5. Complete the circuit wiring as shown in Figure 1.

6. Click the Compiler's **Start** button. When finished, the Compiler Project Compilation window should show zero errors and one warning. Click the **OK** button.

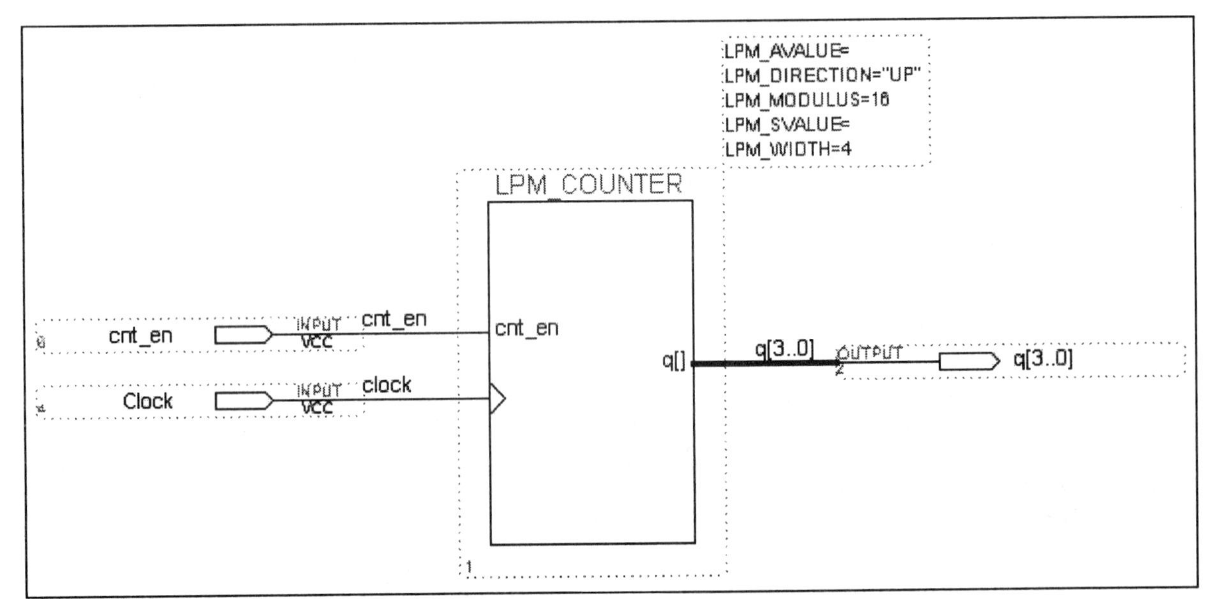

Figure 1

7. Double click on the **RPT** icon that appears below the Fitter box in the Compiler window and answer the following questions.

Question	Numeric Response
How many logic cells were used for this circuit?	a. _______
How many input pins were used?	b. _______
How many output pins were used?	c. _______
What percentage of the IC was used for this circuit?	d. _______
What was the total compilation time for this circuit?	e. _______
How many flip-flops were used for this IC?	f. _______
What name is assigned to unused pins?	g. _______

8. Identify the pin numbers assigned by the software to the following pins.

Pin Name	Pin Number
Clock	h. _______
sload	i. _______
aclr	j. _______
q[0]	k. _______
q[1]	l. _______
q[2]	m. _______
q[3]	n. _______

The software pin assignments may not be correct according to the university board hard-wiring of inputs and outputs. If one is to program the CPLD, then one must reassign pins according to the user manual for the circuit card.

9. Answer the following questions.

What name is assigned to unused pins?	o. ____________________
What pin number is VCCIN?	p. ____________________
Identify all VCCIO pins.	q. ____________________
Identify all the ground pins.	r. ____________________

10. Open the Waveform Editor and create the waveforms shown in Figure 2.

11. Click the Simulate button. Correct all errors before continuing.

Name:	Value:	100.0ns 200.0ns 300.0ns 400.0ns 500
Clock	0	
cnt_en	1	
Q[3..0]	H 0	0
Q3	0	
Q2	0	
Q1	0	
Q0	0	

Figure 2

12. Answer the following questions.

    What is the frequency of the clock? s. ____________________

    What direction does the counter count? t. ____________________

    The counter advances on the ___ edge. u. ____________________

    What is the count sequence? v. ____________________

13. Select **Max+plus II**, then **Timing Analyzer**. Click the **Start** button then click **OK**..

14. The Timing Analyzer shows the propagation delay time of the outputs with respect to the clock input. Record the delay: _______. Close the Timing Analyzer screen.

15. Save all files to Drive A as **binupctr**, then exit the Graphic and Waveform Editors.

Part 2 Procedure

1. Open the Max+plus II software. Assign the project name **updnctr**.

2. Using the LPM_COUNTER megafunction, create an 8-bit up/down counter.

3. Open the Waveform Editor, create clock pulses with a 10 ns period (5 ns pulse width), then verify the counter counts up or down as determined by the LPM_DIRECTION designator. (You will not see the entire count sequence unless you change the End Time.)

4. Obtain hard copies of the Graphic and Waveform Editor files. Label these hard copies **Part 2, Step 4A** and **Part 2, Step 4B**, respectively.

5. Save all files to Drive A as **updnctr** but do not exit the Graphic and Waveform Editors.

6. Modify the parameters box for the LPM_COUNTER in the Graphic Editor to create a Mod 65536 UP counter. Save the Graphic Editor file and change the project name to **mod65536**.

7. Modify your waveforms to reflect the changes in the Graphic Editor file. Save the Waveform Editor file to your disk in Drive A as **mod65536**, then demonstrate the operational circuit to your instructor. Obtain the signature of approval directly on the answer page for this lab.

8. Obtain hard copies of the Graphic and Waveform Editor files. Label these hard copies **Part 2, Step 8A** and **Part 2, Step 8B**, respectively.

9. Write a 1- to 2-page technical summary pertaining to the results obtained for this lab. Include the schematic of the 8-bit binary up/down counter that was created for Step 2 of Part 2.

10. Submit the following for grading:

    - Cover page
    - Typed summary with embedded graphics
    - Completed answer page for this lab
    - Hard copy of the Graphic and Waveform Editor files, **Part 2, Step 4A** and **Part 2, Step 4B**
    - Hard copy of the Graphic and Waveform Editor files, **Part 2, Step 8A** and **Part 2, Step 8B.**

Lab 31: LPM_COUNTER Answer Pages

Name: ______________________________

Part 1

7. 8.

Question	Numeric Response	Pin Name	Pin Number
How many logic cells were used for this circuit?	a. _______	Clock	h. _______
How many input pins were used?	b. _______	sload	i. _______
How many output pins were used?	c. _______	aclr	j. _______
What percentage of the IC was used for this circuit?	d. _______	q[0]	k. _______
What was the total compilation time for this circuit?	e. _______	q[1]	l. _______
How many flip-flops were used for this IC?	f. _______	q[2]	m. _______
What name is assigned to unused pins?	g. _______	q[3]	n. _______

9.

What name is assigned to unused pins? o. ______________________________

What pin number is VCCIN? p. ______________________________

Identify all VCCIO pins. q. ______________________________

Identify all the ground pins. r. ______________________________

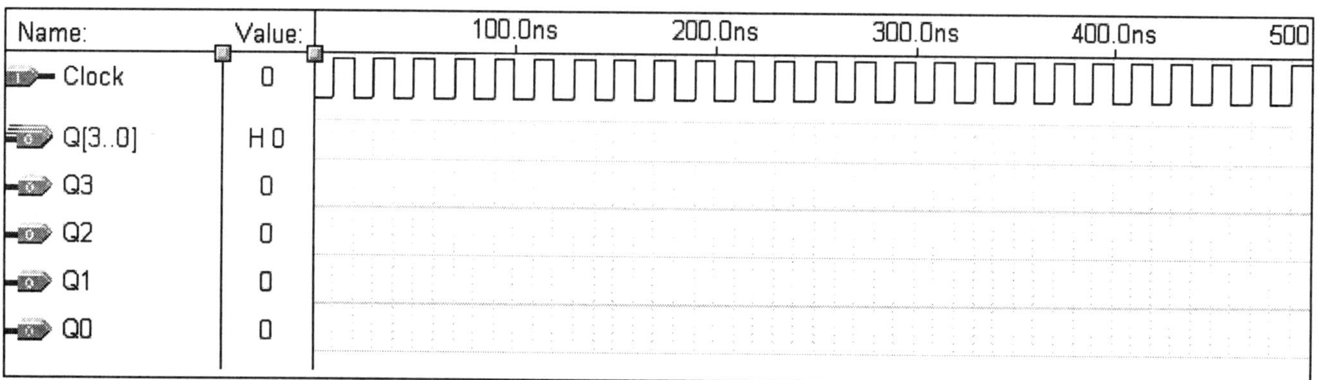

Figure 2

12.

What is the frequency of the clock? s. ______________________________

What direction does the counter count? t. ______________________________

The counter advances on the ___ edge. u. ______________________________

What is the count sequence? v. ______________________________

14. ____________

Part 2

7. Demonstrated to: ______________________________________ Date: _______________

Grade: _____________

Lab 32: LPM_SHIFTREG

Objectives:

1. Creating a 4-bit Si/So shift register using the LPM_SHIFTREG megafunction.
2. Creating a 4-bit Pi/Po shift register using the LPM_SHIFTREG megafunction.
3. Configure the LPM_SHIFTREG as a n-bit ring counter
4. Configure the LPM_SHIFTREG as a n-bit Johnson counter

Materials list:

- Max+plus II software by Altera Corporation
- University Board (optional)
- Computer requirements:
  - Minimum 486/66 with 8 MB Ram
- Floppy disk

Discussion:

The LPM_SHIFTREG is a universal shift register that has numerous synchronous and asynchronous inputs that allow a wide range of circuit configurations. At least one of the inputs: data, aset, aclr, sset, sclr, and/or shiftin ports, must be used in a design. Also, at least one of the outputs, q and/or shiftout, must be used.

The shift pattern of macrofunctions, like the 7491 or 74194, was defined by the conventional drafting techniques of drawing flip-flops on a page. Signal flow was from the left to the right and top to bottom. To show a 4-bit register using flip-flops, Qa would be placed on the left side of the page and Qd on the right. Right shifting would then be from flip-flop A towards flip-flop D. For left shifting, Qd would be drawn on the left side of the page and signal flow from Qd towards Qa.

When microprocessors were introduced, all internal parts were represented by blocks in a programmer's model. Here, the most significant bit was always placed on the left, as with all numbers in mathematics. The direction of data shift was therefore defined by the direction on the page. Right shift was Qd to Qa and left shift was Qa to Qd. This is confusing to beginners. Because the definition for data shift is dependent on the device you are referring to.

The shift pattern of megafunctions matches the definitions for shifting data using software within microcomputers and microcontrollers.

	Macrofunction: 7491 or 74194	**LPM_SHIFTREG and Microprocessors**
Serial Right	So ← Qd ← Qc ← Qb ← Qa ← Si	Si → Qd → Qc → Qb → Qa → So
Serial Left	Si → Qd → Qc → Qb → Qa → So	So ← Qd ← Qc ← Qb ← Qa ← Si

The basic 4-bit (LPM_WIDTH = 4) serial-in serial-out right shift register configuration using the LPM_SHIFTREG is shown in Figure 1. The LPM_DIRECTION options are "Left" or "Right." One could use Q[LPM_WIDTH-1..0] instead of shiftout, but then the output bus must be defined and be sure to take either Q0 or Q3 as the serial output, depending on the direction selected.

Data can be inverted during the shift pattern by setting Inversion to "All" when selecting either the data input or data output name in the port definitions of the Edit Ports/Parameters window. Double clicking on the parameter box (upper right box) of the symbol will open the Edit Ports/Parameters window if additional input or output features need changing.

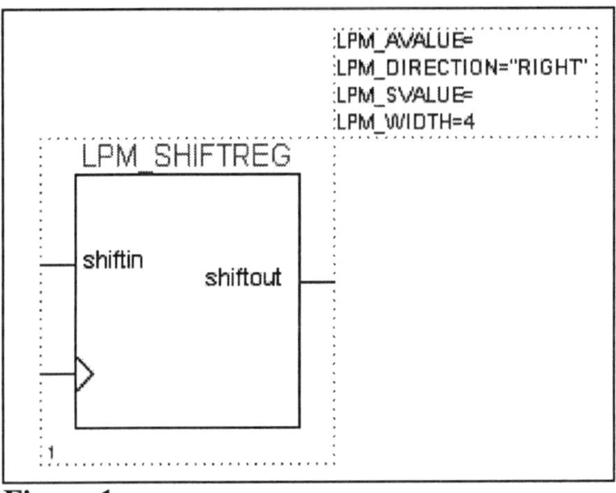

Figure 1

Part 1 Procedure

1. Open the Graphic and Waveform Editors. Assign the project name **siso-lpm-reg**.

2. Select: **Symbol**, then **Enter Symbol**. Double click on the **\Max2lib\mega_lpm** in the Symbol Library. Scroll down Symbol Files and select **lpm_shiftreg**, then click **OK**.

3. An **Edit Ports/Parameters** dialog box should appear on the screen. Identify whether the port is an input or output and if the port is required to create the circuit of Figure 1. Record your responses in Table 1.

Port	I/O	Required	Inversion
Aclr	In/ Out	Yes/No	All/None
Aset	In/ Out	Yes/No	All/None
Clock	In/ Out	Yes/No	All/None
Data[LPM_WIDTH-1..0]	In/ Out	Yes/No	All/None
Enable	In/ Out	Yes/No	All/None
Load	In/ Out	Yes/No	All/None
Q[LPM_WIDTH-1..0]	In/ Out	Yes/No	All/None
Sclr	In/ Out	Yes/No	All/None
Shiftin	In/ Out	Yes/No	All/None
Shiftout	In/ Out	Yes/No	All/None
Sset	In/ Out	Yes/No	All/None

Table 1

4. Add inputs and outputs to the circuit of Figure 1, then open a new Waveform Editor file and create the waveforms shown in Figure 2. Grid Size is set to 50 ns.

5. Identify the data logic level on the rising edge of each clock, then starting with the first data bit, convert this serial string of data to hexadecimal.

Rising edge:	1st	2nd	3rd	4th	:	5th	6th	7th	8th	:	9th	10th
Data bit:	___	___	___	___	:	___	___	___	___	:	___	___
Hexadecimal:				______H	:				______H	:		

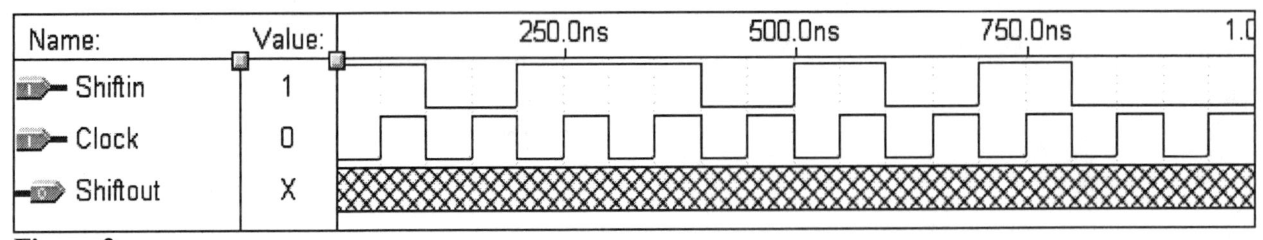

Figure 2

6. Compile and simulate the circuit. Correct all errors before continuing.

7. Sketch the output waveform in Figure 3 for the circuit in Figure 1.

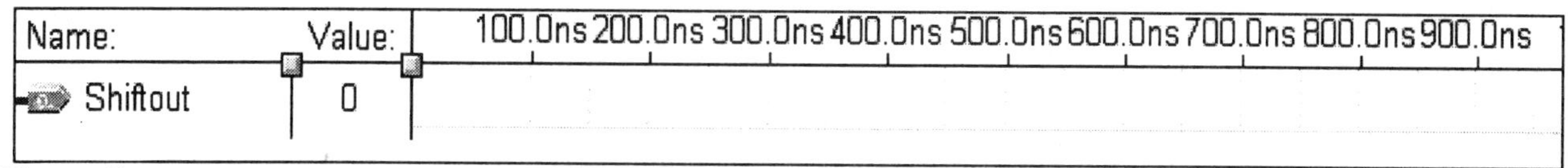

Figure 3

8. How many clock pulses did it take before the shiftin data appeared on shiftout? __________

9. Modify the circuit to make an 8-bit parallel-in serial-out right shift register. Use only inputs and outputs identified in Figure 4. Do mark the status of unnecessary inputs and outputs to unused. Refer to Help on LPM_SHIFTREG for identifying necessary inputs for this function. Save your modifications as **piso-lpm-reg.gdf**.

10. Set Grid Size to 25 ns and create the waveforms shown in Figure 4.

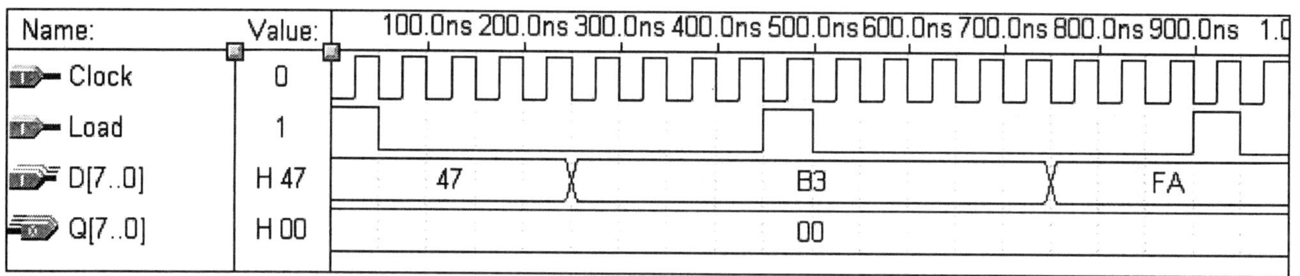

Figure 4

11. Compile and simulate the circuit. Correct all errors before continuing.

12. Complete Table 2 based on the Q[7..0] outputs observed after simulation.

Rising edge of the clock	Q[7..0] becomes:	Convert Q to binary Q_7 Q_6 Q_5 Q_4 Q_3 Q_2 Q_1 Q_0
1st	______H	_ _ _ _ _ _ _ _
2nd	______H	_ _ _ _ _ _ _ _
3rd	______H	_ _ _ _ _ _ _ _
4th	______H	_ _ _ _ _ _ _ _
5th	______H	_ _ _ _ _ _ _ _
6th	______H	_ _ _ _ _ _ _ _
7th	______H	_ _ _ _ _ _ _ _
8th	______H	_ _ _ _ _ _ _ _
9th	______H	_ _ _ _ _ _ _ _

Table 2

13. Based on the recorded results, what data bit is serially inputted as the parallel input data is serially outputted? (a logic-LOW/a logic-HIGH)

14. Explain what modifications to the circuit would be necessary to serial in the opposite logic level from your recorded response to Step 13.

____

15. Based on the simulated results, identify at what times the LPM_SHIFTREG parallel loaded data.
____________, ____________, and ____________

16. Explain what circuit modifications are required if you want to shift left instead of shifting right.

____

17. Obtain hard copies of the Graphic and Waveform Editor files. Label these hard copies **Part 1, Step 17A** and **Part 1, Step 17 B**, respectively.

18. Save all files as **piso-lpm-reg**, then exit the Graphic and Waveform Editors.

Part 2 Procedure

1. Using the LPM_SHIFTREG megafunction, create an 8-bit right-shift ring counter. Add support circuitry to parallel load 01_H into the ring counter on the first clock pulse after a Reset command. After the first clock pulse, the circuit will ring 01_H until the next reset occurs. Your completed circuit will contain only two inputs, clock and reset, and one output, shiftout.

2. Create a set of waveforms showing Clock, Reset, and Shiftout. Apply enough clock pulses between each reset to see the entire ring counter shift pattern.

3. Demonstrate your ring counter to the instructor. Obtain the signature of approval directly on the answer page for this lab.

4. Obtain hard copies of the Graphic and Waveform Editor files. Label these hard copies **Part 2, Step 4A** and **Part 2, Step 4B**, respectively.

5. Exit the Graphic and Waveform Editors.

Part 3 Procedure

1. Using the LPM_SHIFTREG megafunction, create an 8-bit Johnson counter with an asynchronous reset input. Your completed circuit will contain only two inputs, clock and reset, and one output, shiftout.

2. Create a set of waveforms showing Clock, Reset, and Shiftout. Apply enough clock pulses between each reset to see the entire Johnson Counter shift pattern.

3. Demonstrate your Johnson counter to the instructor. Obtain the signature of approval directly on the answer page for this lab.

4. Obtain hard copies of the Graphic and Waveform Editor files. Label these hard copies **Part 3, Step 4A** and **Part 3, Step 4B**, respectively.

5. Exit the Graphic and Waveform Editors.

6. Type a summary with embedded graphics based on the ring and Johnson circuits using the megafunction.

7. Staple the following pages in the sequence listed, then submit the packet to the instructor for grading.

   - Cover page
   - Typed summary
   - Hard copies of the Graphic and Waveform Editor files, **Part 1, Step 17A** and **Part 1, Step 17B**
   - Hard copies of the Graphic and Waveform Editor files, **Part 2, Step 4A** and **Part 2, Step 4B**
   - Hard copies of the Graphic and Waveform Editor files, **Part 3, Step 4A** and **Part 3, Step 4B**

Lab 32: LPM_SHIFTREG Answer Pages

Name: ______________________________

Part 1

3.

Port	I/O	Required	Inversion
Aclr	In/ Out	Yes/No	All/None
Aset	In/ Out	Yes/No	All/None
Clock	In/ Out	Yes/No	All/None
Data[LPM_WIDTH-1..0]	In/ Out	Yes/No	All/None
Enable	In/ Out	Yes/No	All/None
Load	In/ Out	Yes/No	All/None
Q[LPM_WIDTH-1..0]	In/ Out	Yes/No	All/None
Sclr	In/ Out	Yes/No	All/None
Shiftin	In/ Out	Yes/No	All/None
Shiftout	In/ Out	Yes/No	All/None
Sset	In/ Out	Yes/No	All/None

Table 1

5.

Rising edge:	1st	2nd	3rd	4th	:	5th	6th	7th	8th	:	9th	10th
Data bit:	___	___	___	___	:	___	___	___	___	:	___	___
Hexadecimal:			______H		:			______H		:		

Name:	Value:	100.0ns 200.0ns 300.0ns 400.0ns 500.0ns 600.0ns 700.0ns 800.0ns 900.0ns
Shiftout	0	

Figure 3

8. ___________

12.

Rising edge of the clock	Q[7..0] becomes:	Convert Q to binary Q_7 Q_6 Q_5 Q_4 Q_3 Q_2 Q_1 Q_0
1st	______H	__ __ __ __ __ __ __ __
2nd	______H	__ __ __ __ __ __ __ __
3rd	______H	__ __ __ __ __ __ __ __
4th	______H	__ __ __ __ __ __ __ __
5th	______H	__ __ __ __ __ __ __ __
6th	______H	__ __ __ __ __ __ __ __
7th	______H	__ __ __ __ __ __ __ __
8th	______H	__ __ __ __ __ __ __ __
9th	______H	__ __ __ __ __ __ __ __

Table 2

13. (a logic-LOW/a logic-HIGH)

14. ____

15. ___________, ___________, and ___________

16. ____

Part 2

3. Demonstrated to: ____ Date: ________________

Part 3

3. Demonstrated to: ____ Date: ________________

Grade: _________

Appendix A: How do I ...

Assign a Project Name

From the main menu, select **File - Project - Name...** Complete the **Project Name** dialog box (Figure 1).

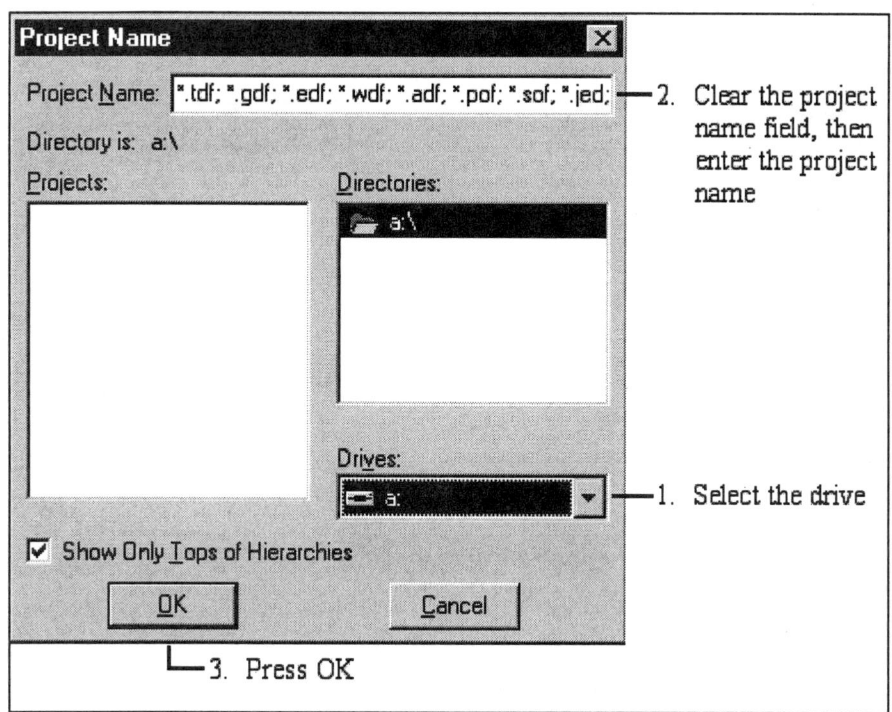

Figure 1

Open New Graphic or Waveform Editor

From the main menu, select **File - New....** Complete the **New** dialog box (Figure 2).

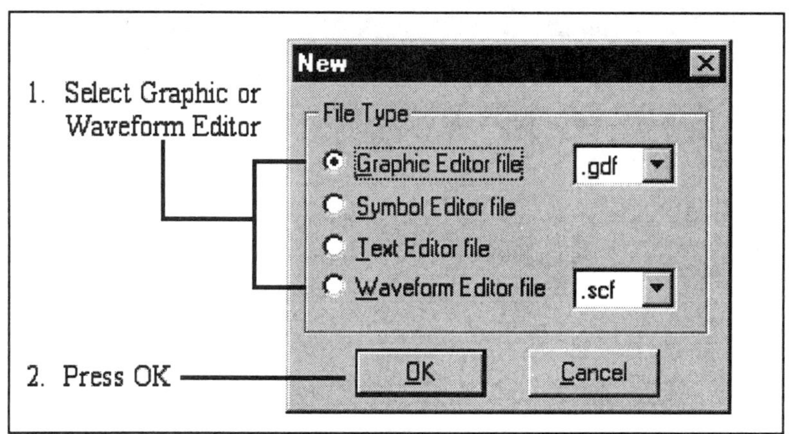

Figure 2

Change the Font or Line Style

From the main menu, select **Option - Font** or **Option - Text Size** (Figure 3). Select the font or line style desired. The Graphic Editor must be in the foreground for this option.

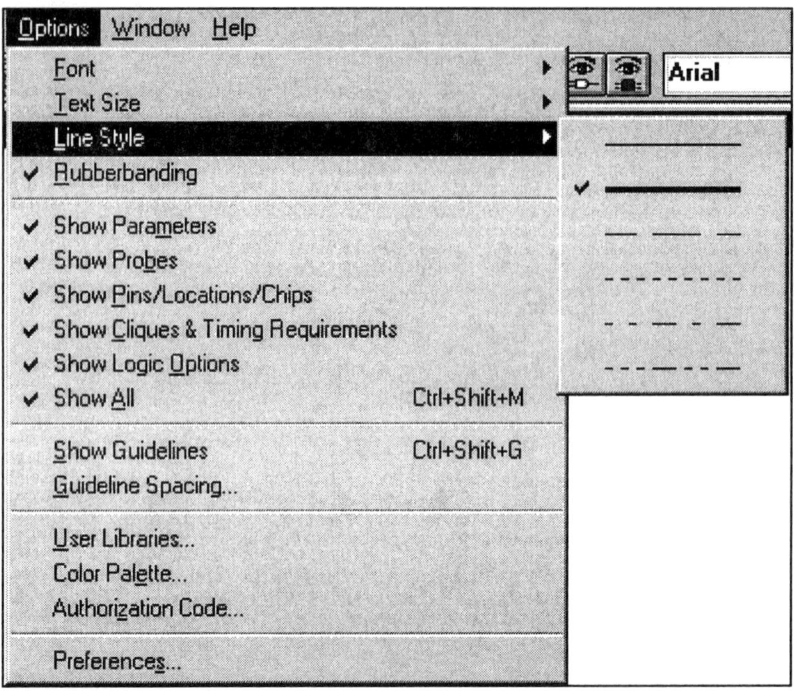

Figure 3

Insert Nodes from the SNF File

When a new Waveform Editor is opened, highlight the first name field, then right click on the mouse button. Select **Enter Nodes from SNF...** (see Figure 4) in the drop-down menu.

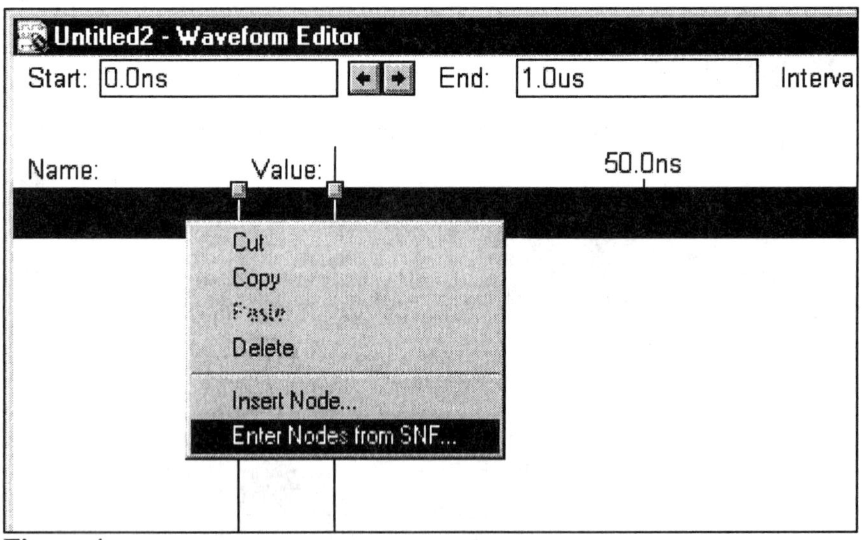

Figure 4

Click List (1) in the top right of the Enter nodes from SNF window (Figure 5), select the nodes (2) then click the Select arrow (3). Click OK. All nodes selected appear in the name fields of the Waveform Editor.

Once created, the radix of a bus can be changed by double clicking on the bus waveform (Figure 7).

Change the Time Base Beyond 1 μs

Bring the Waveform Editor to the foreground. From the main menu, select **File** - **End Time** and enter the end time (Figure 6). Units are necessary, otherwise nanoseconds will be assumed. You will need to redefine the waveforms in the Graphic Editor after changing the time base.

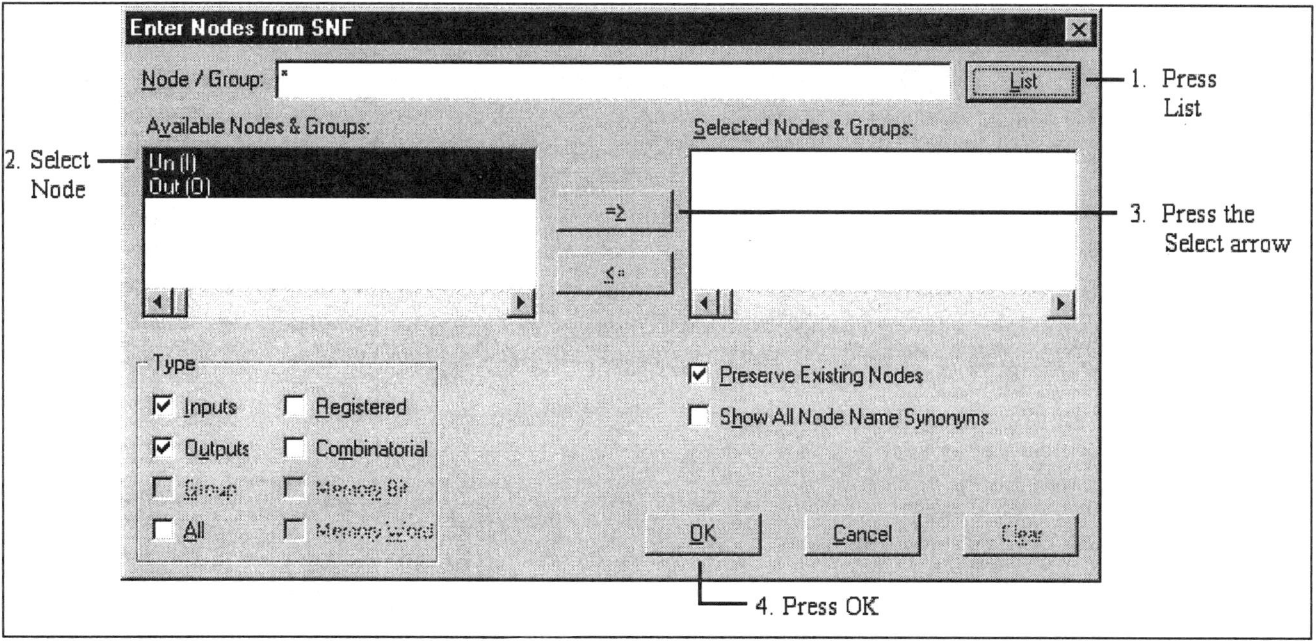

Figure 5

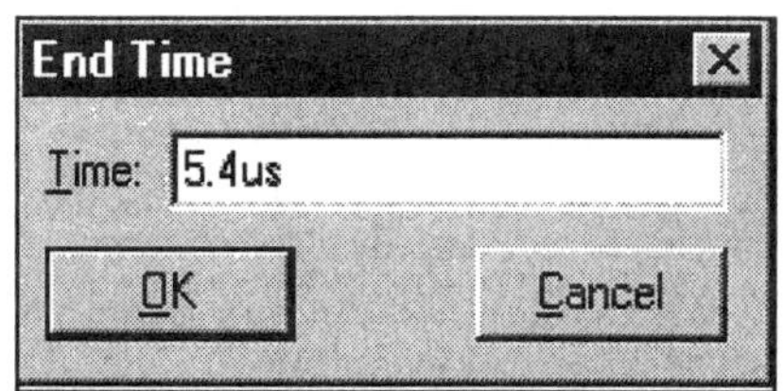

Figure 6

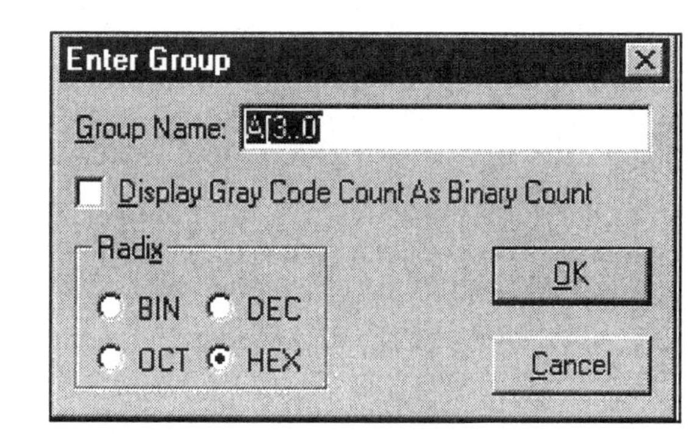

Figure 7

Be sure the End Time in the Simulator window also reflects the time you specified prior to simulating the waveforms (Figure 8).

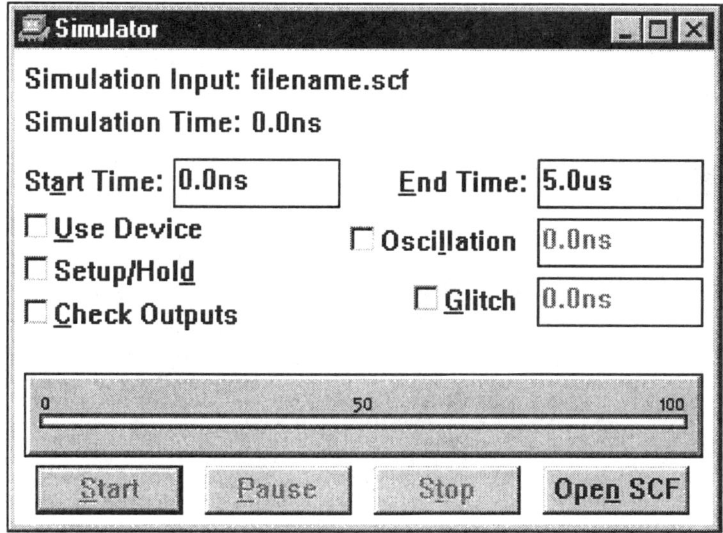

Figure 8

Change the Pin Assignments

Max+plus II automatically assigns pin numbers to inputs and outputs during the compilation of your Graphic Editor file. These pin numbers may be manually changed if necessary. There are eight embedded array blocks (EAB), each containing I/O logic cells in the EPM7128SLC84 chip (see the Floor Plan Editor). The Max+plus II software allocates the logic cells (LC) within each EAB module to "best fit" your design. Some university boards may not have external access to all pins on the EPM7128SLC84 chip so you may find it necessary to reassign some pins for your hardware layout. Refer to the user manual for the university board to determine hardware restrictions for input and output pins.

1. To reassign a pin, select Assign - Pin/Location/Chip in the main menu. Select "Assign Device..." (1) in the Pin/Location/Chip dialog box (Figure 9).

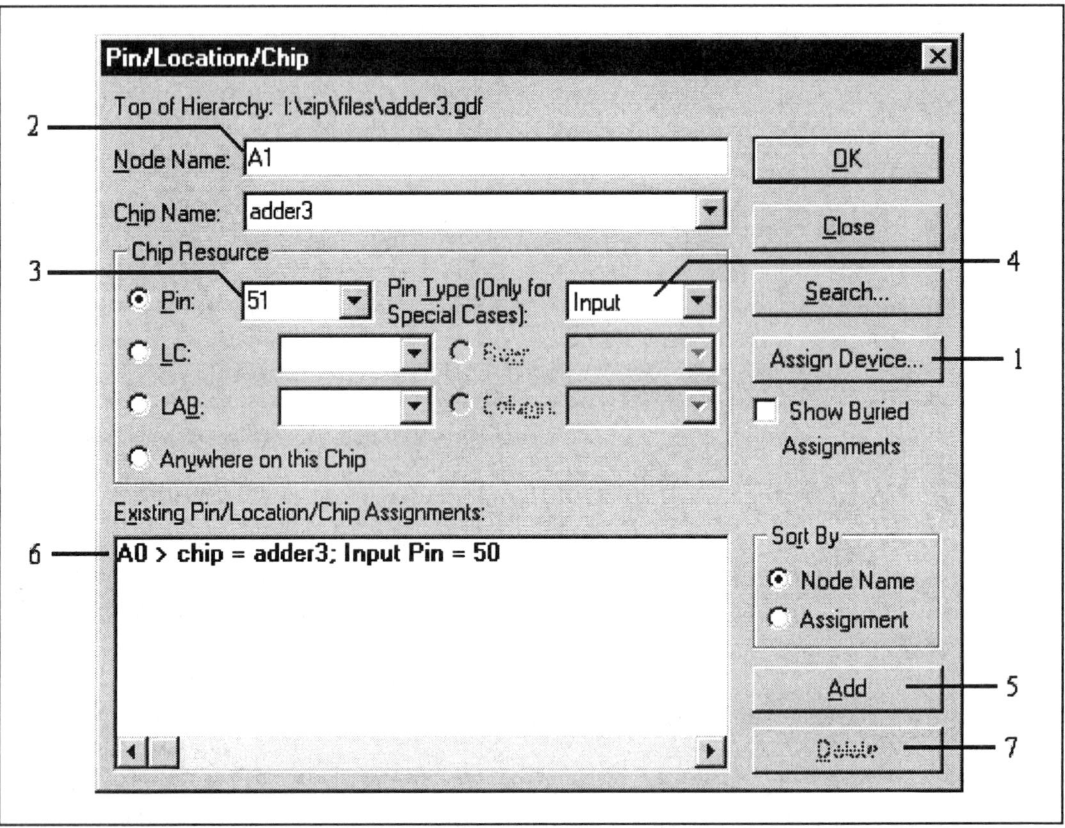

Figure 9

2. Select MAX7000S as the device family and the EPM7128SLC84 as the device when the Device dialog box appears, then click OK (Figure 10).

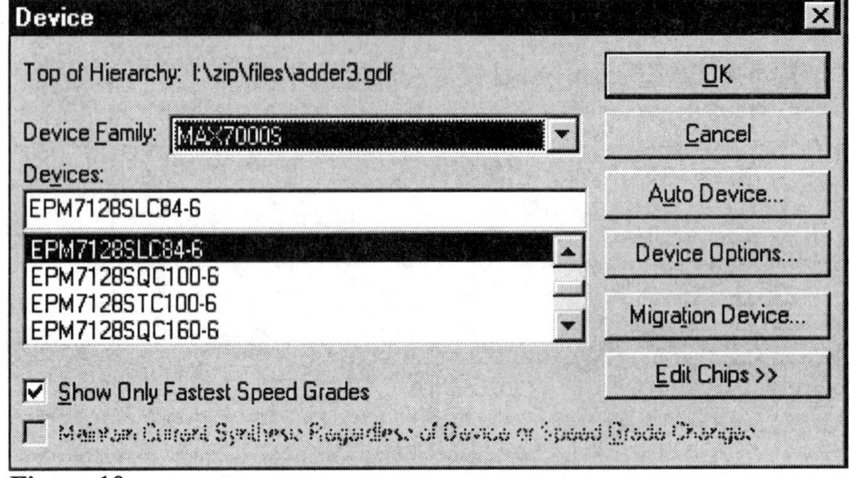

Figure 10

3. Enter the Node Name (2) in the Pin/Location/Chip dialog box.

4. If you don't know the node name, then why are you reassigning pins? To find node names, you can click the "Search" button and choose the node from the list in the resulting dialog box. I assumed that you have already obtained a report (.RPT) listing the pins and have determined which pins needed reassigning. The .RPT is available when compiling the Graphic Editor file or be opened using the Text Editor.

5. Type the pin number (3) that you wish to assign this node name to in the Chip Resource window.

6. Verify the pin type (4) as either an input, output, or bidirectional.

7. Click the "Add" (5) button to add this pin to the "Existing Pin/Location/Chip Assignments" window (6)

8. To add additional pin assignments, repeat Steps 3, 5, and 6.

9. After clicking the "Add" button, the name of the button changes to "Change." If you enter the wrong pin, highlight the pin name in the assignment window (5), click the Change button, make the necessary changes, then click the Add button.

10. To delete a pin, highlight the pin name in the assignment window, then click the delete button (7).

11. Click the OK button to complete the pin assignments.

12. Recompile the Graphic Editor file to update the .RPT file. Save all changes to your disk.

Use the MegaWizard Plug-In Manager

The MegaWizard Plug-In Manager allows you to create or customize megafunction design files that you can use in a design file. These custom megafunction variations are based on Altera-provided megafunctions, including LPM functions. The MegaWizard Plug-In Manager asks questions about the values you want to set for parameters or about which optional ports you want to use.

The wizard generates a Component Declaration file (with the extension .cmp) that can be used in VHDL Design Files and an Include File that can be used in AHDL Text Design Files and Verilog Design Files. The MegaWizard Plug-In Manager also generates a Symbol File that can be used in Graphic Design Files.

1. The MegaWizard Plug-In Manager (Figure 11) can be started in one of three ways:

   A. Select the **MegaWizard Plug-In Manager** command from the **File** menu.

   B. Select the **MegaWizard Plug-In Manager** button in the **Enter Symbol** dialog box.

   C. Start the stand-alone version of the MegaWizard Plug-In Manager by typing the following command at a DOS or UNIX prompt: **megawiz** ↵.

2. Three megafunction libraries are available from which you can choose LPM functions. To view the LPM functions, click the plus sign in the box to the left of the directory. As shown in Figure 12, the "gates" directory contents are now visible.

3. Highlight the megafunction you wish to create or modify (LPM_MUX for this example), select the type of output file, assign the output file name (a:\8by4mux), then click **Next**. Subsequent screens provide by the **MegaWizard Plug-In Manager** will be dependent on the choices you entered.

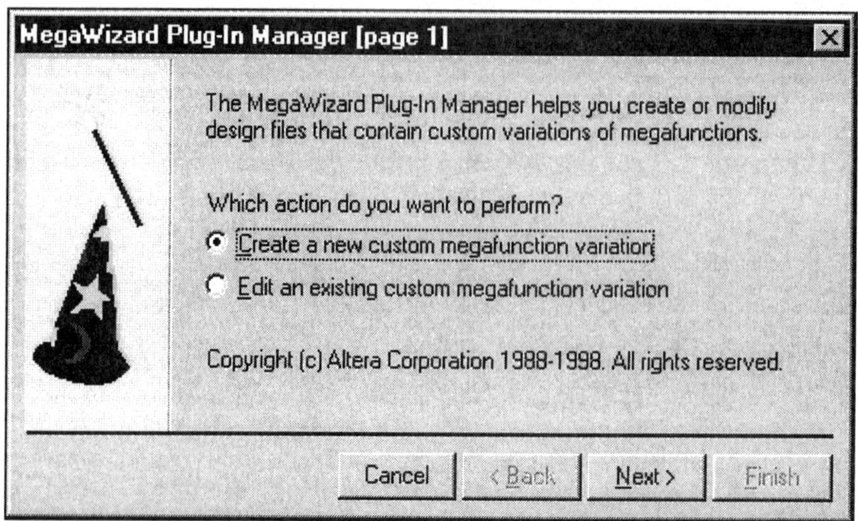

Figure 11

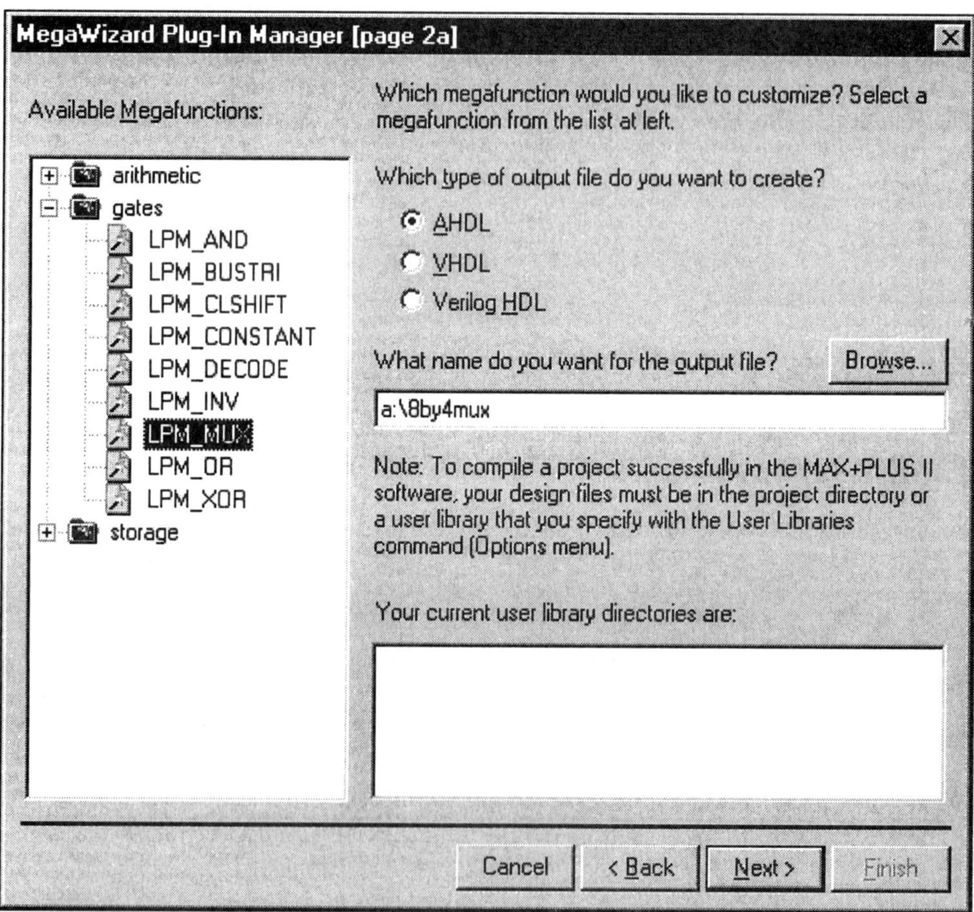

Figure 12

4. Choose the number of data inputs in the **MegaWizard Plug-In Manager** (Figure 13) and the width of each data input. For this example, eight buses are selected, each containing 4 bits. The symbol of the circuit function changes as entries are made in the **MegaWizard Plug-In Manager**.

5. Choose Yes if you want the circuit function to be clock dependent, then select the latency. Latency refers to the number of clock pulses required on the clock input before the output of the circuit function changes. Click **Next**.

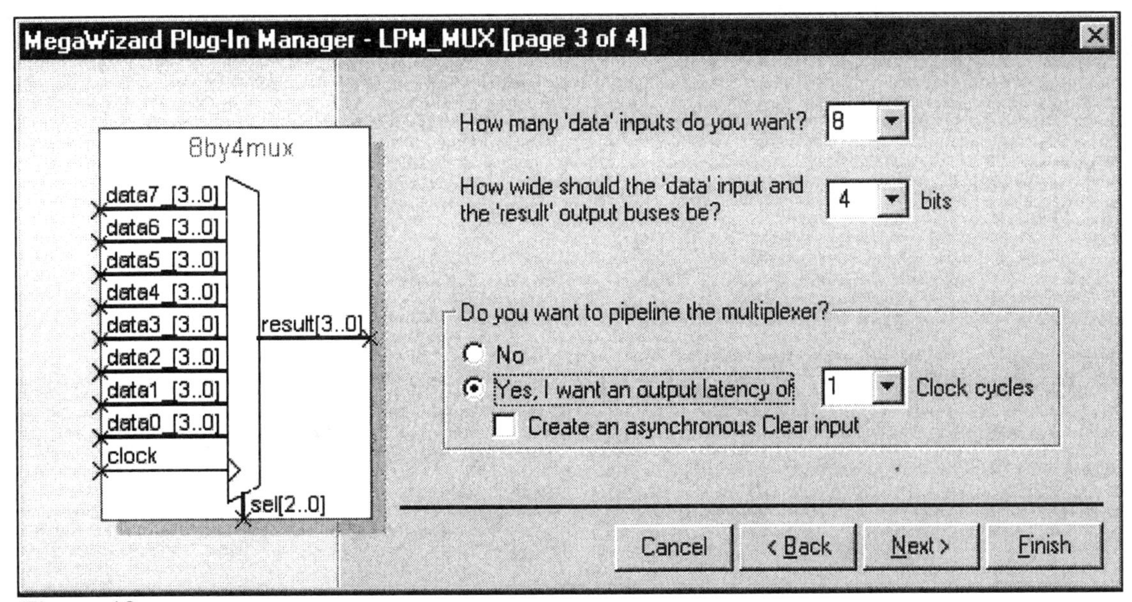

Figure 13

6. The summary of the files created by the **MegaWizard Plug-In Manager** are listed in page 4 of 4 (Figure 14). These files must be present on your disk drive to use the logic symbol created in your design file. Click **Finish**.

7. Think before creating your symbol. The symbol created (Figure 15) will require 40 pins!

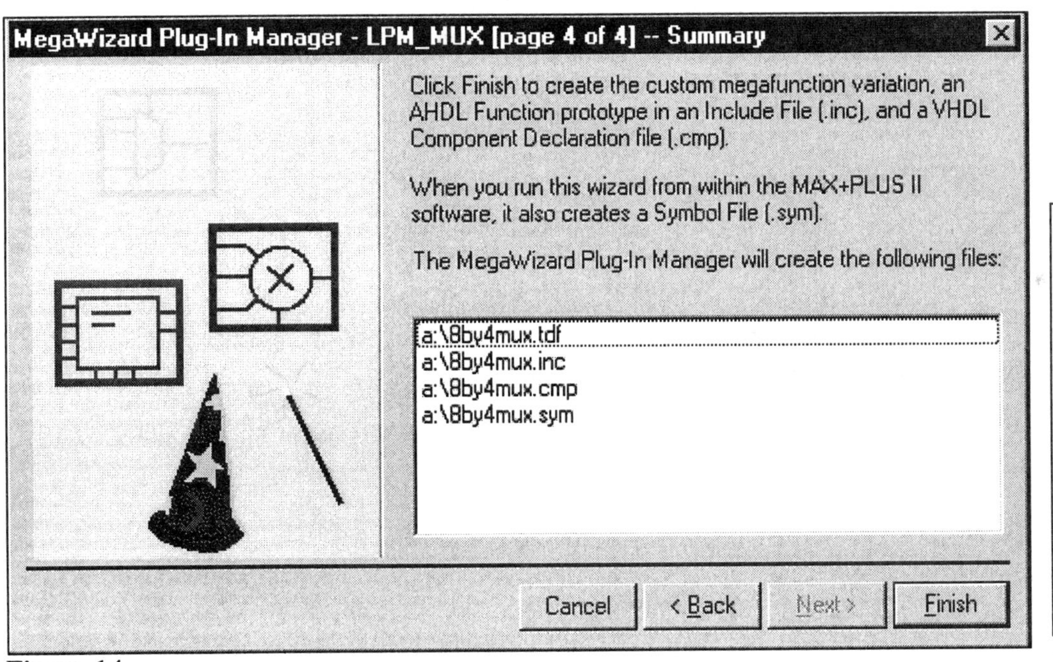

Figure 14

Figure 15

Create your own symbol

1. Open the Max+plus II software and assign a project name to Drive A.

2. Open the Graphic Editor and construct your circuit file including labeled input and output symbols.

3. Open the Waveform Editor and create a set of input and output waveforms.

4. Compile and simulate your circuit file to verify the circuit behaves according to your design specifications.

5. Select **File - Create Default Symbol.**

6. Your symbol is now available on your disk in Drive A. This disk must be in Drive A for you to use the symbol in other design files.

The Compiler Window

Figure 16 shows the Compiler window when the Compiler button in the main menu is clicked. Assuming zero errors and zero warnings, click the Start button to finish the compilation. The following tasks are completed by the compiler.

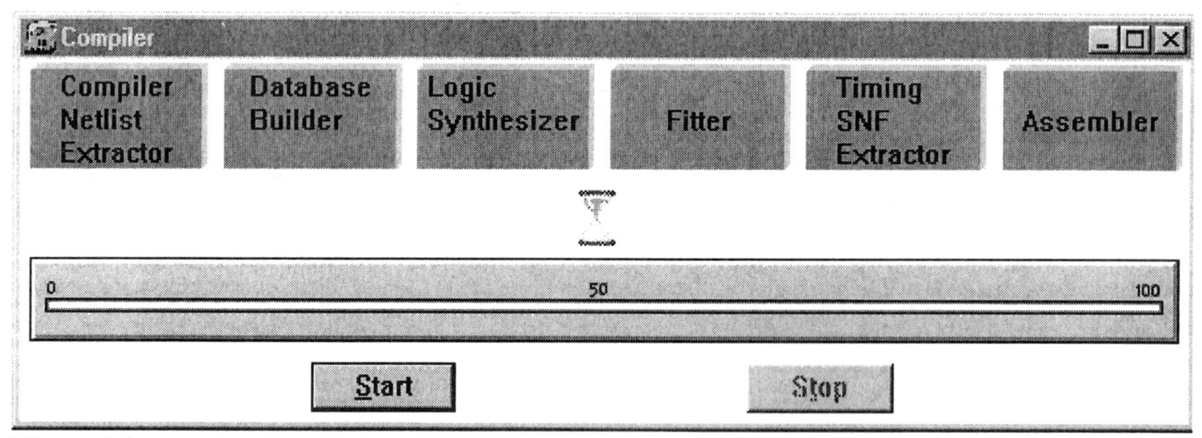

Figure 16

Compiler Netlist Extractor[1]

The Compiler module converts each design file in a project into a separate binary CNF. The file names of the CNF(s) are based on the project name.

The Compiler Netlist Extractor also creates a single HIF that documents the hierarchical connections between design files.

This module contains a built-in EDIF Netlist Reader, VHDL Netlist Reader, and XNF Netlist Reader and converters that translate ADFs and SMFs for use with Max+plus II.

During netlist extraction, this module checks each design file for problems such as duplicate node names, [1] Max+plus II Help window for the Compiler Netlist Extractor missing inputs and outputs, and outputs that are tied together.

Database Builder[2]

The Database Builder uses the HIF to link the CNFs that describe the project. Based on the HIF data, the Database Builder copies each CNF into the project database. Each CNF is inserted into the database as many times as it is used within the original hierarchical project. The database thus preserves the electrical connectivity of the project.

The Compiler uses this database for the remainder of project processing. Each subsequent Compiler module updates the database until it contains the fully optimized project. In the beginning, the database contains only the original netlists; at the end, it contains a fully minimized, fitted project, which the Assembler uses to create one or more files for device programming.

---

[1] Max+plus II Help window for the Compiler Netlist Extractor

[2] Max+plus II Help window for the Database Builder

As it creates the database, the Database Builder examines the logical completeness and consistency of the project, and checks for boundary connectivity and syntactical errors (e.g., a node without a source or destination). Most errors are detected and can be easily corrected at this stage of project processing.

Using the database created by the Database Builder, the Logic Synthesizer calculates Boolean equations for each input to a primitive and minimizes the logic according to your specifications.

Logic Synthesizer[3]

For projects that use JK or SR flip-flops, the Logic Synthesizer checks each case to determine whether a D or T flip-flop will implement the project more efficiently. D or T flip-flops are substituted where appropriate, and the resulting equations are minimized accordingly.

The Logic Synthesizer also synthesizes equations for flip-flops to implement state registers of state machines. An equation for each state bit is optimally implemented with either a D or T flip-flop. If no state bit assignments have been made, or if an incomplete set of state bit assignments has been created, the Logic Synthesizer automatically creates a set of state bits to encode the state machine. These encodings are chosen to minimize the resources used.

The Logic Synthesizer then uses a rule-based expert system to group the project requirements for I/O pins, tristate buffers, and register resources into individual logic cell-sized units. In addition, in device architectures with shareable expander product terms, the number of logic cells and expander product terms are balanced to use resources efficiently. As part of the logic minimization and optimization process, logic and nodes in the project may be changed or removed.

Throughout logic synthesis, the Logic Synthesizer detects and reports errors such as illegal combinatorial feedback and tristate buffer outputs wired together ("wired ORs").

Fitter[4]

Using the database updated by the Partitioner, the Fitter matches the logic requirements of the project with the available resources of one or more devices. It assigns each logic function to the best logic cell location and selects appropriate interconnection paths and pin assignments.

The Fitter attempts to match any resource assignments made for the project with the resources on the device. If it cannot find a fit, the Fitter allows you to override some or all of your assignments or terminate the compilation.

The Fitter module generates a Fit File that documents pin, buried logic cell, chip, clique, and device assignments made by the Fitter module in the last successful compilation. Each time the project compiles successfully, the Fit File is overwritten. You can back-annotate the assignments in the file to preserve them in future compilations.

Regardless of whether a fit is achieved, the Fitter generates a Report File that shows how the project is implemented in one or more devices. You can specify whether to include optional sections in the Report File with the Report File Settings command (Processing menu).

Timing SNF Extractor[5]

The Timing SNF Extractor is turned on with the Timing SNF Extractor command (Processing menu). It is also turned on automatically when you turn on the EDIF Netlist Writer, Verilog Netlist Writer, or VHDL

[3] Max+plus II Help window for the Logic Synthesizer

[4] Max+plus II Help window for the Fitter

[5] Max+plus II Help window for the Timing SNF Extractor

Netlist Writer command (Interfaces menu). The Timing SNF Extractor cannot be turned on at the same time as the Functional SNF Extractor or the Linked SNF Extractor.

A timing SNF describes the fully optimized circuit after all logic synthesis and fitting have been completed. Regardless of whether a project is partitioned into multiple devices, the timing SNF describes a project as a whole. Therefore, timing simulation and timing analysis (including delay prediction) are available only for the project as a whole. Neither timing simulation nor functional testing is available for individual devices in a multi-device project. Functional testing is available only for a single-device project.

A timing SNF is not created until a project compiles without any errors. The Timing SNF Extractor can run repeatedly during compilation. The Compilation Settings & Times Section of the Report File records the time spent during processing in this module during project compilation.

Assembler[6]

The Assembler module completes project processing by converting the Fitter's device, logic cell, and pin assignments into a programming image for the device(s), in the form of one or more POFs, SOFs, Hex Files, TTFs, Jam Files, JBC Files, and/or JEDEC Files. POFs and JEDEC Files are always generated; SOFs, Hex Files, and TTFs are always generated if the project uses FLEX 6000, FLEX 8000 or FLEX 10K devices; and Jam Files and JBC Files are always generated for MAX 9000 or MAX 7000AE projects. If you turn on the Enable JTAG Support option in the Classic & MAX Global Project Device Options dialog box (Assign menu) or the Classic & MAX Individual Device Options dialog box, the Assembler will also generate Jam Files and JBC Files for MAX 7000A or MAX 7000S projects. After compilation, you can also use SOFs to create different types of files for configuring FLEX 6000, FLEX 8000 and FLEX 10K devices with Convert SRAM Object Files (File menu).

The programming files can then be processed by the Max+Plus II Programmer and the MPU hardware to produce working devices. Several other programming hardware manufacturers also provide programming support for Altera devices.

The Assembler can run repeatedly during compilation to generate each programming file. Programming files are not created until a project compiles without errors. The Compilation Settings & Times section of the Report File records the time spent during processing in this module during project compilation.

Get Help on a Component

The help screens are very useful and informative. Altera also provides a Help button in the Tool Bar. First click on the Help button (see Figure 17) and the mouse pointer turns into an arrow with a question mark. Then click on the component you want help with and either an explanatory message or the function table for the macrofunction appears on the monitor.

Figure 17

[6] Max+plus II Help window for the Assembler

Appendix B: Error Messages

I cannot predict nor provide solutions to all possible errors students will make when using the Max+plus II software but here are a few common errors/solutions that my students have encountered. One thing you must remember when using a computer is to **save often**!!! Nothing like spending a couple hours working on a project when the software decides to lock up the computer and refuses to respond to your commands, short of hitting the Reset button. Everything is lost? No problem, you did save your work, didn't you?

Error 1: **Cannot find the SCF file.**

Occurs when trying to simulate the file.

Solution: Save the Waveform Editor file to your disk in Drive A. Also, be sure both Graphic and Waveform Editor files **AND** the Project Name have been assigned to Drive A (Figure 1). You must have a current copy of the Waveform Editor opened (and on disk).

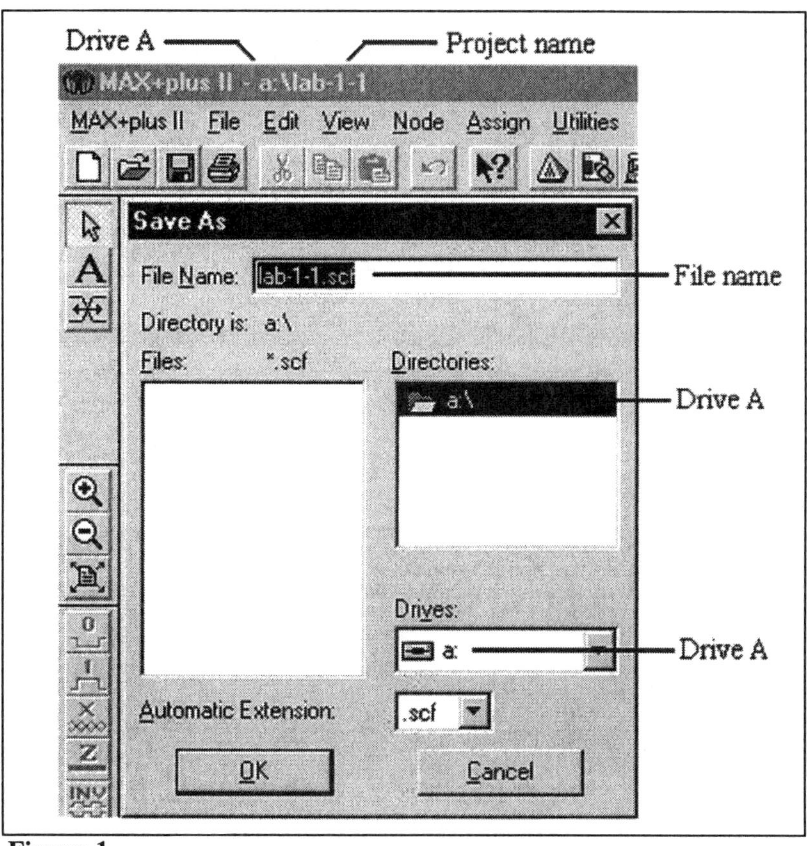

Figure 1

Error 2: **Cannot Simulate – No SCF is loaded for the project.**

You forgot to save the Waveform Editor file.

Solution: Bring the Waveform Editor to the foreground and save the file.

Error 3: **The software won't let you save files or assign a project name to Drive A.**

Bad disk in Drive A or the disk is full.

Solution: Use the Windows Explorer to verify disk space, read/write ability, or replace disk.

Error 4: **The software displays an internal error message.**

Max+plus II had a problem with the operating system.

Solution: Exit Max+plus II and re-open the software. Hopefully, your file was saved to the disk so you may reopen it. If you have to recreate your file, then this is a good practice session.

Error 5: **The software automatically boots you right out of the working area, you're now looking at the Windows^R desktop screen.**

Oops, this typically is an "Operator Error" that causes the software to behave irrationally.

Solution 1: Many times, I have noticed that students like to minimize each screen instead of reducing the work area size to see what is in the background. As a result, when you compile, you don't see the Start button to finish the compilation, then try to continue to simulate. Hence, students miss many steps in the process and the software "gives up." Follow the step-by-step procedure and maximize windows before compiling or simulating. You miss the details of each screen left in minimize form, and hence, skip steps in the process.

The compiler box (Figure 2) will disappear when you click the OK button and you may think you're done with the compiler, but you're not. Maximize the Compiler icon to click the Start button in the Compiler window.

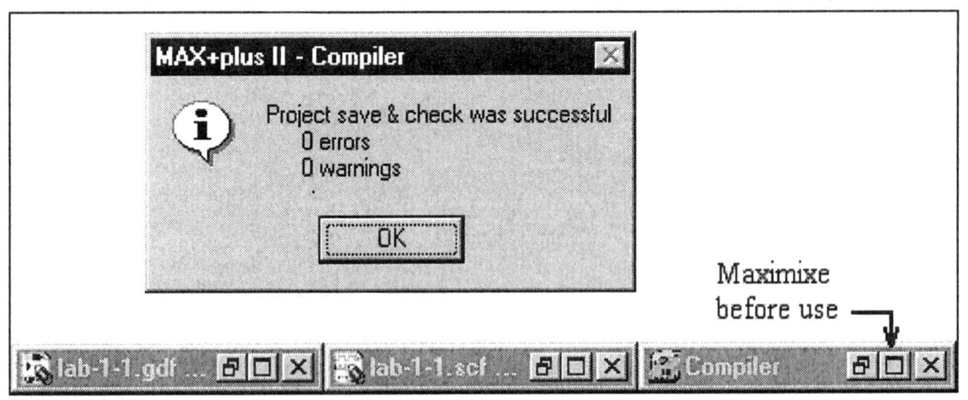

Figure 2

Solution 2: Your directory on the floppy disk may be full. Directories on diskettes are limited to 127 file names. Create a different directory on your disk to place files or obtain a new diskette. You may also delete all but .GDF and .SCF files to free up disk space.

Error 6: **Compiled, simulated, and the output waveform did not change at all.**

You assigned the output as an input in the Waveform Editor.

Solution: Double click on the waveform and reassign the wave as an output. Then re-simulate the circuit. Pay particular attention to the And2 labels in Figure 3.

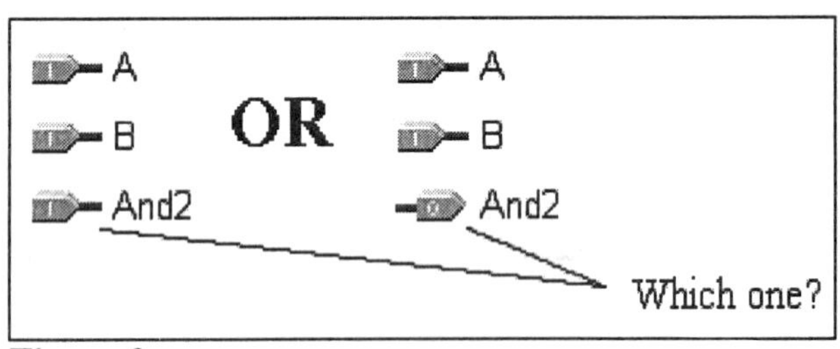

Figure 3

Error 7: **Programmer cannot detect hardware.**

Unrecognized device or socket is empty. (Figure 4)

Solution 1: Apply voltage to the circuit card being programmed.

Solution 2: Select the proper Parallel Port under Hardware Setup in the Options menu.

Solution 3: Be sure the interface cable between the circuit card containing the EPM7128SLC84 and the computer printer port is connected.

Solution 4: If this is an NT station, be sure the Byte Blaster driver (located on the Max+plus II CD-ROM) was installed. Installation of the drive will require administrative rights if this is a networked machine.

Solution 5: For versions before 9.xx, download the self-extracting **univers.exe** file from Altera's university support page at *http://www.altera.com.* Place this file on the Maxplus2 directory, then execute the file.

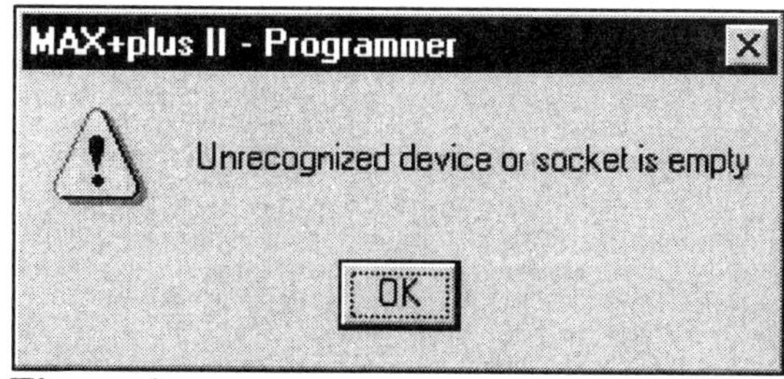

Figure 4

Error 8: **Unrecognized device or socket is empty.**
ByteBlaster is not present - - check power and cables. (Figure 5)

Solution: Connect the ByteBlaster (or interface cable) to the computer and board containing the EPM7128SLC84 chip.

Apply 7- to 12-VDC to the board containing the EPM7128SLC84 chip.

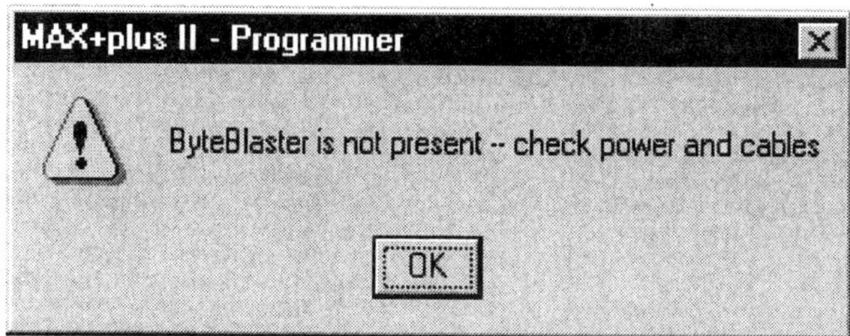

Figure 5

Error 9: **JTAG chain information specified in this dialog box does not match.** (Figure 6)

Solution: Apply power to the circuit card containing the EPM7128LCM84 chip.

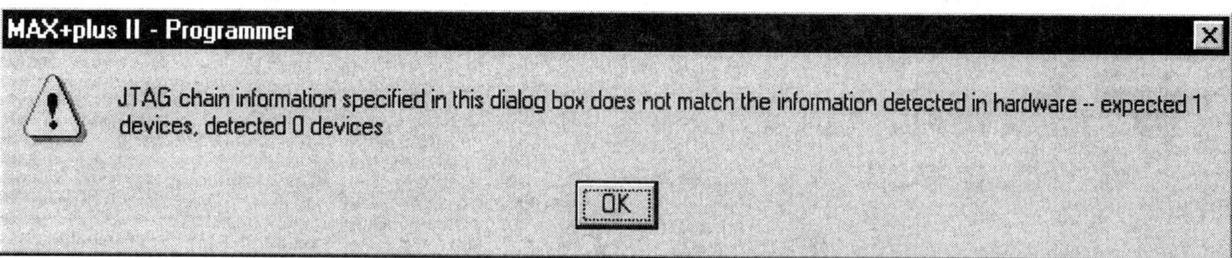

Figure 6

Error 10: **Not in Multi-device chain mode.**

Solution: Click the OK button in the error window. (Figure 7)

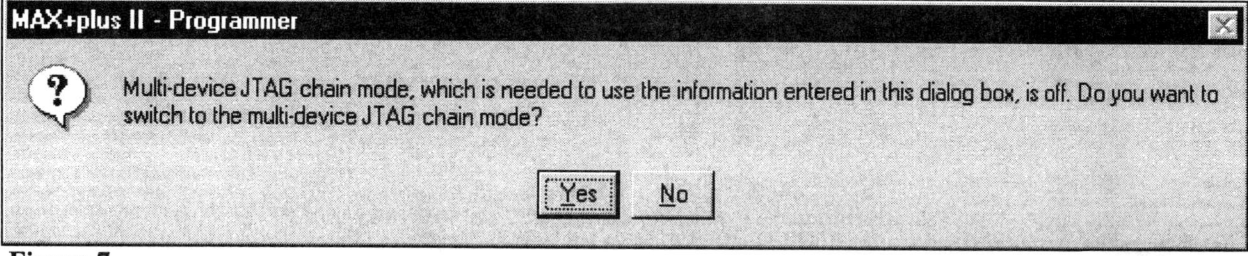

Figure 7

Error 11: **Internal error message appears when opening the software.** (Figure 8).

Solution: Change the year on the clock, **Start - Control Panel**. Non Y2K (year 2000) compliant computers do not display the current year (or date). As such, the computer may default to a year that is well beyond the software expiration date.

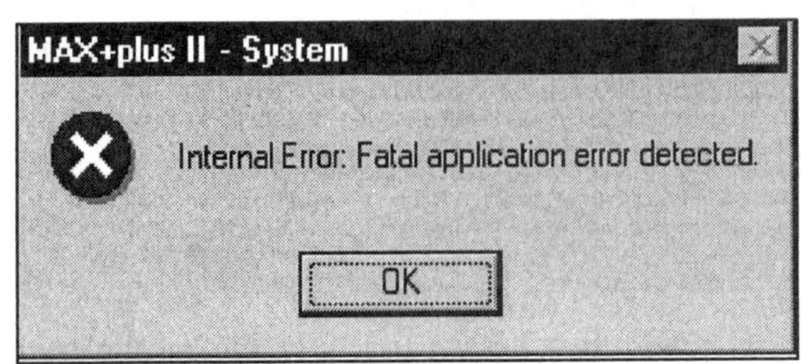

Figure 8

Error 12: **Current license file support does not include the 'Graphic Editor' application or feature.** (Figure 9)

Solution: You either have an invalid license or don't have a license for this computer. Obtain a license.dat file from Altera's web site: ***www.altera.com*** then follow the instructions sent by Altera to license the software.

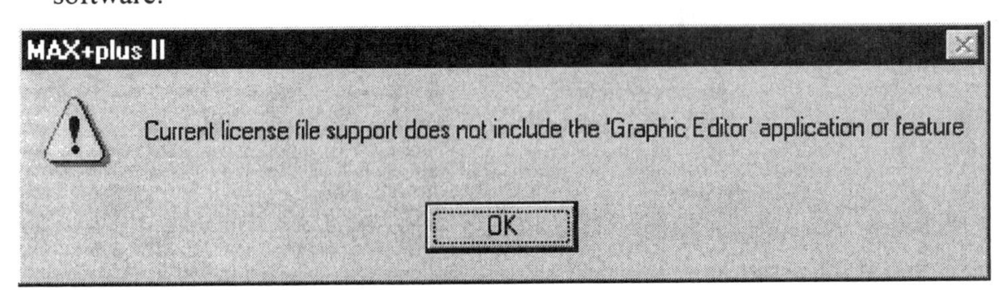

Figure 9

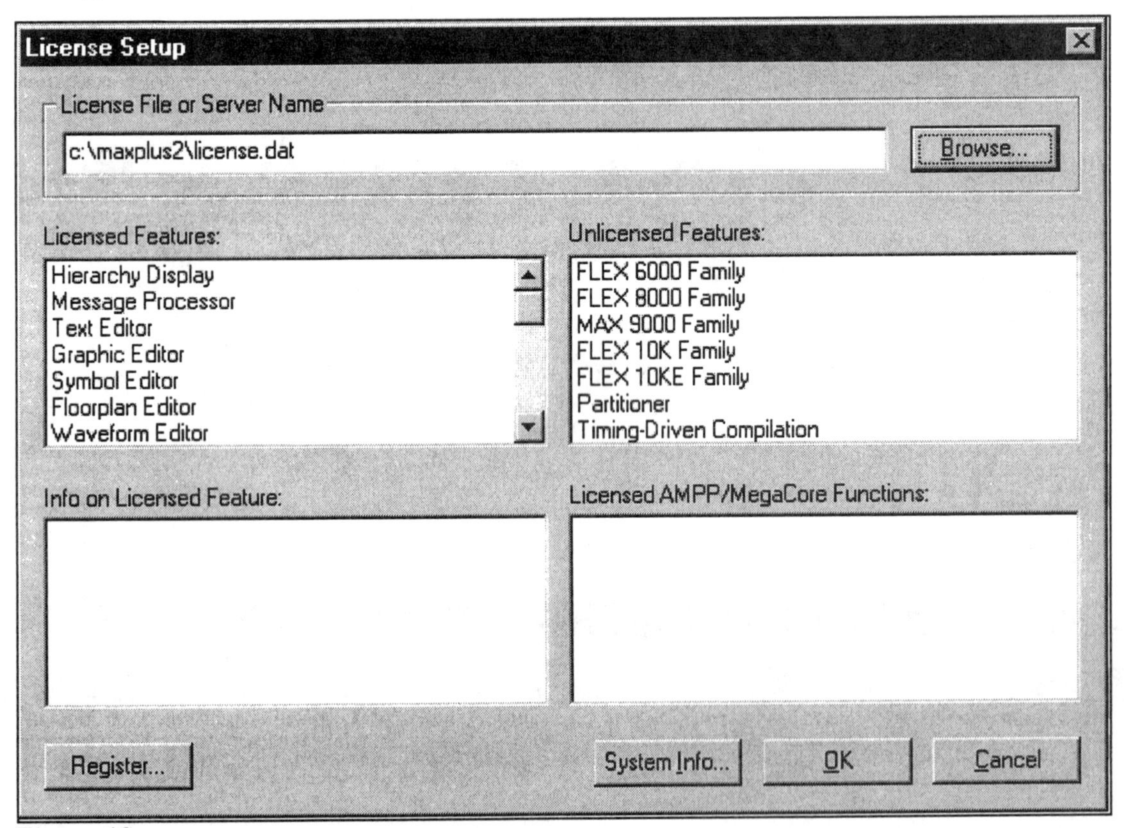

Figure 10

Be sure the drive, path, and license.dat file name are identified in the License File or Server Name window in the License Setup box as shown in Figure 10. Use the Browse button and directory to locate the license file. Be sure the file does not contain multiple extensions. You should create the file using Notepad since other word processors embed additional hidden code into the body of the text and add additional extensions, making the file unrecognizable to Max+plus II.

Error 13: **Cannot use the Compile or Simulate buttons, they both are a light grey shade.** (Figure 11)

Solution: Assign the project name to match the Graphic and Waveform Editor files.

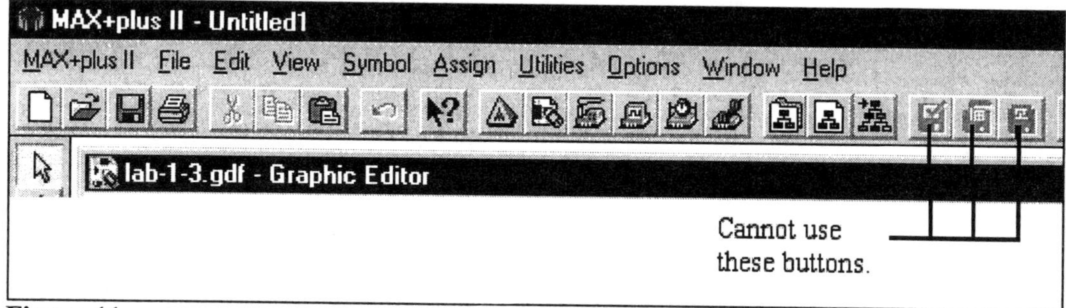

Figure 11

Appendix C: Programming the 7128S

Programming university boards containing in-system programmable CPLDs is straightforward if using Max+plus II Version 9.xx or 10.xx. It is well worth the time to upgrade if you have earlier versions of the software.

1. Most University boards are sent with an AC/DC power adapter that delivers 7- to 12-volts DC to the input of an on-board 5-volt regulator. Plug in the AC/DC adapter.

2. Connect the interface cable between the hardware and the parallel printer port on the computer, LPT1. If you are using the Byte Blaster from Altera, connect the Byte Blaster in the JTAG port on the circuit board to the printer port on the computer.

3. Open the Max+plus II software.

4. Open the Graphic (.gdf) and Waveform (.scf) files of the circuit you want to program to the hardware.

5. Run the Compiler and Simulator. Correct all errors before continuing.

6. Each University board supports various inputs and outputs that are hard-wired to the CPLD. It is necessary to assign pins according to the University board user manual. Refer to the section titled "Change the Pin Assignments" in Appendix A for step-by-step instructions on assigning pins, then recompile the circuit.

7. Select the Max+plus II option in the main menu (Figure 1).

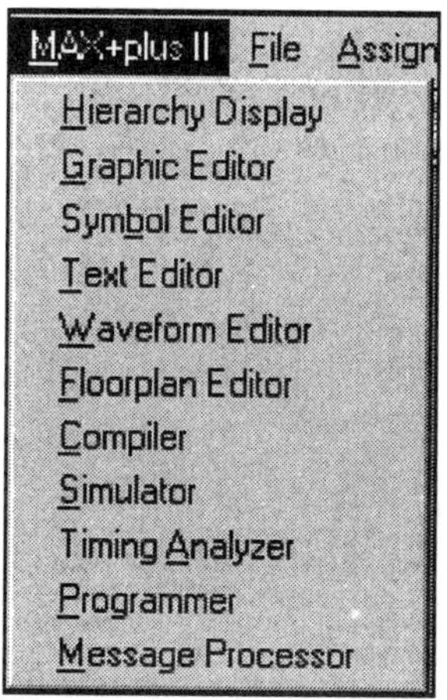

Figure 1

8. Select the **Programmer** option in the Max+plus II drop-down menu. The Programmer dialog box (Figure 2) will appear on the screen.

   If you have successfully compiled the project after assigning pins, a program object file (.pof) is created. Your project name with .pof extension and the CPLD device should appear in the name fields of the programmer window. Do verify that this is the project you want to program into the CPLD.

   You should also see that four of the six buttons have black lettering. When the Programmer window is opened, the software tries to locate the hardware, and when found, these buttons have dark lettering. If for some reason the software is unable to detect the hardware, an error message is displayed (see Appendix B) and these buttons will be a grey shade.

9. Click the Program button. You should see a red bar in the Progress Bar and the Module boxes turn dark blue as the software examines (including blank-checking), configures, programs, tests, and verifies that the CPLD is programmed. When complete, click OK.

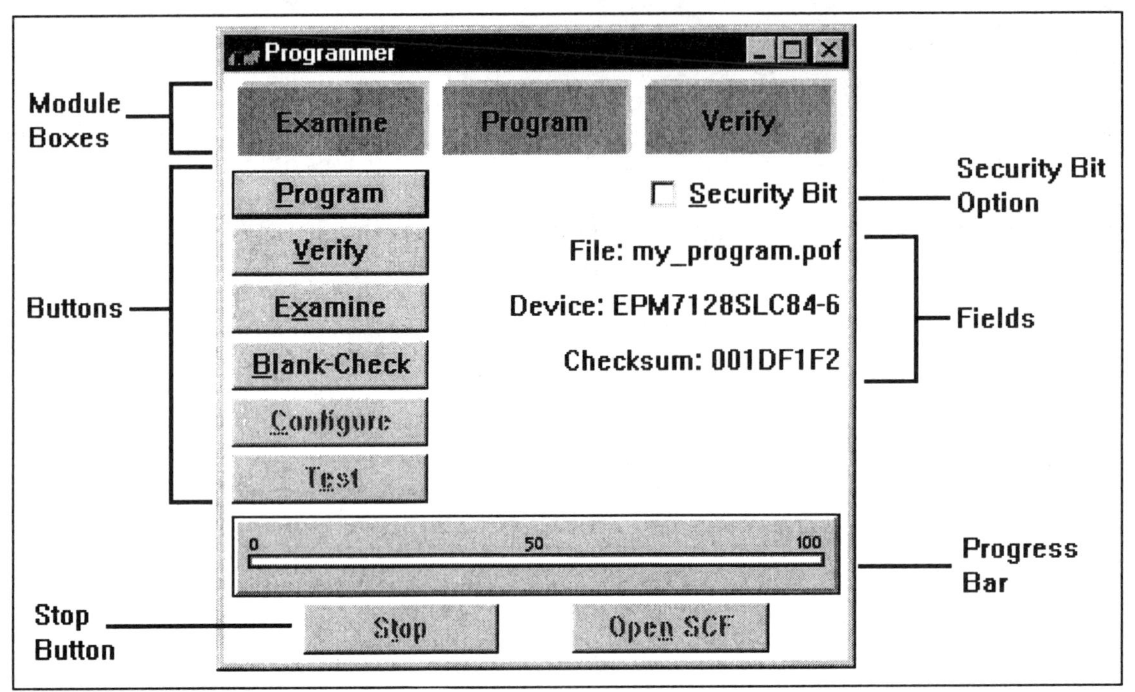

Figure 2

10. The CPLD is now programmed. Complete the necessary hardware interfacing as necessary to external devices to complete your project or lab.

Appendix D: Wiring Circuits

The Breadboard

All breadboards have numbers and letters creating a grid shown in Figure 1. Row X and Row Y contain two solid strips of metal isolated from all other rows. Each numeric column of holes will contain two isolated strips of metal with 5 holes associated with each strip. For instance A1, B1, C1, D1, and E1 all share the same metal strip whereas F1, G1, H1, I1, and J1 all share another metal strip. Breadboards come in different sizes, the larger they are, the more they cost.

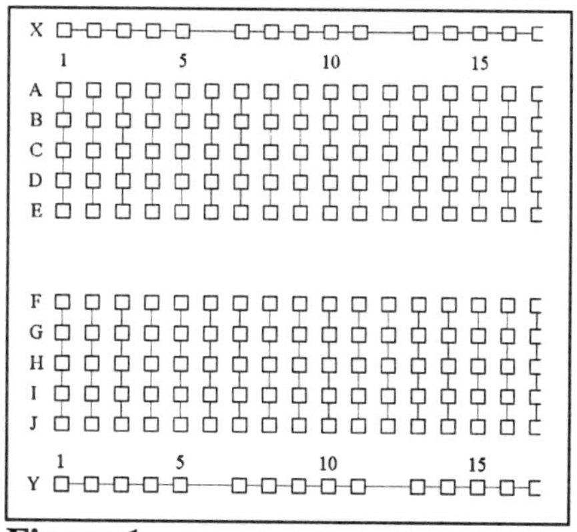

Figure 1

It is best to use 22 gauge or 24 gauge insulated wire on breadboards. Prepare the wire by stripping off 1/4" insulation from both ends. Estimate and cut the length of wire required when making jumper wires. If the distance between two holes is 2", then start with a 2 1/2" jumper wire. The shortest distance between two holes is a straight line unless there are components in the way. So take this in account when determining the length of wire to use. Avoid using excessive wire lengths and be certain that exposed metal leads above the surface of the board do not touch each other.

When placing components on a breadboard, make sure that each lead of the component is inserted into a different set of holes. Resistor R_1 inserted into holes A4 and E4 shown in Figure 2 is shorted by the common metal strip associated with these holes. Regardless of the resistor value, zero ohms will be measured across this resistor. This is not a good wiring plan.

Place resistors and other components to bridge the gap between different columns of holes. The resistor R_2 lead at D8 and the resistor R_3 lead at E8 form a common node making a series circuit. The jumper wire between X13 and A13 completes the continuity from the right side of resistor R_2 plugged into D13 and any hole in Row X. By the same token, the jumper between J8 and Y8 completes continuity between the bottom of resistor R_3 connected to G8 to any hole on Row Y. R_2 could have been placed between D8 and X8, eliminating the jumper wire between X13 and A13.

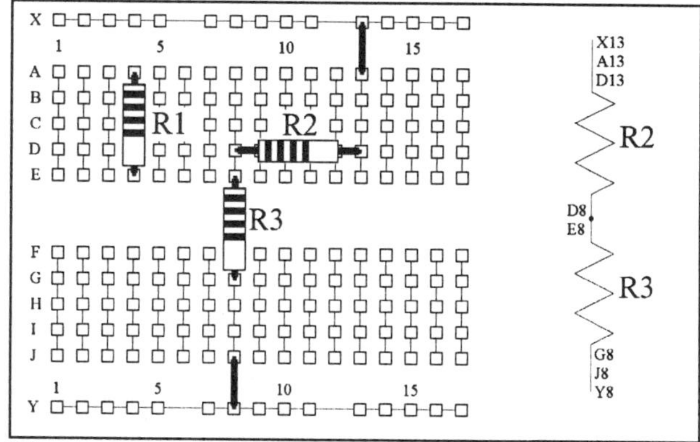

Figure 2

Wiring SPDT Switches

Figure 3 shows the placement and wiring of a Quad In-line SPDT switch producing a logic-LOW or logic-HIGH on the center contact of each switch, SW1, SW2, SW3, and SW4, assuming +Vcc is wired to a point on Row X and ground is connected to a point on Row Y.

1. The SPDT switches are designed to be mounted and soldered on a circuit board. When used on a breadboard, the four plastic stand-off tabs should be cut so the body of the switch mold will be flush on the breadboard surface. If these plastic standoffs are not cut, the switch body has a tendency to pop off the breadboard when the switches are toggled.

2. Insert the pins of the switch body into holes E1, E2, E3, E5, E6, E7, F1, F2, F3, F5, F6, and F7 on the breadboard.

3. Place short (3/4") jumper wires between X1 and A1, X3 and A3, X5 and A5, X7 and A7, J1 and Y1, J3 and Y3, J5 and Y5, and J7 and Y7.

4. Connect jumper wires from the center contact of switch 1, C1, the center contact of switch 2, C2, the center contact of switch 3, C3, and the center contact of switch 4, C4, to the destination points as dictated by the schematic you are wiring.

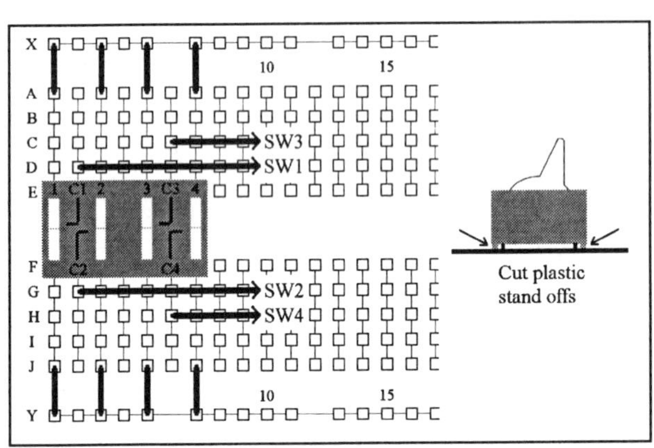

Figure 3

Wiring an Oscillator

When doing experiments at home, you may need a low frequency oscillator as a signal source. An inexpensive oscillator may be constructed using a 555 timer integrated circuit (IC) shown in Figure 4. Refer to Lab 14 for a technical discussion on the circuit operation.

ICs will have a notch or indentation on one end of the plastic mold containing the circuit (chip) within the IC. Locate the chip with numeric designators 555 and position the chip with the notch (or indentation) to the left and pins pointing away from you. Pin 1 is on the lower left and the remaining pins are sequentially numbered in a counter-clockwise direction.

1. Insert the 555 timer IC into holes E1, E2, E3, E4, F1, F2, F3, and F4 with pin 1 inserted into pin F1 on the breadboard. (See Figure 5)

2. Cut two 1" jumper wires then strip 1/4" of the insulation off from each end of the jumper. Insert one jumper between X1 and A1. This will apply +Vcc to pin 8 of the chip once 5 volts is applied to one of the holes in Row X.

3. Insert the second 1" jumper between J1 and Y1. This will ground pin 1 of the IC.

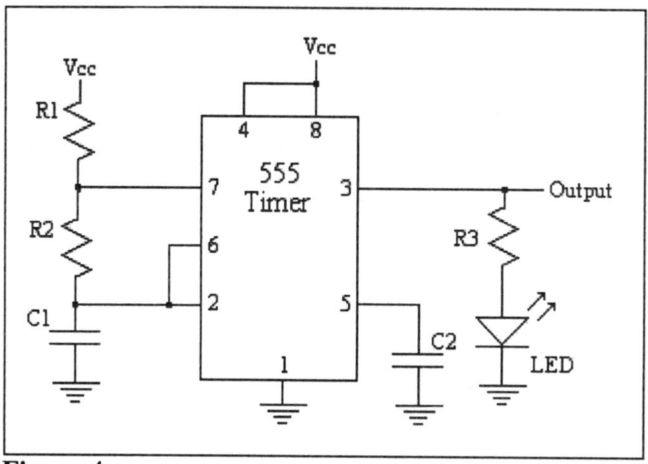

Figure 4

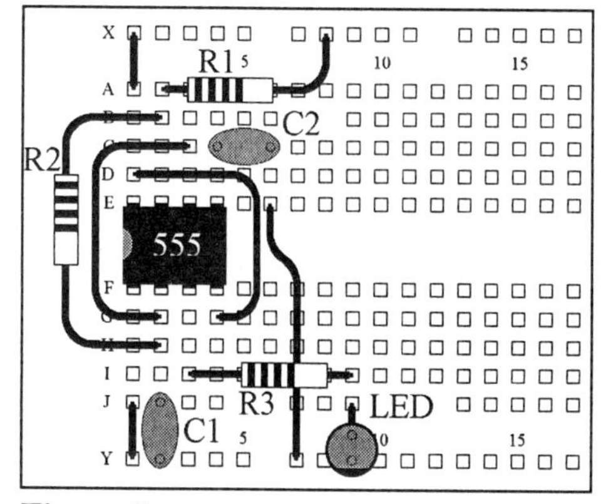

Figure 5

4. Cut three 2" jumper wires then strip 1/4" of the insulation off from each end of the jumpers. Insert one jumper between C3 and G2 on the breadboard. This jumper will short pin 2 of the IC to pin 6 of the IC.

5. Insert another 2" jumper between D1 and G4 on the breadboard. This jumper will short pin 4 of the IC to pin 8 of the IC. Pin 4 is an active-LOW reset that is disabled when pin 4 is wired to pin 8 (or +Vcc).

6. Insert the last 2" jumper between E6 and Y7 on the breadboard.

7. Locate two 33 KΩ resistors (orange, orange, orange). Bend the resistor leads 90 degrees and insert resistor R_1 between A1 and X8.

8. Connect the other 33 KΩ resistor, R_2, between B2 and H2.

9. Locate a 22 μF capacitor, C_1. Insert the + capacitor lead into hole J3 and the – capacitor lead into hole Y3 on the breadboard.

10. Locate a 1 KΩ resistor (black, brown, red). Bend the resistor leads 90 degrees and insert resistor R_3 between I3 and I9.

11. Locate a 0.001 μF capacitor. Insert capacitor C_2 between C4 and C6.

12. Locate a light emitting diode (LED). The flat side of the diode glass base is closest to the cathode lead. Insert the cathode led into Y9 and the other lead into J9.

13. Apply +5 volts to a hole in Row X and the power supply ground to a hole in Row Y. The LED should turn on and off about once a second. Decreasing the value of C_2 will increase the frequency of the oscillator.